Bogdan M. Popescu (Ed.)

Hydrocarbons of Eastern Central Europe

Habitat, Exploration and Production History

With 206 Figures and 26 Tables

Springer-Verlag
Berlin Heidelberg New York
London Paris Tokyo
Hong Kong Barcelona
Budapest

BOGDAN M. POPESCU
Petroconsultants S.A.
24, Chemin de la Mairie
1258 Perly-Geneva
Switzerland

ISBN 3-540-55014-3 Springer-Verlag Berlin Heidelberg New York
ISBN 0-387-55014-3 Springer-Verlag New York Berlin Heidelberg

Library of Congress Cataloging-in-Publication Data. Hydrocarbons of Eastern Central Europe: habitat, exploration and production history/Bogdan M. Popescu, ed. p. cm. "This volume arises out of the Symposium on the Hydrocarbon Exploration Opportunities in Central-Eastern Europe and the USSR, organized by Petroconsultants and held in Geneva, Switzerland, between 22 and 24 October 1990"—Pref. Includes bibliographical references and index. ISBN 3-540-55014-3 (alk. paper).—ISBN 0-387-55014-3 (alk. paper: U.S.). 1. Petroleum—Prospecting—Central Europe—History—Congresses. 2. Petroleum—Geology—Central Europe—Congresses. I. Popescu, Bogdan M., 1944– . II. Petroconsultants S.A. III. Symposium on the Hydrocarbon Exploration Opportunities in Central-Eastern Europe and the USSR (1990: Geneva, Switzerland) TN271.P4H93 1993 553.2′8′0943—dc20 93-26022

Printed in Germany

Typesetting: Macmillan India Ltd., Bangalore 25
32/3130/SPS – 5 4 3 2 1 0 – Printed on acid-free paper.

Preface

This volume arises out of the Symposium on the Hydrocarbon Exploration Opportunities in Eastern-Central Europe and the USSR organized by Petroconsultants and held in Geneva, Switzerland, between 22 and 24 October 1990. This was the first time that officials and experts from the newly emerging, post-communism democracies released to representatives of the Western oil industry first-hand information on what had been for the previous 40 years, one of the most uncommunicative industries in these countries. In addition, reserve and production figures were classified state secrets, hence, it was not surprising that the area appeared to the free world oil industry as obscure, albeit reportedly hydrocarbon-rich.

The achievements of the more than 100-year-old petroleum industries of Eastern Central Europe (ECE) obviously resulted in an exceptionally great amount of information difficult to describe, even in a succinct manner, in the allowed editorial space of this volume. The authors have used mostly unpublished statistical data from the archives of the national state companies. Therefore, the bibliographical lists are not intended to be complete: they only highlight a few of the more important or recently published papers. A list of "Suggested Reading", which could supplement the information released in this volume, was appended by some authors. Moreover, the rapidly evolving political situation in the area and the high pace of the economic legislation enactment since late 1990 led us to eliminate the fiscal, legal and administrative aspects presented in the original Symposium papers.

This book is intended to be useful to specialists in both industry and academia who have an interest in ECE petroliferous provinces. The emphasis has been on the historical evolution of activities, but an important element is petroleum geology. The authors have made a flurry of statistical data available, especially of the post-World War II period, of which only very little had been released before. These data are supplemented by geological summaries which show a remarkably good understanding of the tectono-stratigraphic framework of hydrocarbon accumulations in Albania, Bulgaria, the Czech Republic and Slovakia, eastern Germany, Hungary, Poland and Romania, i.e. in the Carpathian, Dinaride and Balkan alpine foldbelt and its foreland (see figure, p. VI).

If questions related to political risk are omitted, the reader will perhaps try to find an answer to a few other, major questions: how much undiscovered resources remain to be found in the apparently extensively explored ECE petroleum provinces?; are there any frontier areas which could be tapped by the latest exploration technology and concepts of Western industries?; if hydrocarbons are found, are there any incentives to produce them? A number of Western oil companies think they can give a positive answer to these questions,

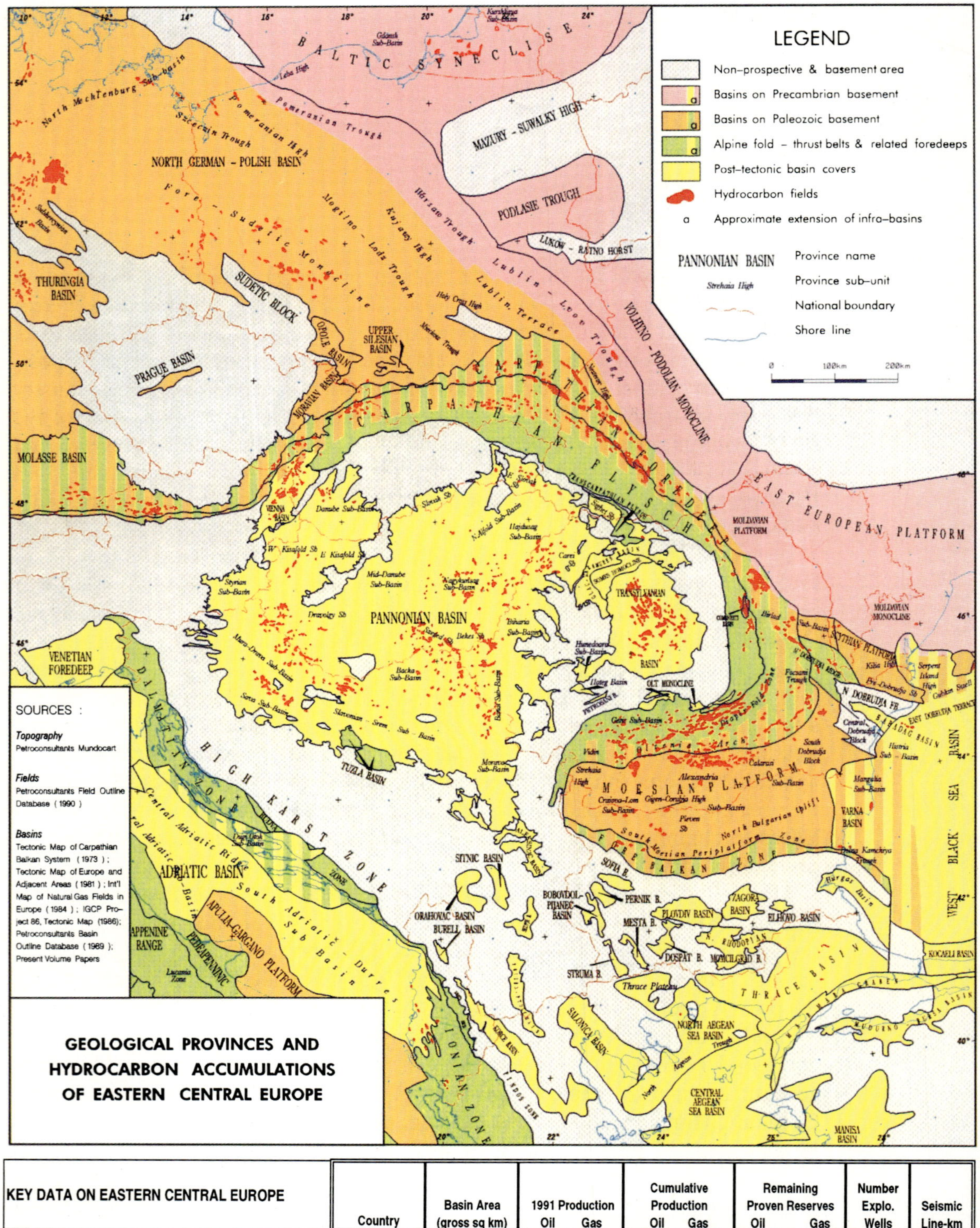

GEOLOGICAL PROVINCES AND HYDROCARBON ACCUMULATIONS OF EASTERN CENTRAL EUROPE

KEY DATA ON EASTERN CENTRAL EUROPE

NOTES: Figures, as of 1.01.1992, came from author's updates, state agencies releases, Petroconsultants' World Production and Reserves Statistics and World Petroleum Trends reports. Production and reserves are in million tonnes oil and billion cubic meters gas. Oil includes NGLs. Gas excludes non-hydrocarbon gases. Exploratory well count includes an unidentifiable amount of extension wells; it is estimated that up to 25% might have been true wildcats. Seismic includes 3D surveys line-km.

^ = estimated

Country	Basin Area (gross sq km)	1991 Production Oil	1991 Production Gas	Cumulative Production Oil	Cumulative Production Gas	Remaining Proven Reserves Oil	Remaining Proven Reserves Gas	Number Explo. Wells	Seismic Line-km
Albania	28,750	.75	.14	45 ^	5.0	25.5	1.5^	800^	26,370
Bulgaria	77,300	.06	.01	7.9	3.9	5^	.5	400^	62,000^
Czech/Slovak	28,400	.14	.39	6.4	30.8	2^	10^	2,000^	38,800
E. Germany	70,000	.63	1.2	2.4	60	1.9	40	2,030	80,000^
Hungary	72,500	2.2	5.4	72.2	144.7	21.5	104.4	5,257	112,000^
Poland	257,000	.14	.37	18.2	142	12.9	159.1	6,554	260,230
Romania	150,000	6.8	29	646	1,043	206	563	20,000^	276,650
TOTAL	**684,000**	**10.7**	**36.5**	**795^**	**1,430**	**275^**	**880^**	**37,000^**	**856,050^**

and chapters dealing with a specific country will certainly give further valuable insights and convincing arguments in favour of increased exploration in the petroliferous basins of ECE.

At the time of writing this Preface, western oil companies have secured acreage in Albania, Bulgaria, Czech Republic, eastern Germany, Poland, Romania and Slovakia. Additional bidding rounds were underway or will be opened in Albania, Bulgaria, Hungary, Poland and Romania. Drilling and seismic surveys were completed or are planned for the next 3 to 5 years. A promising Tertiary gas discovery has already been made offshore Bulgaria in 1993.

Acknowledgements. I am grateful to Gerald Dixon, Tom Jameson and Gerard Messin for their helpful reviews of the original typescripts. Petroconsultants gave access to various proprietary databases and generously supported the typing, manual and computer drafting. I wish to acknowledge the endeavour of Eden Bobrowski, Liz Lador, Philippe Chessel, Joseph Simantov and Ely Schneeberger, who made the difficult editing task easier. Theodor Orasianu kindly supplied additional information on Romania. Finally, I thank Wolfgang Engel, Monika Huch and Andrea Weber of Springer-Verlag who gracefully accepted delays and late changes.

Geneva, October 1993

BOGDAN M. POPESCU

Contents

List of Contributors

A. Atanasov
Exploration and Production of Oil and Gas Co., 8, V. Levsky 5800 Pleven, Bulgaria

M. Blizkovsky
Geofyzika, Jecna 29a, 61246 Brno, Czech Republic

P. Bokov
Scientific Research Institute of Mineral Resources, H. Kabakchiev Blvd 23, 1505 Sofia, Bulgaria

P. Constantinescu
Petroconsultants S.A., 24, Chemin de la Mairie, 1258 Perly-Geneva, Switzerland

J. Francu
Geological Survey, Leitnerova 22, 60200 Brno, Czech Republic

B. Gaza
Moravian Oil Company, Uprkova 6, 69530 Hodonin, Czech Republic

V. Hlavaty
Nafta, 90845 Gbely, Slovakia

N. Ionescu
109 Str. Isbiceni, Bucharest, Romania

A. Kocak
Geofyzika, Jecna 29a, 61246 Brno, Czech Republic

J. Kokai
Ministry of Industry and Trade, P.O.B. 22, Budapest H-1518, Hungary

P. Kostelnicek
Moravian Oil Company, Uprkova 6, 69530 Hodonin, Czech Republic

S. Lunga
Nafta, 90845 Gbely, Slovakia

B. Monov
Scientific Research Institute of Mineral Resources, H. Kabakchiev Blvd 23, 1505 Sofia, Bulgaria

M. Morkovsky
Geofyzika, Jecna 29a, 61246 Brno, Czech Republic

E.P. Müller
Erdöl-Erdgas Gommern GmbH, Postfach 21, 3304 Gommern, Germany

P. Mueller
Geological Survey, Leitnerova 22, 60200 Brno, Czech Republic

A. Novotny
Geofyzika, Jecna 29a, 61246 Brno, Czech Republic

R. Ognyanov
Committee of Geology and Mineral Resources, 23, Princess Marie Louise Blvd., 1000 Sofia, Bulgaria

T. Piperi
Petroconsultants S.A., 24, Chemin de la Mairie, 1258 Perly-Geneva, Switzerland

B. Sejdini
Albpetrol, DPIM House, Durresi Street 83, Tirana, Albania

D. Tochkov
Exploration and Production of Oil and Gas Co., Varna Branch, 9000 Varna, Bulgaria

D. Vass
Geological Institute of Dionyz Stur, Mlynska dolina 2, 81704 Bratislava, Slovakia

V. Vuchev
Geological Institute, Bulgarian Academy of Sciences, Acad. G. Bonchevstr, b/k 24, 1113 Sofia, Bulgaria

J. Zagorski
Geonafta. ul. Krucza 36, 00-921 Warsaw, Poland

1 Petroleum Exploration in Albania

B. SEJDINI[1], P. CONSTANTINESCU[2], and T. PIPERI[2]

CONTENTS

1 Introduction

Situated in the western part of the Balkan Peninsula, with a surface of 28,748 km^2, Albania includes portions of the petroliferous Durres and Ionian sedimentary provinces.

As an element of the southern branch of the Alpine folded belt and of the Apulia Plate, this area has witnessed the Mesozoic rifting and the Cenozoic collision of the European and African Plates.

All the hydrocarbon discoveries made so far in Albania are located in the Ionian Zone and in the Durres Basin, also called the Peri Adriatic Depression (PAD). For this reason, the present study will concentrate only on the description of these two major basins, and will briefly describe some geological characteristics of potentially prospective areas little or not yet explored.

2 Geological Framework

Reports of oil discovery in Albania at the beginning of this century triggered the interest of geologists who started carrying out some field studies. They focussed on the folded and thrusted Albanides whose structure was described by Austrian and German geologists in the period between the two world wars.

In the early 1950s, Albanian geologists resumed geological research with Soviet technical and financial assistance. The first geological map at scale 1:200,000 accompanied by an explanatory notice and numerous cross-sections was published in 1970. Later, a small scale (1:2,500,000) tectonic map was published.

Part of the southeast Adriatic region, the territory of Albania is characterized by the progressive advancement from north to south, of overthrusting of the more external parts of the Dinaride-Hellenide (locally called Albanides) chains onto the Apulian-Adriatic foreland (Flores et al. 1991).

The Adria Plate, or the Apulian-Adriatic foreland, is made up of Permo-Triassic clastics and evaporites, Mesozoic carbonates overlaid by Tertiary carbonates and clastics. In Italy, it is characterized by gentle folding and systems of normal faults.

[1] Albpetrol (New Ventures Dept.), DPIM House, Durresi Street 83, Tirana, Albania

[2] Petroconsultants S.A., 24, Chemin de la Mairie, 1258 Perly-Geneva, Switzerland

The Apulian–Adriatic foreland extends eastwards to Albania and may represent a rather deep prospective area offshore Albania.

2.1 Structure

The thrust belt of the Albanides comprises the Inner and Outer Albanides. They are uncomformably covered by the post-tectonic Durres, Korca and Burrel basins (Figs. 1, 2, 3).

Deformation of the Inner Albanides took place by the end of the Jurassic and the beginning of the Cretaceous, whereas for the Outer Albanides this process started in the Late Eocene and lasted to the Pliocene. During the Neogene, large volumes of clastics were transported in the foredeep situated on the eastern margin of the Apulia Plate.

From the petroleum point of view, the Outer Albanides and the Neogene molasse basins represent the main interest.

The outer part of the Albanides comprises the Krasta, Kruja, Ionian and Sazani Zones and is characterized by successive folds thrusted westerly over the Adriatic (Apulia) foreland. The main thrusting phase that affected the Kruja Zone took place during Middle/Late Oligocene, and for the Ionian Zone during Middle Miocene.

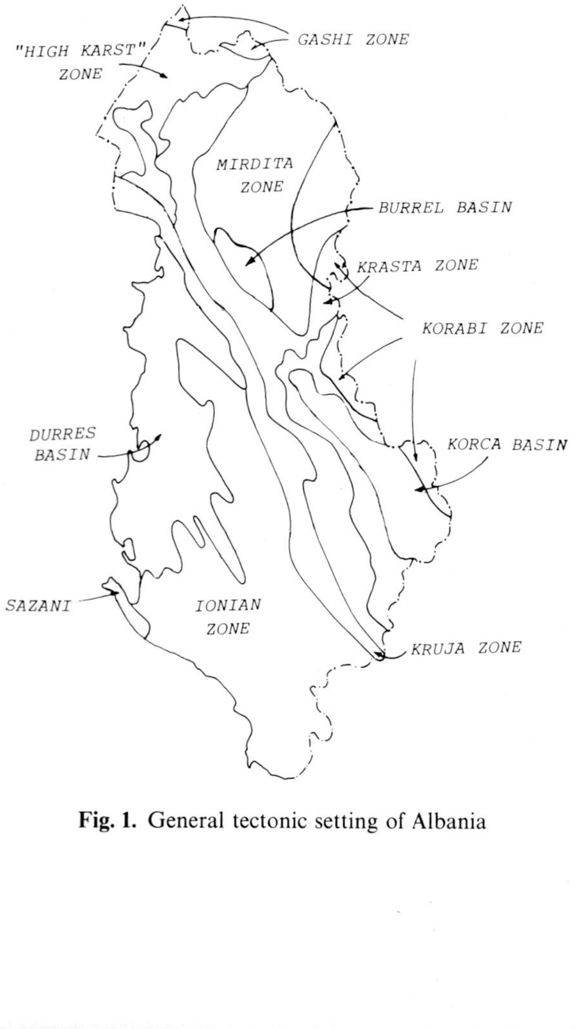

Fig. 1. General tectonic setting of Albania

The molasse foredeep is made up of a 8000-m-thick Neogene sequences of the Durres Basin and the external, overthrusted part of the Ionian Zone. The thrust front migration towards the west formed gentle folds into the foredeep.

2.1.1 Sazani Zone

This uplifted foreland block is located in the southwestern part of Albania and outcrops in Karaburun Peninsula and Sazani Island (Figs. 1, 3).

The contact with the Durres Basin and Ionian Zone is a reverse fault with the Sazani Zone in the upper position (back thrusting?) in the Sazani Island and in the Dukati gulf area. This contact turns to an overthrust of the Ionian Zone over the Sazani Zone south of Llogara village. The area was in an extensional tectonic regime during most of the Mesozoic, which resulted in extensive normal faulting of the platform carbonates.

The Sazani Zone is made up of Triassic to Upper Oligocene carbonates which were deposited in a shallow marine environment. This carbonate facies extends northwards below the Vlora Gulf. During the Paleocene, the Sazani platform was exposed subaerially. The carbonate depositional cycle is unconformably overlain by the Lower Miocene Premolasse clastics. The uplift of the Sazani Zone and reverse faulting is of an Early Aquitanian–Pliocene age.

2.1.2 Ionian Zone

This is a major oil and gas productive area in Albania (Figs. 1, 3). It crops out in the central southwestern parts of the country and extends south onshore towards the western part of Greece. In the west-central part of Albania, the Ionian Zone is overlaid by the post-tectonic Durres Basin.

Early Syn-Rift Unit. The reconstruction of the basin history shows that a large area, which includes Albania, was an extension-type basin (Matavelli et al. 1991) with a fault-controlled subsidence during the Triassic and Early Jurassic. A number of uplifted and tilted blocks, loci of carbonate sedimentation, appeared through differential sinking. In the sunken starved basin area, large

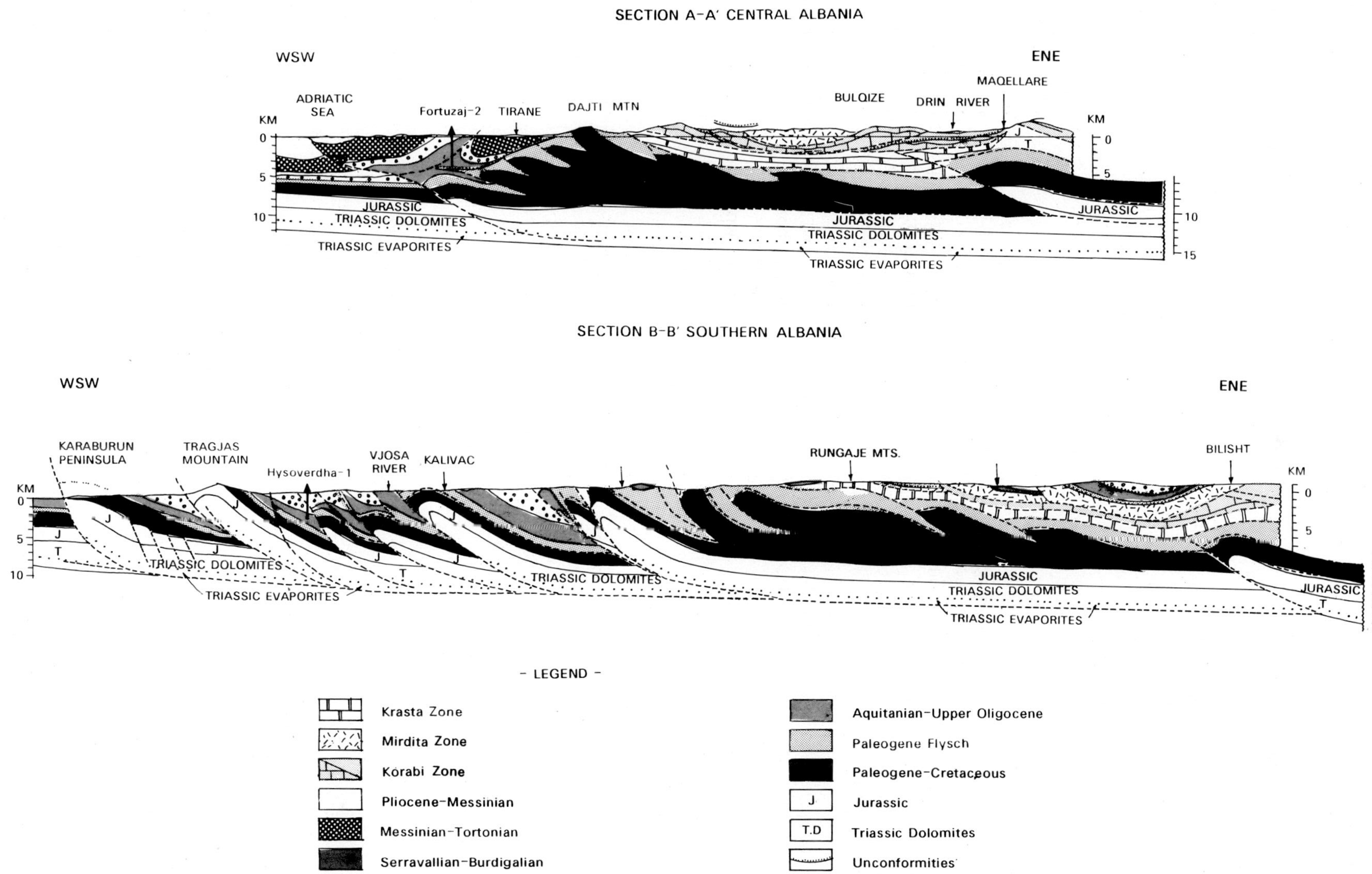

Fig. 2. Interpretative general cross-sections through the Albanides and their foreland (see Fig. 3 for location) (After H Bakia, et al., 1992)

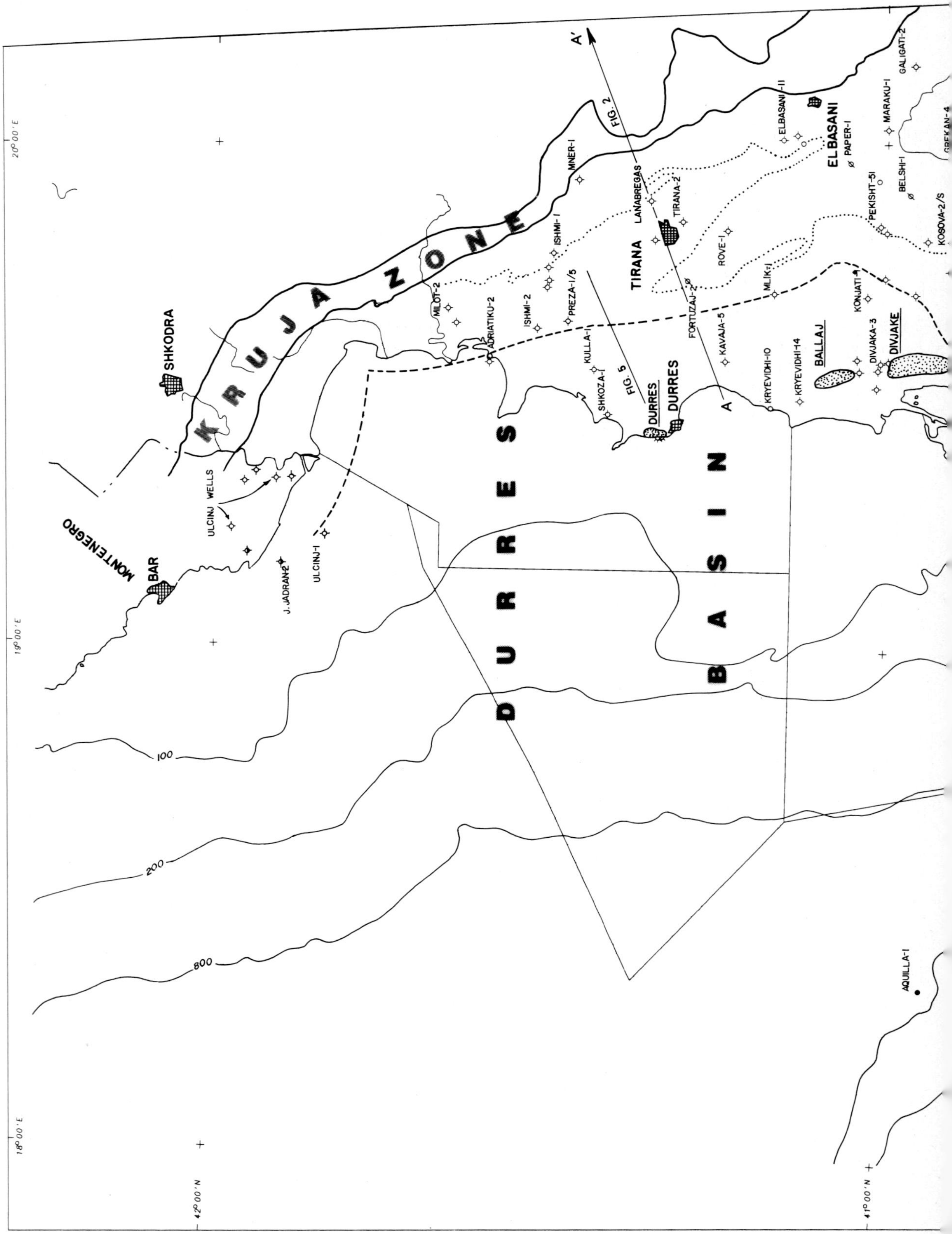
KRUJA ZONE
DURRES BASIN
SHKODRA
MONTENEGRO
BAR
ULCINJ WELLS
ULCINJ-1
J. JADRAN-2
MILOT-2
ADRIATIKU-2
ISHMI-2
ISHMI-1
PREZA-1/5
MNER-1
KULLA-1
SHKOZA-1
FIG. 5
DURRES
DURRES
TIRANA
LANABREGAS
TIRANA-2
FORTUZAJ-2
KAVAJA-5
ROVE-1
MLIK-1
KRYEVIDHI-10
KRYEVIDHI-14
BALLAJ
DIVJAKA-3
DIVJAKE
KONJATI-1
FIG. 2
A
A'
ELBASANI-11
ELBASANI
PAPER-1
PEKISHT-51
BELSHI-1
KOSOVA-2/S
MARAKU-1
GALIGATI-2
AQUILLA-1
100
200
800
20°00' E
19°00' E
18°00' E
42°00' N
41°00' N

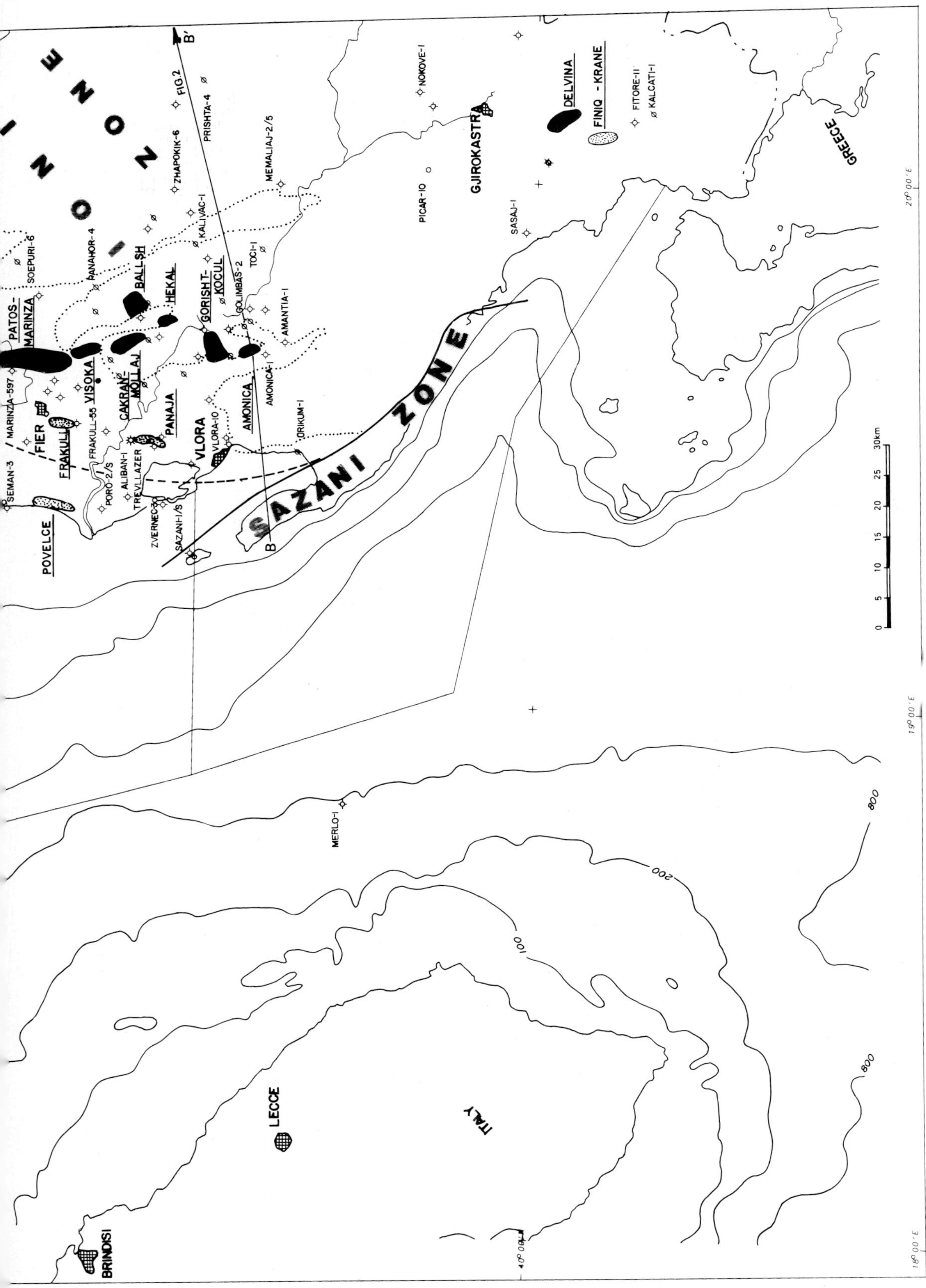

Fig. 3. Oil and Gas field map (After Petroconsultants FSS map)

– EXPLORATORY WELL SYMBOLS (of Fig. 3, pp. 4 and 5) –

Gas well | Oil shows
Dry well | Target not reached

– FIELD SYMBOLS –

Gas or gas and condensate field | Oil field

– OTHER SYMBOLS –

Limit of offshore blocks | Concealed front of the Ionian Zone
Eastward extension of the Durres basin | Isobath in meters

volumes of evaporites and anoxic sediments were deposited.

Late Syn-Rift Unit. In the Early Jurassic, the Ionian Zone started to increase in subsidence and was connected with open sea areas. From the Upper Jurassic until the Neocomian, the sedimentation was mostly pelagic.

Extensive normal faulting, having a westward and southwestern trend, was very active from Middle Jurassic to Lower Cretaceous. It was associated with the widening of the Tethys. A result was stronger halokinesis and the occurrence of basic intrusives.

Compressional Unit. The Ionian Zone was subject to compressional movements from Late Cretaceous to Quaternary. The transition from the extensional to the compressional regimes is likely to have occurred initially through the activation and the inversion of the listric faults (Flores et al. 1991).

From Oligocene to Pliocene, stronger compressional events gave rise to asymmetrical anticlines and thrusts (Figs. 4, 5). During the Miocene, Styrian phase, the subsidence was associated with thrusting, uplift of intermediate and external blocks and Triassic evaporite diapirism.

2.1.3 Durres Basin

The basin is located in the west-central part of Albania, but extends mainly offshore in the Adriatic Sea. About 80% of the whole basin lies offshore and is undrilled in Albanian waters (Figs. 1, 3). The onshore portion has so far proved to be the most productive gas province of Albania.

The Durres Basin is a young basin which comprises sediments of Neogene to Quaternary age and is a part of the larger South Adriatic "Clastic" Basin. The basin is wedge-shaped and towards the southeast overlies the Ionian Zone (Figs. 4, 5) and farther east the more internal Kruja Zone.

The first compressional events that occurred in the basin are very likely of Middle Miocene age. The Middle Miocene thrusting and uplift of the Outer Albanides was probably very intense and closely tied with a process of erosion, especially in the frontal part of many compressional structures. The Upper Miocene folds of the Durres basin that appeared as a result of the intense tectonic activity, have a north-south, en-echelon trend with a progressive shift towards north-northeast.

These folds were again affected by the Upper Pliocene and Pleistocene compressional events that gave rise to thrusting and asymmetric folding, settled conformably or independent of the main, Styrian phase, overthrusts (Dalipi 1985).

By the end of Messinian and during the Early Pliocene, the post-tectonic deposits of the Durres Basin accumulated on the moving Ionian and Kruja thrust zones.

2.1.4 Korca and Burrel Basins

These post-tectonic basins are elongated depressions filled with Tertiary clastic. They lie over the Inner Albanide ophiolite nappes in eastern Albania (Fig. 1).

While the 20-km-long Burrel Basin has only a thin Tortonian sedimentary fill, Korca is a 50-km-long basin filled with up to 3500 m of Eocene to Quaternary clastics. Both basins have a simple, faulted, synclinal section. In the Korca Basin extensive block faulting took place to the Messinian. Some Mesozoic half grabens were preserved below the Tertiary cover.

Two major sedimentation cycles have been distinguished in the Korca Basin: a Lower Paleogene Flysch and an Upper Neogene Molasse. The Paleogene flysch-type depositional sequence covers unconformably the Cretaceous or the ophiolite basement. Compressive Styrian movements created a series of uplifted reverse faulted blocks (e.g. Berzeshte Area) towards the front of the ophiolite nappe.

The transgressive Mio-Pliocene molasse was preserved only in the central narrow part of the Korca Basin and represents the whole sedimentary fill of the Burrel Basin. It may reach up to 3 km in thickness, in southeast Korca, the deepest part of the basin. Towards the northwest, the Tortonian clastics unconformably cover the ophiolites, the Mesozoic and Paleocene sediments.

2.2 Stratigraphy

Three main sedimentation sequences (cycles in Albanian literature) were identified in the geological evolution history of the Outer Albanides and their foreland (Dalipi 1985). The general stratigraphy and related events are presented in Fig. 6.

Upper Triassic-Aquitanian Cycle. This comprises a transgressive series deposited between Upper Triassic and Lower Cretaceous (Neocomian) and a regressive series which lasted from Hauterivian to Aquitanian.

Albeit older deposits may exist in the subcrop of the studied areas, the first transgressive series known is represented by Triassic evaporite deposits (Halogene Formation) which can exceed a thickness of 3500 m. They consist of gypsum, anhydrite, halite and dolomite.

The Halogene Formation is overlaid by a series of predominantly carbonate deposits (Carbonate Formation) covering the Upper Triassic–Eocene chronostratigraphic interval. The Carbonate Formation includes a wide variety of carbonate facies ranging from subaerian to deep water. Slope and deep water facies have a higher associated clastic component.

The Flysch and Flyschoidal Formation is Oligocene to Lower Burdigalian in age. It apparently overlays conformably the Eocene members of the Carbonate Formation. This major change in the depositional regime was controlled by the sediment supply from an eastern, Albanide, source area and was the result of the increased Alpine tectonic activity. During the period of flysch deposition, the facies changed to shallow marine and continental towards east and south.

Burdigalian-Lower Tortonian Cycle. This comprises a transgressive and a regressive series, grouped into the Premolasse or Shliri (Schlier) Formation.

The Premolasse Formation is made up of an upward fining sequence of clays, marls and siltstones over the basal conglomerates and sands. Local olistolite, shallow water limestones or deep water slump structures have been noted.

Upper Tortonian-Pliocene Cycle. This is also called the Molasse Formation. This sedimentation stage is divided into two depositional sequences: Early Molasse and Late Molasse.

The Early Molasse (Upper Tortonian-Messinian) deposits are associated with the regional transgression Tortonian Sea. They were divided into: Bubullima, Marinza, Driza, Gorami, Kucova and Polovina members.

The Messinian regressive event was marked by the presence of gypsum bearing clastics only in the west and northwest Durres Basin. Elsewhere, typical molasse clastics of mixed marine and brackish water facies developed.

The late Molasse (Pliocene) is made up of two formations well developed in the Durres Basin: Helmesi and Rrogozhina.

The transgressive Helmesi Formation began with basal conglomerate passing upward to siltstones, marls and clays, locally with sandstone intercalations. In the regressive Rrogozhina Formation the proportion of coarse clastics increases.

3 Hydrocarbon Habitat

Due to the differences in the tectonic and sedimentary evolution, the resulting conditions of hydrocarbon accumulation will be presented by major productive sedimentary province and prospective area.

3.1 Sazani Zone

More detailed data concerning the hydrocarbon potential of the carbonate sequences are available for the Upper Cretaceous-Paleocene rocks which outcrop in the area of Mali-Kanalit hills. The geochemical analysis performed on the samples from this area showed a low content of organic matter (less than 0.2%).

The only well drilled in the area, Sazan 1s, which reached TD of 3932 m was abandoned dry in 1982. Some 3400 m of carbonates (dolomites and limestones) were encountered, ranging in age from Upper Triassic to Lower Cretaceous. The transgressive Lower Miocene (Burdigalian) shales account for some 580 m. A well drilled in the Durres

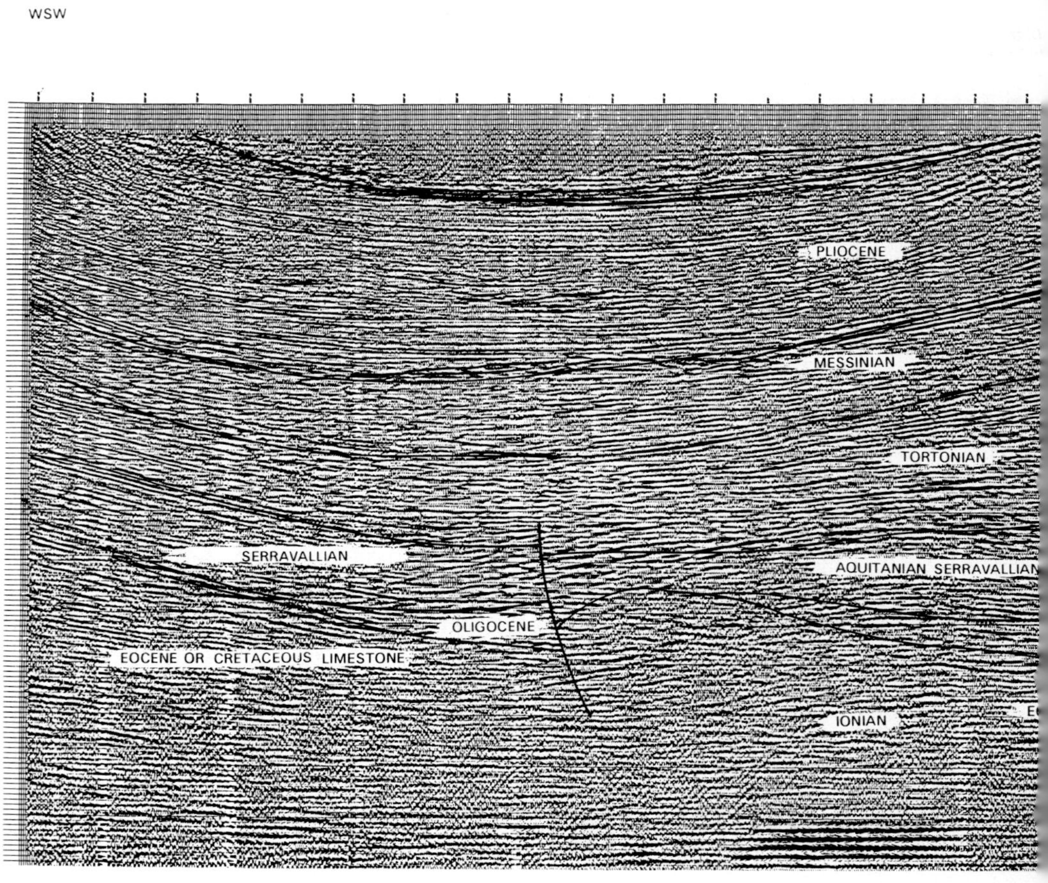

Fig. 4. Seismic profile 127/89. Speculative structural interpretation (see Fig. 3 for location)

Basin, Zvernec-3, encountered some shows in the basin floor carbonates belonging to the Sazani Zone.

The source rocks are likely to be located in Triassic and Early Jurassic. Bitumen shows in the outcrops led to the conclusion of a migration process through the carbonate sequences. The maturation of the organic matter is assumed to be recent (Pliocene to Pleistocene), and should be due to the overloading of the Neogene deposits and to the ongoing overthrusting of the Outer Albanides over the Inner Albanides.

3.2 Ionian Zone

This is the major petroliferous basin of the country. Within the Ionian Basin the following carbonate Mesozoic source rocks can be distinguished (Curi and Stamuli 1990):

1. Upper Triassic dolmitic shale and bituminous shale.
2. Lower Jurassic (Lower Lias) dolomitic shale.
3. Toarcian calcareous shale.
4. Lower Cherty Member (Dogger).
5. Upper Cherty Member (Malm).
6. Clayey micritic limestone (Lower Cretaceous).

Geochemical studies indicated that the hydrocarbons generated from these source rocks are of algal and mixed marine origin (types I and II). All the source rocks have a relatively high level of maturation and the highest values are encountered in the northwestern part of the basin. A schematic graph showing the depth versus age distribution of the source rocks in this basin is shown in Fig. 7.

The source rocks contain significant amounts of oil-prone kerogen, composed largely of amorphous alginite and leptinite material, and the Total Organic Carbon (TOC) content varies from 1% in Toarcian calcareous shale to 26% in bituminous shales (Table 1).

Presentday average geothermal gradient within the basin is estimated at 2 °C/100 m (Flores et al. 1991), and is likely to be higher in the northwestern

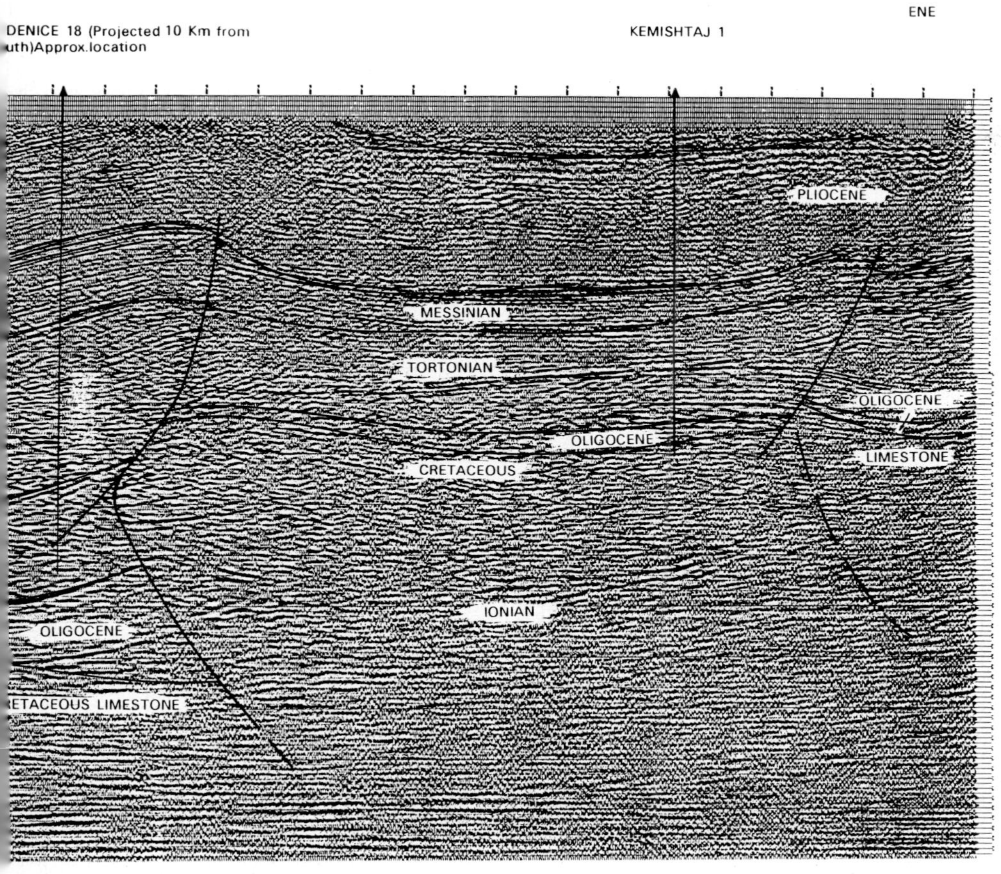

part, at the contact with the Durres Basin. The heat flow variations are thought to reflect the shape of the overlaying Durres Basin on the Ionian Zone and the depth of the basement. The vertical migration of oil has occurred along the main fractures and the faulting systems that developed in the limestones. The lateral migration in the eastern marginal areas of the Ionian Zone was facilitated by the low-amplitude fault pattern.

The maturation of the organic matter, which very likely started during the Late Cretaceous, presents a relatively non-uniform depth zonation throughout the Ionian Zone.

It was possible to distinguish three oil generation and migration phases: the first phase originates in Burdigalian, the second in Tortonian, and the third phase coincides with the overlapping of the Ionian Zone by the Durres Basin sediments which led to the oil generation and migration from Pliocene to Early Pleistocene.

Reservoirs are the Upper Cretaceous–Eocene carbonates sealed by the shales of the Flysch and Flyschoid Formation or by the Tortonian and Messinian shales.

3.3 Durres Basin

The molasse deposits of Upper Miocene (Tortonian) and Pliocene include large amounts of shales, which are considered to be the source rocks of the gas fields in the Durres Basin. The TOC varies from 0.2 to 0.4%. The present potential (S_2) can reach a value of 0.15 and the Hydrogen Index averages 50.

The organic matter is of a continental lacustrine origin (Type III) and so far, has proved to be gas-prone only. The low values of the parameters mentioned above and of the vitrinite reflectance suggest that the gas is of biogenic nature.

Geochemical analyses run on well samples offered the following data concerning the organic matter composition: amorphous materials 4%; phytoplanton 0.6%; herbaceous origin 12%; high plants and woody material 46%; inertinite 36%.

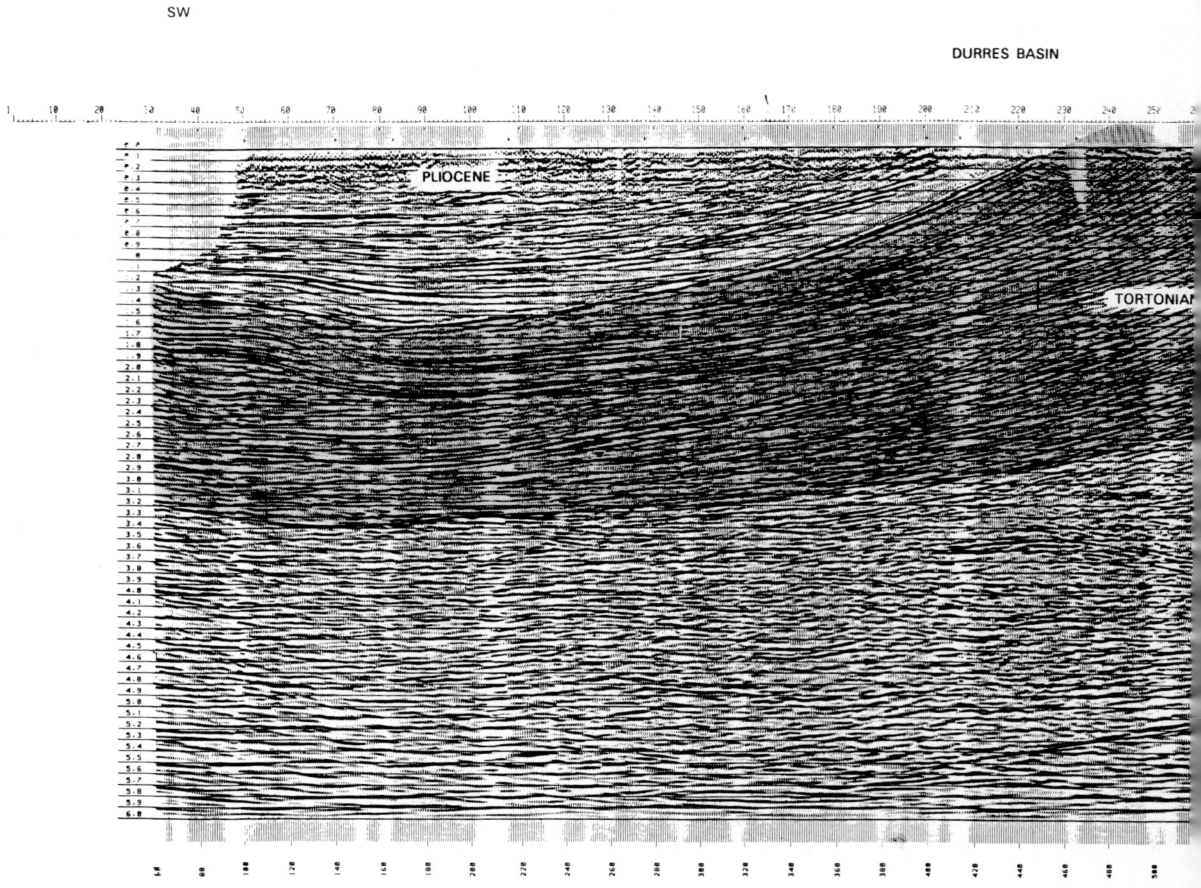

Fig. 5. Seismic profile 15/88. Speculative structural interpretation (see Fig. 3 for location)

The gas produced from six fields is dry and the methane content is over 96%. The ratio of $\delta^{13}C$ methane ranges from 55 to 70‰.

Durres Basin oil fields were most probably sourced by lateral or vertical migration from formations belonging to the Ionian Zone, underlying the reservoirs of the molasse cycle. Different genetic types of gas show an uneven depth distribution in the basin (Fig. 8). A very good correlation between the oils from sandstone reservoirs of the Durres Basin and those of carbonate reservoirs of the Ionian Zone have been noted (Curi 1991).

Hydrocarbons accumulated into the Tortonian and Messinian sands and sandstones. They were sealed by the finer clays and siltstones of the same age.

3.4 Kruja Zone

The regional studies carried out in the area disclosed a moderate to good prospect for hydrocarbons. However, all the data concerning the Kruja Zone have to be regarded with a certain reserve, taking into account the fact that the area is practically unexplored as yet. A positive aspect is provided by the oil seeps in the Tortonian, deposited transgressively over the eroded limestrones of the Kruja Zone. Moreoever, the oil seeps give a certain assumption of oil maturation and generation.

In the outcrops three main potential source rocks were differentiated:

- Lower Oil Shale Member (Bituminous shales) of the Carbonate Formation (Upper Cretaceous).
- Upper Oil Shale Member of the top of the Carbonate Formation.
- Shales and the shaly limestones of Eocene age.

The TOC varies from 0.5 to 6%, and the organic matter is both algal and mixed marine (types I and II). The content of the amorphous material is up to 80%.

Geochemical studies carried out on samples from

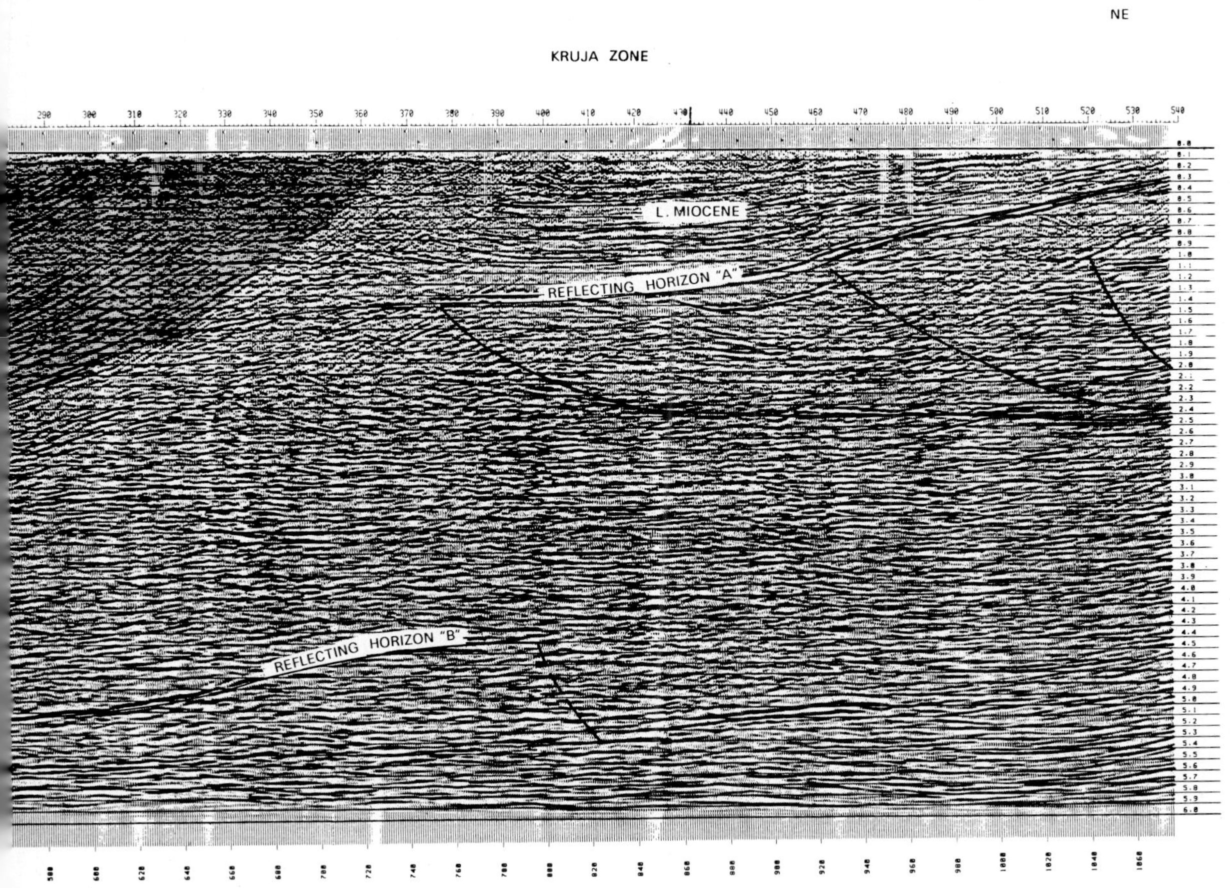

outcrops show that the maturity of source rocks is typically in the range $R_o = 0.3–0.5\%$.

It is obvious that at greater depths, where the organic matter was not subjected to biodegradation, the maturation is likely to be higher. Adequate reservoirs and seals can be found in the Carbonate and Flysch formations.

4 Main Albanian Oil and Gas Fields

According to geochemical characteristics, the lithology and age of their reservoirs, two groups of fields have been differentiated in Albania: Cretaceous-Paleogene and Neogene.

4.1 Major Plays

Most of the country's fields are located in structural and combination traps. Two significant plays, each typical of one of the productive provinces of Albania, have been described so far.

4.1.1 Upper Miocene-Pliocene Clastic Play

The clastic play is characteristic of the oil and gas fields from the Durres Basin. It could be one of the main targets in the Korca Basin. Proven reservoirs range in age from Tortonian to Messinian with the exception of one field that produces from the Pliocene clastics. The stratigraphic traps are in sand pinch-outs and sand lenses. Combination traps were also reported. Seals are the intrafomational shales.

Reservoirs are deltaic and alluvial plain channel clastics charged with hydrocarbons generated by the Ionian Zone source rocks or biogenic gas.

4.1.2 Upper Cretaceous-Eocene Carbonate Play

The carbonate play comprises a wide range of carbonate reservoirs, mainly limestones with fracture porosity. It is characteristic of the Ionian Zone and it is likely to be found in the Kruja and Sazani zones also.

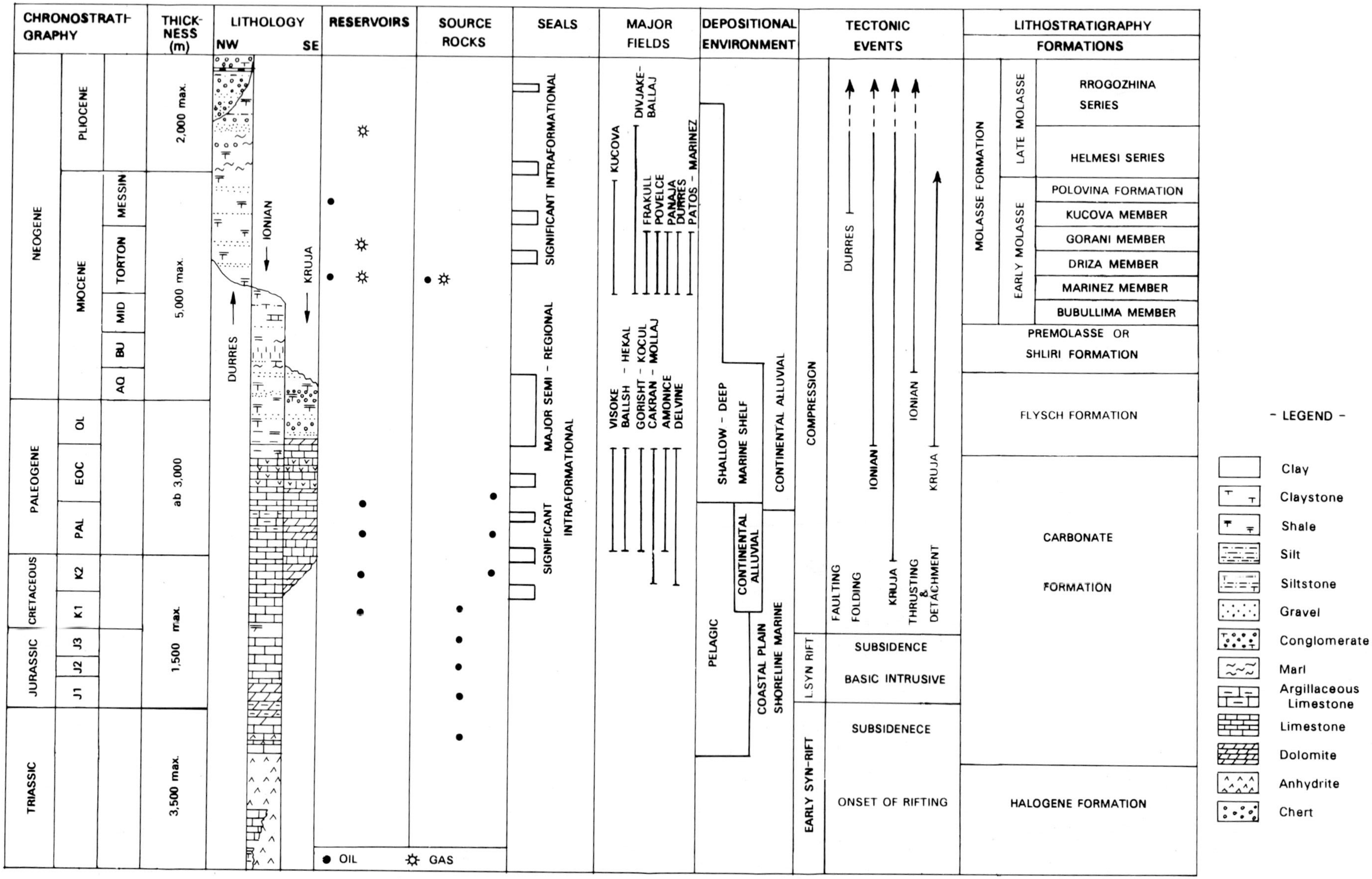

Fig. 6. Generalized stratigraphy and main tectonic events of Durres Basin, Ionian and Kruja zones. (Partial data from Gjoka et al. 1990)

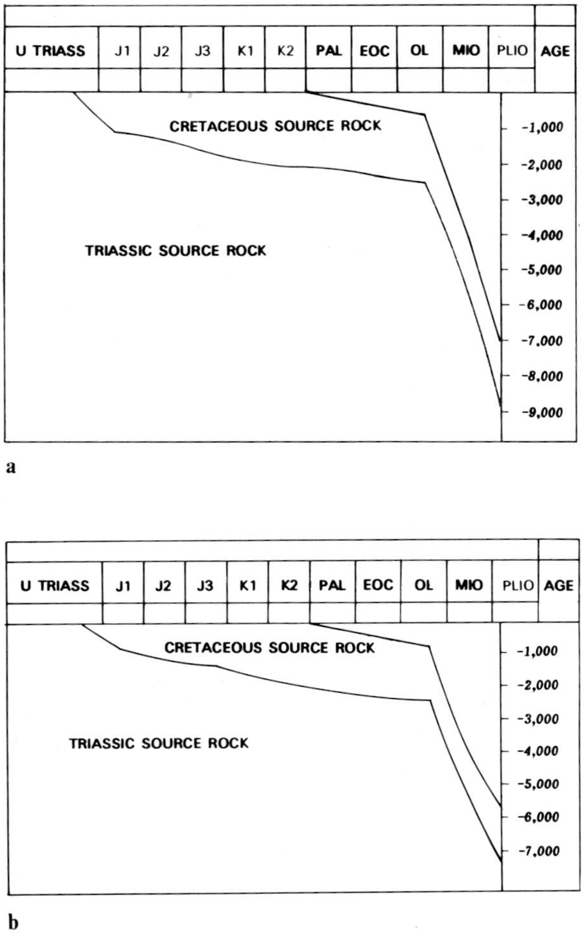

Fig. 7a,b. Age versus depth graph showing the source rocks burial in the Ionian Zone. **a** SE Ionian Zone **b** NW Ionian Zone. (Curi 1991)

Traps consist of fault-controlled anticlines, and the reservoirs (mainly of massive type) are usually sealed by the fine clastics of the Flysch Formation. In some cases the trap is of hydrodynamic type or lithological-hydrodynamic.

Some of the uplifted features were subsequently eroded by marine transgression and channelling. The younger Neogene transgressive sequences provided the seal for the reservoirs.

A brief description of the main fields together with their geological sections is given below, while a few main characteristics are presented in Table 2. The areal distribution of the fields within the Durres Basin and the Ionian Zone is shown in Fig. 3.

4.2 Neogene Oil and Gas Fields

Neogene oil and gas fields were found so far only in the Durres Basin. The Neogene (Tortonian-Pliocene) play occurs in seven fields (two oil and five gas) and accounts for most of the gas and over 50% of the total oil production of the country. All fields are in sandstone and sand reservoirs.

Exploration and production of the Neogene gas fields are confronted with serious problems due to high and differential pressures.

4.2.1 Patos–Marinza

This field is situated in the southeastern part of the Durres Basin and is represented by a monocline structure (southeast updip), with sandstone layer pinch-outs (Fig. 9).

The depositional sequence is of onlap–toplap type, and shows a moderate to high sedimentation rate.

Within the field three main oil and gas bearing formations were separated from bottom to top: Marinza-2, Marinza-1, and Driza. The oil/gas contact is at about 1500 m in Marinza-2 and 1340 m in Marinza-1. The field was discovered in 1928, has in-place reserves of 240 million tonnes and a cumulative production at the end of 1990 of some 15.35 million tonnes.

4.2.2 Kucova-Arza

These fields are located in the central eastern part of the Durres Basin and covers some 14 km^2. The reservoirs are represented by sand lenses, sealed by shales (Fig. 10). The length of sand lenses ranges between 100 to 1500 m, and they dip towards the centre of the field.

The following productive pay zones were identified: Driza, Gorani, Kucova and Polovina. As of end 1990, the field produced 3.15 million tonnes of oil and 62 million m^3 of gas. In-place reserves are estimated at 68 million tonnes oil and 680 million m^3 gas.

4.3 Cretaceous–Paleogene Oil and Gas Fields

These fields are situated only in the Ionian Zone, and the reservoirs are in the carbonate play. The oil in the Ionian Zone is heavy, viscous and sulphurous, with the exception of the Cakran field.

Table 1. Characteristics of organic matter from potential source rocks

Average values of the geochemical parameters of the potential source rocks

Structures	Age	Lithology	TOC (%)	Hydrogen index (HI)	Current potential (s_2)	T_{max}	Vitrinite refl. (R_o)	λ Max (nm)	TAI
		Bituminous shale	26.01	700	181.5				
	Cr1–Cr2	Argillaceous lst., shale	0.8	652	5.2	413	0.48	560	0
		Argillaceous shale	1.5	520	7.81				
	J3	Argillacous lst.	0.57	400	2.1	430	0.51	570	2.5 to 3
		Bituminous shale	5.25	508	26.5				
Gjere Mountain	J2	Marl, alligaceous lst.	1.07	411	4.4	432	0.52	590	2.5 to 3
		Bituminous argillaceous shale	4.9	588	28.8				
	J1–3	Marl, clayey shale	1.1	380	4.2	432	0.55	550	2.5 to 3
		Marly clay shale	1.3	417	5.51				
	J1–2	Micritic lst.	0.14	401	0.41	432	0.53	582	2.5 to 3
	J1–1	Bituminous dolomitic shale	15.66	450	90	434	0.55	600	2.5 to 3
		Argillaceous shale	1.14	443	5.03				
	Cr1–Cr2	Micritic lst.	0.3	300	0.12	432	0.3	577	2.5 to 3
	J3	Argillaceous lst.	0.49	126	0.62	430	0	0	0
Ftere	J2	Marly lst.	0.49	343	1.51	430	0	605	2.5 to 3
		Marly shale	1.3	380	4.52				
	J3–1	Marly lst.	0.8	546	7.14	428	0.46	600	2.5 to 3
		Bituminous dolomitic shale	10.84	513	55.71				
	J1–1	Dolomite	0.22	170	0.4	435	0	602	2.5 to 3

Kurvelesh		Bituminous shale	25.95	615	159.2				
	Cr1–Cr2	Marly shale	1.67	354	6.23	424	0.42	541	2
		Argillaceous shale	5.22	527	27.6				
	J3	Argillaceous lst.	0.82	431	3.51	430	0.47	580	2.5 to 3
		Bituminous shale	5.32	552	29.35				
	J2	Shale lst.	1.1	451	4.97	432	0.53	580	2.5 to 3
Kremenare	Cr1–Cr2	Argillaceous shale	0.83	0	0	0	0.5	597	2.5
	J3	Argillaceous micritic lst.	0.12	555	0.63	422	0	0	0
	J13	Marl	1.39	523	7.27	415	0	0	0
Ballsh		Argillaceous bituminous shale	6.33	274	1.74				
	Cr1	Shale, argillaceous lst.	1.01	522	5.24	424	0.42	0	0
	J3	Dolomitic marl	7.08	630	26.8	420	0	0	0
	J2	Dolomite	1.96	571	11.21	420	0.35	0	0
Patos/Verbas	Cr1	Marl	4.84	586	16.82	414	0	0	0
	J3	Dolomite	1.01	510	5.21	420	0	0	0
	J2	Dolomite	2.6	620	16.1	420	0.37	571	0
Cakran	Cr1	Shale, arillaceous lst.	5.72	590	33.17	430	0.42	0	0
Gorisht/	Cr1	Argillaceous shale	8	569	45.5	407	0.41	0	0
Kocul	J13	Dolomitic marl	1.86	546	10.17	417	0.34	590	0

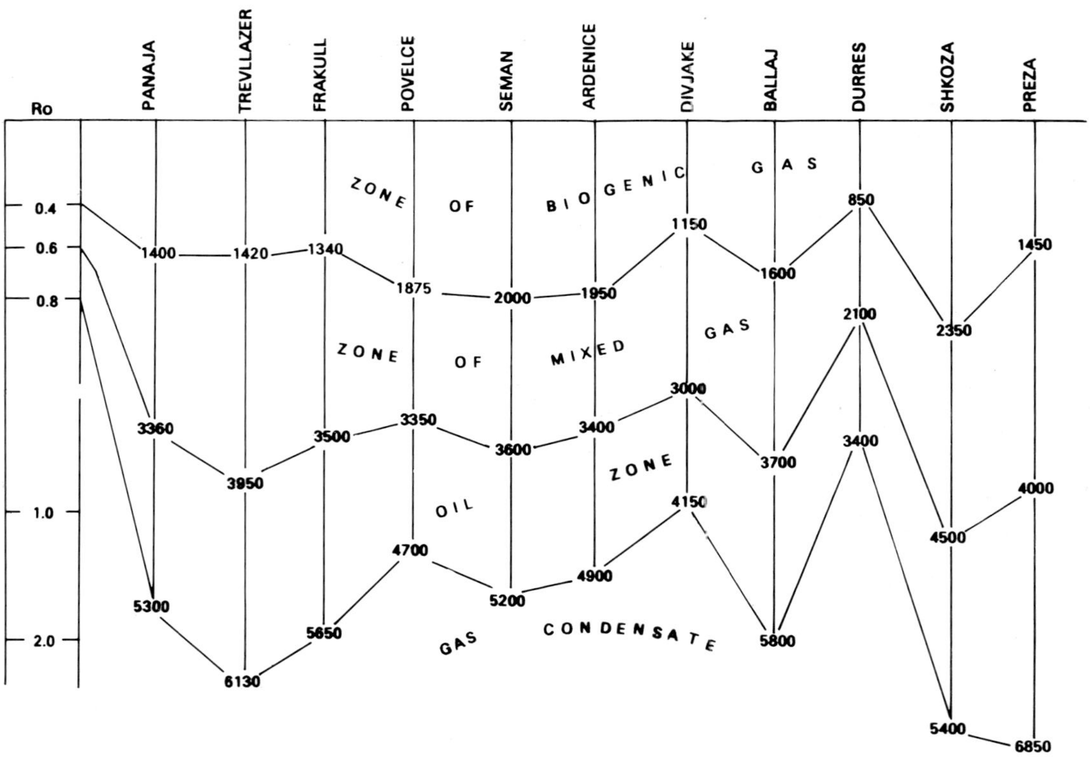

Fig. 8. Depth zonation of hydrocarbons in the Durres Basin. (Curi 1991)

Table 2. Characteristics of main Albanian oil and gas fields

a) OIL FIELDS

Reservoirs	Discovery year	Reservoir age	Reservoir lithology	Perm. (md)	Porosity (%)	API (degrees)	Cumulative production MMbbl[a]
Driza	1928	Tortonian	Sandstone	800	22.5	9 to 12	64.6
Marinza 1	1957	Tortonian	Sandstone	550	24.2	9 to 24	28
Marinza 2	1957	Tortonian	Sandstone	350	25.5	16.5 to 35.5	19.3
Kucova	1929	Tort.-Messin	Sandstone	70	22.8	14.5 to 22	23.3
Visoka	1963	K-Pg	Limestone	200	4 to 5	5 to 16	37.6
Gorisht-Kocul	1965	K-Pg	Limestone	300	4 to 5	13 to 16	78
Ballsh-Hekal	1966	K-Pg	Limestone	40–300	4 to 5	13	32.8
Cakran-Mollaj	1977	K-Pg	Limestone	2–600	4 to 5	11.5 to 37	19.3
Delvina	1987	K-Pg	Limestone	0.2	3 to 5	23.5 to 26	2.5

b) GAS FIELDS

Reservoirs	Discovery year	Reservoir age	Reservoir lithology	Depth to top pay	Cumulative production m^3[b]	Remarks
Divjake	1963	Tortonian	Sandstone	2200–3100	1,371,420,340	Abnormal pressure
Frakull	1965	Tortonian	Sandstone	700–2500	240,297,804	
Divjake	~1970	Pliocene	Sandstone	500–1300	371,461,455	
Finiq	1970	Paleogene	Limestone	200–2000	220,114,621	Gas and condensate
Ballaj	1983	Pliocene	Sandstone	500–1300	331,227,778	
Povelce	1987	Tortonian	Sandstone	1700–3500	159,625,120	Abnormal pressure
Panaja	1989	Tortonian	Sandstone	2500	16,368,846	
Durres	1989	Tortonian	Sandstone	2200	6,649,490	

[a] 1.1.1991. [b] 1.1.1990.

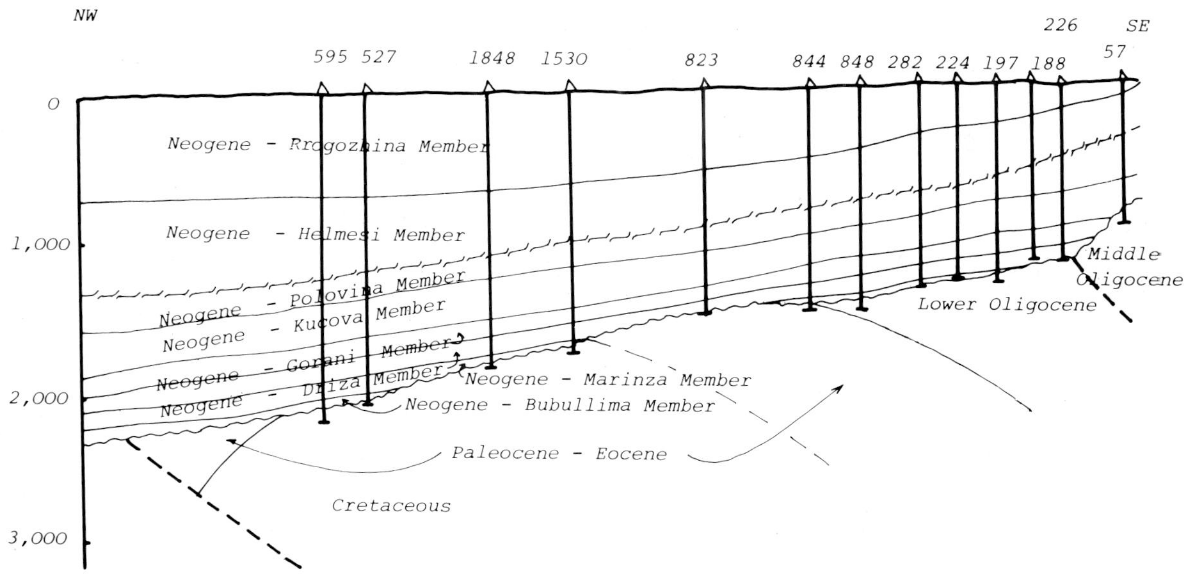

Fig. 9. Simplified geological section through the Patos-Marinza oil field.

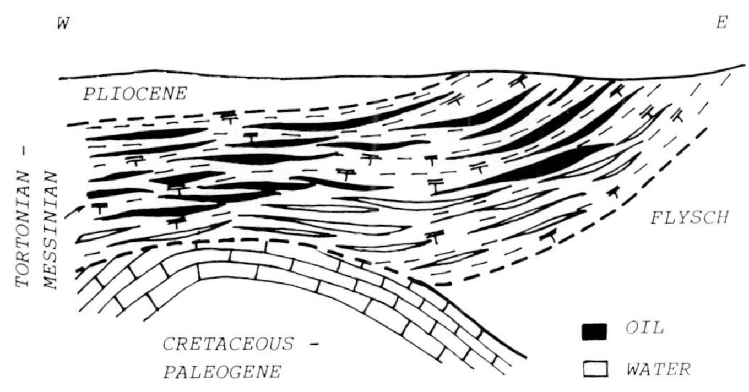

Fig. 10. Simplified geological section through the Kucova oil field

The largest fields are located inthe northwestern part of the Ionian Zone, i.e. Visoka, Ballsh, Gorisht-Kocul and Cakran.

4.3.1 Visoka

The field, which is in a mature stage of exploitation, is situated 13 km southeast of Fieri on the Patos-Verbas structure (Figs. 11, 12). The trap is represented by an anticline sealed by the Oligocene Flysch Formation. Laterally the trap is of hydrodynamic type. The Visoka field has a tilted oil-water contact: 1060 m in the southern part, 1980 m in the northern part. Cumulative production as of end 1990 amounted to 5.15 million tonnes of oil and 193 million m^3 of gas with in-place reserves of 73 million tonnes oil and 255 million m^3 gas.

4.3.2 Ballsh and Hekal

Located 29 km southeast of Fieri, the fields are represented by two superposed faulted anticlines (Figs. 12, 13). The Ballsh field is rather shallow with a gas/oil contact at about 730 m depth. The oil column is over 550 m thick. As of the beginning of 1991 the cumulative production was 4.5 million tonnes of oil and 220 million m^3 of gas. The production is very low because of the high water content (especially in the southern part). The only solution that would make an increase in production possible appears to be the application of re-

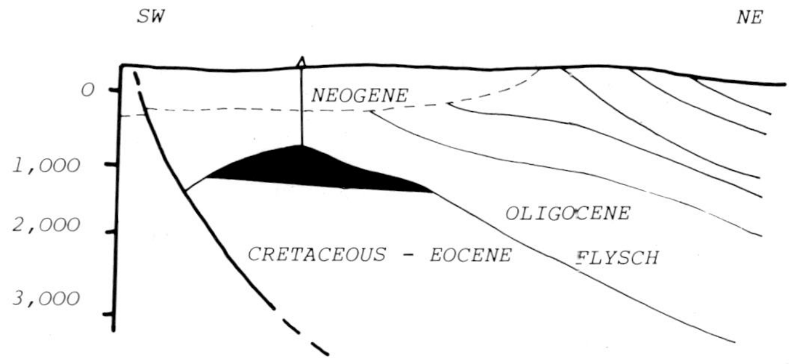

Fig. 11. Simplified geological section through the Visoka oil field

servoir stimulation with solvents and polymers. In-place reserves are estimated at 63.5 million tonnes of oil and 700 million m^3 of gas.

4.3.3 Gorisht-Kocul

This field is situated 20 km east of Vlora (Figs. 14, 15). The reserve estimates have been revised considerably since the producing area actually covers only 4.5 km^2 against the initial estimation of 9.7 km^2. They are now estimated at 154 million tonnes of oil and 635 million m^3 of gas in-place reserves. Exploitation of this field is characterized by the high density of wells.

4.3.4 Cakran-Mollaj

Located 6 km to the west of Ballsh, the field produced so far some 2.7 million tonnes of light, sulphurous and asphaltic oil and 8600 million m^3 of associated gas. It also provided a fairly good gas condensate production, which decreased dramatically after the blow-out of the well Cakran-37 seriously damaged the reservoir. The field is represented by an asymmetric anticline, controlled by a major fault to the west (Figs. 16, 17).

5 Exploration History

The presence of bitumen in Albania has been known since Roman times. Various occurrences have been reported from the coastal plain from Durres to Vlora. Asphalt was mined from the Pliocene clastics in the vicinity of Selenica and Romesi. First production of up to 7000 tonnes/year was recorded at the beginning of the century (Macovei 1938). The first wells were drilled after World War I and seismic work was carried out only after World War II.

5.1 Geophysical Exploration

5.1.1 Onshore Seismic

Three main periods can be distinguished in the evolution of the seismic surveys:

1951–1975. A total of 3920 line-km were shot in the Durres Basin, mainly in the Vlora-Ishem area. The operations were carried out by two to four seismic crews using Soviet-made oscillograph recording stations with 24–60 channels, one-fold coverage. In the field off-end shooting was used; the shot holes averaged 20 to 30 m. The seismic data are of poor quality and have only historical value.

First experiments with analogue recording stations were carried out between 1970 and 1975. In this period, the volume of the acquisition was very low.

1976–1980. Multifold coverage was the main feature of this period when 5500 line-km were recorded using Chinese-made analogue equipment DZ-663 type with 24 channels, with 24 geophones/channel (six-fold coverage).

Six seismic crews operated every year, about 66 crew-months/year; the depth of the shot holes drilled in the flysch area for the dynamite source averaged 20 to 30 m. In the flat areas, the wells were shallower.

Characteristic for this period are the extensive surveys carried out in rugged areas. Each seismic program was accompanied by experimental studies

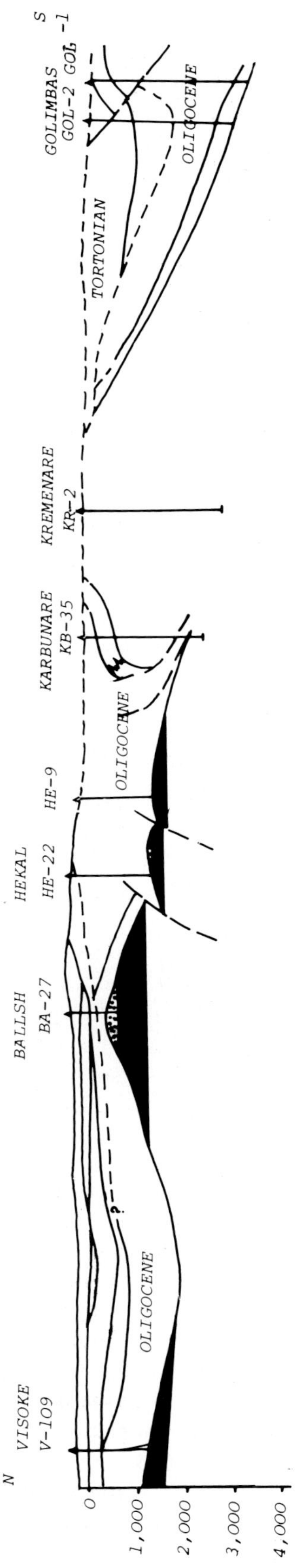

Fig. 12. Simplified geological section through the Visoka–Ballsh–Hekal–Golimbas structures

(noise pattern of geophone array, the pattern of sources).

1981–1990. First experiments using digital recording were carried out during the 1980s. About 10,000 line-km were shot during this period and about half of these recordings were digital. Activity came to a standstill in late 1991.

There were five seismic crews operating (about 55 crew-months/year), two of them using digital recording. Digital equipment is of SERCEL type (SN-338 HR), with 48 channels. Seismic operations were concentrated mainly in the prospective areas such as the Ionian Zone and Durres Basin and less in the Kruja Zone.

The cumulative, post-Second World War seismic work resulted in a very dense grid in the Durres Basin, usually from 500 × 500 m to 1 × 1 km. In the Ionian Zone and partly in the Kruja Zone, the seismic grid is less dense, from 1 × 1 to 2 × 2 km.

5.1.2 *Offshore Seismic*

Regional to semi-regional seismic surveys were carried out offshore on the coast line, in an attempt to improve the understanding of the deep structure. Another purpose was the delineation of the Neogene basin and its morphology towards the offshore area.

First seismic testing offshore was initiated by the Seismic-Gravimetric Enterprise in 1976 near the Lalzi Gulf, and was followed by the first survey started in 1980.

About 1200 line-km were shot between 1980 and 1991, of which 50% have a good signal-to-noise ratio. The *Naftetari* vessel used a 48-channel analogue recording station with 12 hydrophones per channel (24-fold coverage) and had a dynamite energy source. The results of the survey indicated prospective structures in the Vlora, Durres and Rodon areas, where a number of leads have been delineated.

Foreign companies ran about 8,200 line-km in their offshore holdings during the late 1991–early 1992 period.

5.1.3 *Gravity*

First gravity measurements were carried out before 1944 by the Italian company AIPA near Kucova and Drashovice. The gravity surveys resumed after

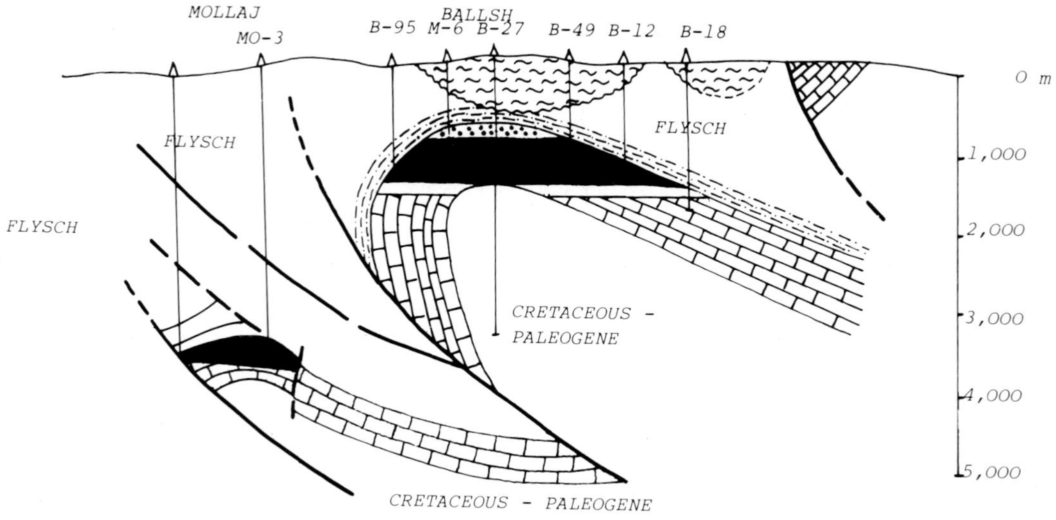

Fig. 13. Simplified geological section through the Ballsh and Mollaj oil fields

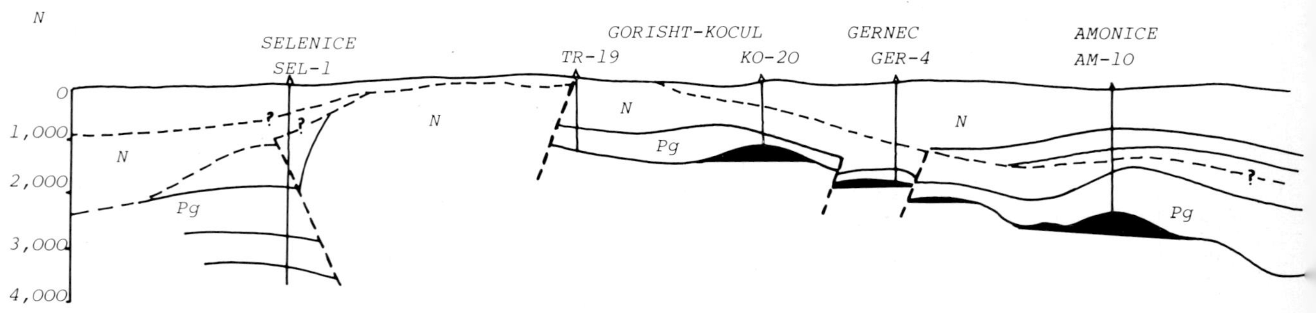

Fig. 14. Simplified geological section between Selenica (*N*) and Amonica (*S*) structures

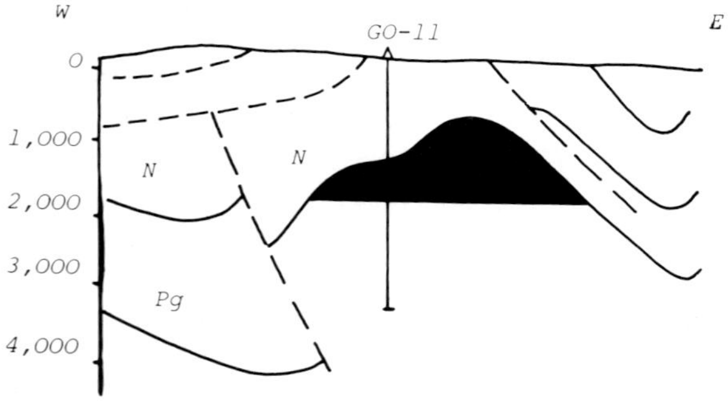

Fig. 15. Simplified geological section through the Gorisht–Kocul field

the World War II and two periods can be distinguished.

1951–1962. Gravity operations started with Russian technical assistance in 1951 and covered the Durres Basin mainly the area between Vlora and Durres. The results were summed up in a chart of the Bouguer anomaly, at a scale of 1:100,000. The gravimeters used were Soviet-made GAK-3M.

1963–1991. About 22,000 line-km of such surveys were carried out during this period, covering all of the Outer Albanides, Burrel and Korca Depressions at a scale of 1:50,000.

In addition, various areas in the Durres Basin and in the Ionian Zone were covered using detailed survey, such as 1:25,000 or 1:10,000. Different detailed surveys were carried out in Inner Albanides with 1:10,000 and 1:50,000 surveying. Chinese CG-2 and Canadian SZ-2 gravimeters were used.

No offshore gravity surveys have been carried out so far, but some 27,000 km of satelite gravimetric line is available.

All surveys were summed up in the State Gravity Network, and a Bouguer map for Outer Albanides at a 1:50,000 scale covers all sedimentary basin areas.

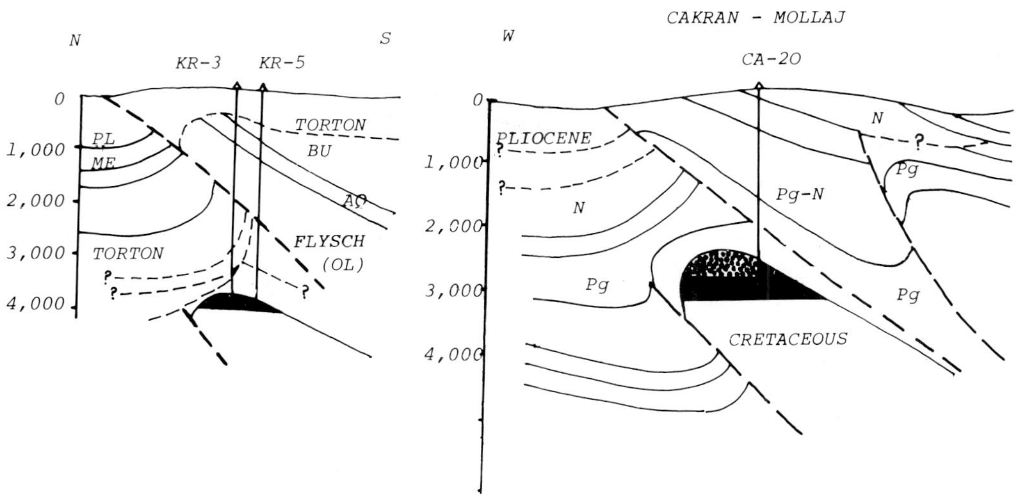

Fig. 16. Simplified geological section through the Cakran-Mollaj field

Foreign companies run about 6200 line-km in their offshore holdings during the late 1991–1992 period.

5.1.4 Electric Methods

A limited volume of data was gathered in the period from 1976–1989 using resistivity methods. Electrical methods were used to identify shallow features near the salt-flysch contacts, shallow carbonate targets between 1000 and 2000 m and sand/shale sequences from the Pliocene deposits.

5.2 Drilling and Production

As a result of a few geological studies carried out by the Ministry of the Italian Navy in Southern Albania, between the rivers Vjosa and Vlora, the first well was spudded at Drashovice in 1918. The well was drilled to a total depth 200 m, encountered oil at different levels, and yielded up to 3.5 tonnes/day of heavy oil (Lazzari 1964).

After this first success, a number of companies secured exploration permits. Between 1926 and 1930, several exploratory wells were drilled by the state-controlled Italian company Azienda Italiane Petroli Albania (AIPA) on anticlinal structures of the Durres Basin, such as Peshtan-1 (1926), Selevec-1 (1929) and Trevllazer-1 (1930).

Other wells were dirlled close to oil seeps: in the Patos Monocline Zone, by the British company Anglo-Persian Oil Co. (APOC); in the Penkove Zone by the Italian companies Societa Italiana Miniere de Seleniza (SIMSA) and AIPA; in the Korca Basin by the French company Syndicat Franco-Albanais (SFA) and in the Pekisht-Karthnek-Kucova Zone by AIPA.

In 1928 the first oil field was discovered in the Driza reservoir of the Patos-Marinza field. Here, a first well drilled to 273 m TD blew out and some 60 t/day flowed during 2 years (Macovei 1938). Further appraisal found only smaller production flows.

The first steady commercial flow was, however, recorded in the Kucova field from 1929 onwards. In 1940 the oil production from Kucova and Driza fields increased to 154,000 tonnes and fell to 25,000 tonnes in 1944 as a result of the Second World War.

After World War II, exploration resumed with the drilling of wildcats Kruja-1 and 2, Karbunare-1 and Selisht-1. In the late 1950s, exploration drilling focussed on the Neogene monocline of the Patos-Marinza area. As a result, the Marinza 1 and 2 reservoirs were discovered in Tortonian sandstones in 1957. Deeper structures identified in Ardenica, Frakull and Pekisht areas were the targets of an extensive drilling programme.

In the early 1960s, the exploration of the carbonate structures in the Ionian Zone began. It led to the discovery of the following fields: Visoka oil field, discovered by the well G-22 in May 1963, followed 2 years later by the Gorisht oil field (discovery well Gorisht-2 in March 1965), the Verri oil pool (discovery well Verri 645), Ballsh oil and gas field (1966, discovery well Ballsh-14), Finiq-Krane gas and condensate field discovered in 1970

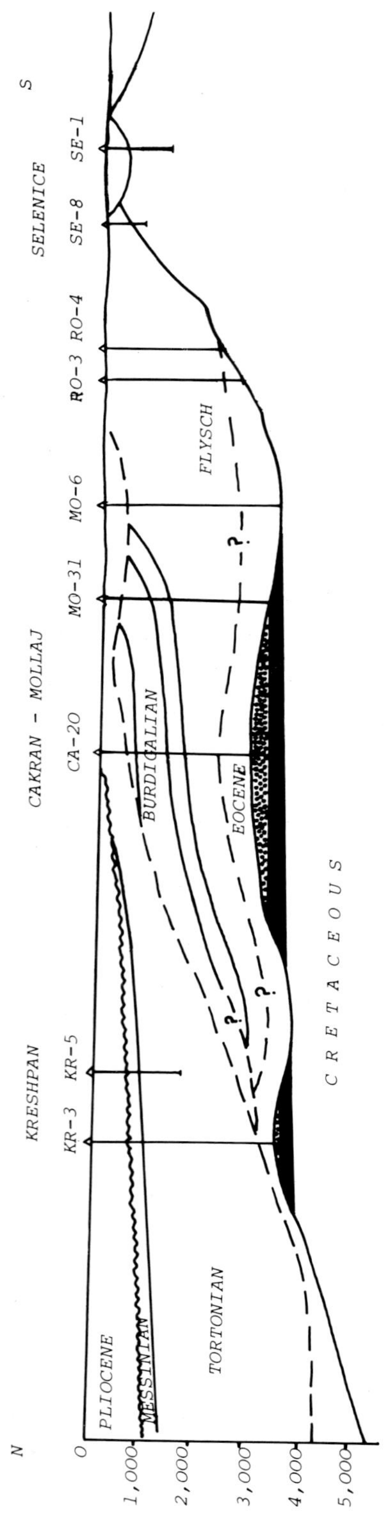

Fig. 17. Simplified geological section through Keshpan Cakran-Mollaj and Selenice structures

(Finiq) and in 1973 (Vurgu 6), Cakran-Mollaj gas, condensate and oil field (discovery well Cakran-12, August 1977), Hekal oil and gas field (discovery well Hekal-5, 1978), Amonica oil field (discovery well Amonice-7, 1980) and Delvina gas, condensate and oil field (discovery well Delvina 4, 1987).

After the Second World War, with Soviet and Chinese aid, the oil production increased gradually, and in 1960 reached some 725,000 tonnes. After 1961, operations were conducted solely by the Albanian state companies, which further increased the amount of oil produced to over 1,000,000 tonnes in 1968, the year when production from carbonate reservoirs was for the first time larger than those from the traditional clastic reservoirs. Oil production reached the peak level of 2,225,000 tonnes in 1974 (Fig. 18).

Until the late 1960s, gas production was associated gas only. Exploratory drilling carried out since 1960 in the Durres Basin led to the discovery of the following Tortonian gas fields: Divjake (discovery well Divjake-2, completed in 1963); Frakull (discovery well Frakull-11, 1965); Povelce (discovery well Povelce-1, 1987); Panaja (discovery well Panaja-10, 1989) and the Durres gas field (discovery well Durres-15, 1989). The Pliocene clastic play was discovered in the late 1960s. At the end of 1990, the cumulative non-associated gas production was 2.75 billion m^3 of gas (Table 4).

The development of the non-associated gas reserves in the Durres Basin resulted in a boost of production starting in 1968 (Table 4, Fig. 19). It reached 909 million m^3/year an all time record high, in 1982 mainly due to the blowout of well Cakran 37, and stabilized to over 200 million m^3/year in the late 1980s.

Reserves in-place are estimated at 650 million tonnes of oil and 37 billion m^3 of gas. Recovery by natural drive averages some 10% for oil, i.e. some 71.5 million tonnes and up to 80% for gas i.e. some 30 billion m^3. Considering the in-place oil reserves, even a small increase in recovery factors would bring significant increase in oil production.

Over 1000 exploratory wells and some 3500 oil and gas development wells have been drilled up to now in Albania. The evolution of drilling meterage during a recent 6-year period is given in Table 5. The deepest well in the country was completed in 1988 on the Ardenice structure and reached a total depth of 6700 m (Ardenice 18).

Since October 1991, drilling activity has practically ceased because of financial, spare-part, mud and power supply problems.

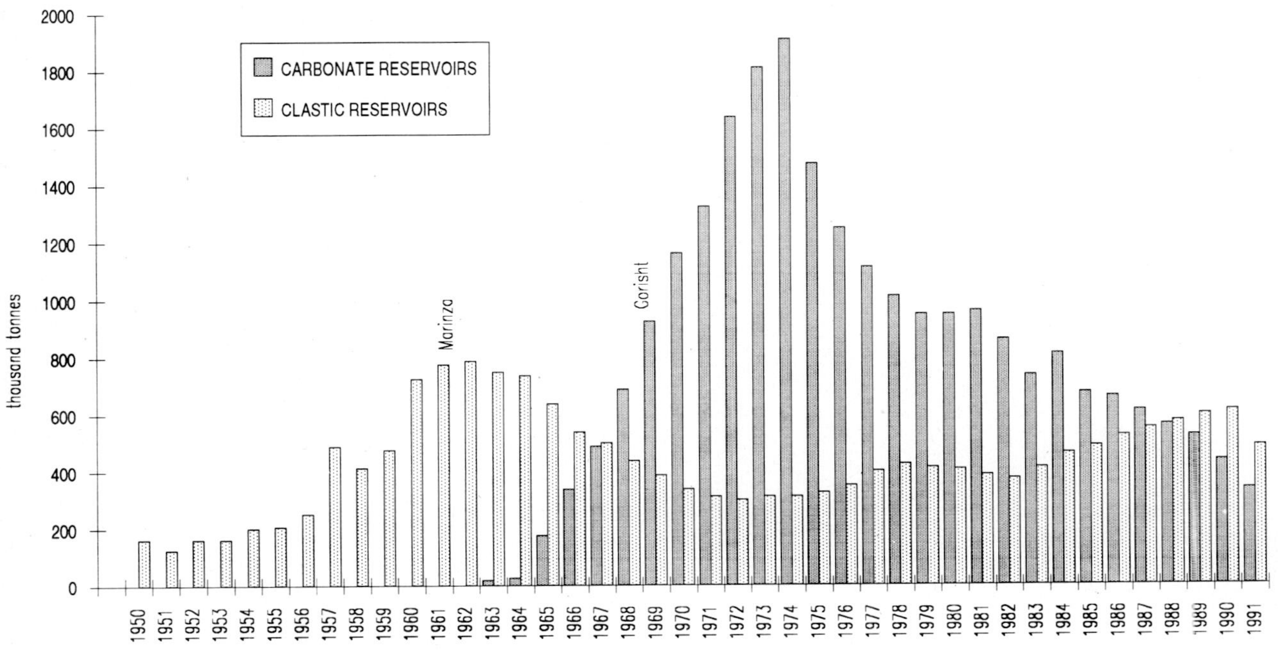

Fig. 18. Oil production per reservoir type (1950–1991)

Table 3. Results of drilling for non-associated gas during 1981–1990

Year	Structure	Wells drilled	Completed gas
1981	Divjake	6	3
	Frakull	4	1
1982	Divjake	3	2
	Frakull	1	1
1983	Divjake	3	1
	Fakull	2	1
1984	Divjake	16	12
	Ballaj	1	0
	Frakull	3	0
1985	Divjake	28	20
	Ballaj	4	3
	Frakull	2	0
1986	Divjake	31	19
	Ballaj	26	19
1987	Divjake	24	18
	Ballaj	27	15
	Frakull	8	4
	Povelca	1	1
1988	Divjake	18	9
	Ballaj	18	8
	Frakull	9	7
	Povelca	7	3
1989	Divjake	17	8
	Ballaj	23	11
	Frakull	10	1
	Povelca	8	5
	Durres	2	1
	Panaja	1	1
1990	Divjake	11	4
	Ballaj	13	2
	Frakull	2	1
	Polvelca	8	5
	Durres	1	1
	Panaja	4	0
	Total	342	187

Post-Second World War, all drilling activity was conducted by the state-owned General Directorate of Oil and Gas (D.P.N.G). In September 1990, the government authorized D.P.N.G. to accept foreign participation in several offshore blocks. In 1991 contracts were signed with Deminex, Chevron, Occidental, Hamilton and AGIP. At least three of these companies are planning to drill in 1993 or early 1994.

In 1992, Albpetrol was formed as a state oil corporation, taking over all DPNG activities, and responsibility for all of the Albanian oil industry. Albpetrol has invited foreign participation in onshore fields, and is expected to offer onshore exploration contracts soon.

Table 4. Non-associated gas production from main reservoirs (1966–1990)

Field	Reservoir	1966	1967	1968	1969	1970	1966–1970
Divjake	Tortonian	0	0	69,084,954	80,726,356	73,654,457	223,463,767
	Pliocene	0	0	0	0	0	0
	Total	0	0	69,084,954	80,726,356	73,654,457	223,465,767
	Cumulative total	0	0	69,084,954	149,811,310	223,465,767	
Field	**Reservoir**	**1971**	**1972**	**1973**	**1974**	**1975**	**1971–1975**
Divjake	Tortonian	94,155,879	99,182,646	140,269,345	163,688,982	188,938,981	686,235,813
	Pliocene	6,488,925	7,331,968	2,205,090	2,411,834	1,455,835	19,893,657
	Sub-total	100,644,804	106,514,614	142,474,435	166,100,816	190,394,816	706,129,470
Frakull	Tortonian	355,722	7,050,006	15,237,284	5,386,702	7,952,554	35,982,268
Finiq	Pg2-3	0	11,388	43,981	1,002,200	50,639,606	51,697,175
	Total	101,000,526	113,576,008	157,755,700	172,489,718	248,986,976	793,808,913
	Cumulative total	324,466,293	438,042,301	595,798,001	768,287,719	1,017,274,695	
Field	**Reservoir**	**1976**	**1977**	**1978**	**1979**	**1980**	**1976–1980**
Divjake	Tortonian	116,090,484	97,084,323	86,818,582	48,830,153	24,351,863	373,175,405
	Pliocene	2,416,709	2,162,729	4,608,976	5,736,702	5,433,866	20,358,982
	Sub-total	118,507,193	99,247,052	91,427,558	54,566,855	29,785,729	393,534,387
Frakull	Tortonian	5,988,415	7,328,540	14,784,467	22,987,746	25,618,686	76,707,854
Finiq	Pg2-3	38,828,442	64,064,856	43,817,872	17,104,403	2,677,308	166,492,881
B.Palle	Tortonian	0	0	0	80,000	166,000	246,000
	Total	163,324,050	170,640,448	150,029,897	94,739,004	58,247,723	636,981,122
	Cumulative total	1,180,598,745	1,351,239,193	1,501,269,090	1,596,008,094	1,654,255,817	

Field	Reservoir	1981	1982	1983	1984	1985	1981–1985
Divjake	Tortonian	16,599,978	7,625,636	7,211,912	7,612,112	13,984,287	53,033,925
	Pliocene	3,860,259	7,748,460	12,287,250	28,262,659	76,640,266	128,798,894
	Sub-total	20,460,237	15,374,096	19,499,162	35,874,771	90,624,553	181,832,819
Frakull	Tortonian	14,411,327	13,691,895	12,409,884	12,810,805	11,368,933	64,692,844
Finiq	Pg2	1,077,228	540,227	14,800	67,510	88,100	1,787,865
B.Palle	Tortonian	128,000	140,500	137,300	166,000	147,600	719,400
Ballaj	Pliocene	0	0	0	0	6,943,345	6,943,345
Trevllazer		0	0	0	0	196,000	196,000
	Total	36,076,792	29,746,718	32,061,146	48,919,086	109,368,531	256,172,273
	Cumulative total	1,690,332,609	1,720,079,327	1,752,140,473	1,801,059,559	1,910,428,090	
Field	Reservoir	1986	1987	1988	1989	1990	1986–1990
Divjake	Tortonian	5,499,411	4,247,627	6,288,586	13,301,136	6,174,670	35,511,430
	Pliocene	85,515,189	45,404,637	23,214,872	28,358,837	19,916,387	202,409,922
	Sub-total	91,014,600	49,652,264	29,503,458	41,659,973	26,091,057	237,921,352
Frakull	Tortonian	8,629,827	13,063,206	18,792,225	13,140,009	9,289,571	62,914,838
Finiq	Pg2	77,200	54,000	5,500	0	0	136,700
B. Palle	Tortonian	107,600	73,500	72,000	80,400	97,000	430,500
Ballaj	Pliocene	89,232,143	90,358,751	47,218,236	47,936,722	49,538,581	324,284,433
Trevllazer	Tortonian	35,000	0	0	0	0	35,000
Delvina	Paleogene	0	377,350	1,429,500	2,595,364	11,224,153	15,626,367
Povelce	Tortonian	0	5,264,425	23,310,673	54,062,417	76,987,605	159,625,120
Durres	Tortonian	0	0	0	3,759,120	2,890,370	6,649,490
Panaja	Tortonian	0	0	0	10,295,123	16,358,551	26,653,674
	Total	189,096,370	158,843,496	120,331,592	173,529,128	192,476,888	834,277,474
	Cumulative total	2,099,524,460	2,258,367,956	2,378,699,548	2,552,228,676	2,774,705,564	

Table 5. Evolution of drilling meterage in the 1986–1991 period. (After Shehu and Johnston 1991)

Year	Exploration	Development	Total
1986	158,000	208,725	366,725
1987	148,997	131,797	280,794
1988	151,000	214,121	365,121
1989	198,475	205,027	403,502
1990	154,807	233,159	387,966
1991	6,932	100,213	107,145

6 Future Exploration and Production Opportunities

6.1 Durres Basin

Although this basin is apparently at a mature stage of exploration, the geophysical surveys carried out so far showed that there are still very promising deeper targets not drilled due to technological reasons. For instance, structures like Divjake, Durres, Povelce were tested only up to the upper part of the overpressured zone.

Due to the poor quality of the equipment, lack of spares (geophones, cables), inadequate processing capacity and obsolete hardware and software, seismic shooting has been confined mainly to the identification of structural traps.

The northern part of the Durres Basin, near the shore line (where the molasse lies transgressively on the carbonates from the Ionian Zone), can still offer good opportunities, the generation potential for hydrocarbons being considered as "good"; the probability of identifying stratigraphic traps in the area seems to be rather high.

Other post-tectonic basins, Korca and Burrel, have not yet been assessed. Preliminary evaluation by Albanian geologists suggest that at least the Korca Basin may have met the geological requirements for the formation of petroleum accumulations.

6.2 Ionian Zone

In addition to known rather shallow hydrocarbon accumulations, there are also good prospects in deep carbonate structures (4000–6000 m) overlaid by the Neogene deposits of the Durres Basin. Such structures were identified on the regional and semi-regional seismic profiles, but further detailed survey and deep drilling was never carried out for technical and financial reasons.

In central-eastern Albania, the frequent occurrences of oil seeps west of the Krasta-Kucali Zone, in the Kruja and Ionian Zones, and the existence of

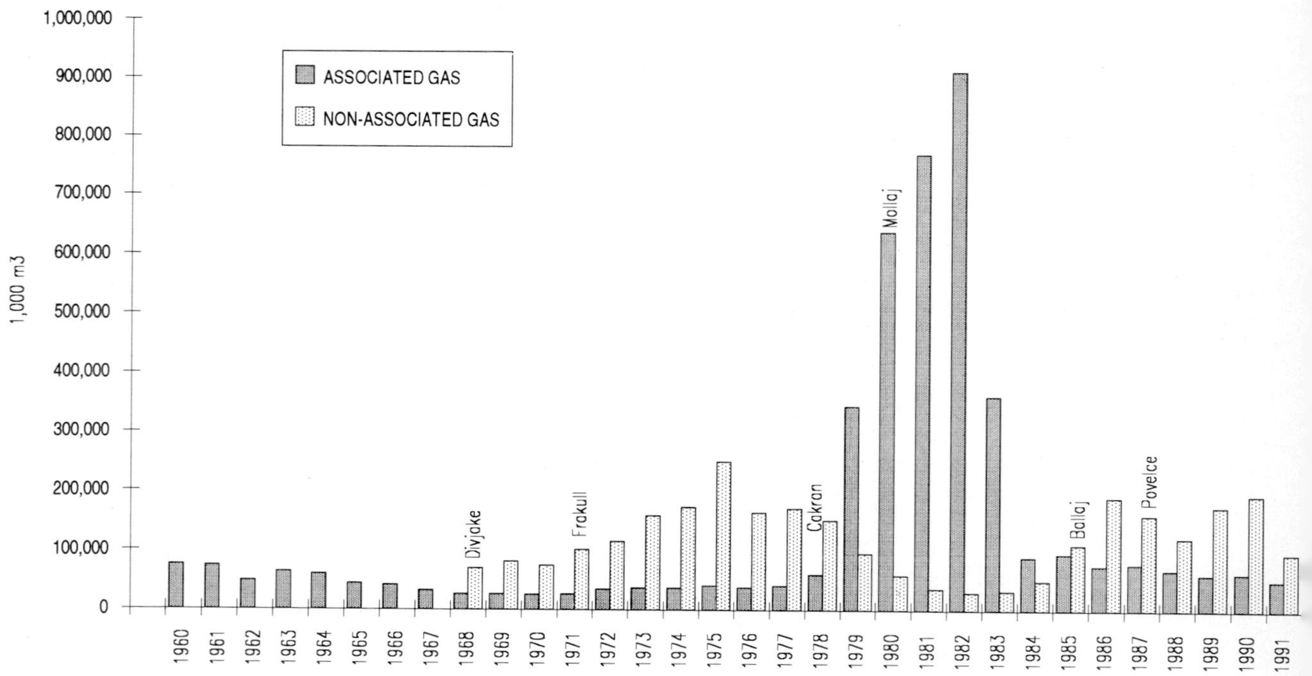

Fig. 19. Gas production (1960–1991)

potential source rocks correlated with the good values of the vitrinite reflectance of the samples from the wells drilled in the area are good arguments for further exploration.

In southern Albania, the recent discoveries of oil – both light and heavy – and of condensate at Delvina and the oil shows from several wells (Tepelene, Bence, Proshte, Golimbas) show an optimistic prospect for the exploration of the southern part of the Ionian Zone.

6.3 Offshore Albania

Offshore Albania, exploration and production blocks were granted to Deminex, Chevron, Occidental, Hamilton and Agip (Fig. 3). Semi-regional seismic surveys began in 1991 and drilling is expected to begin in 1993. Here, in addition to Durres Basin plays, the deeper Mesozoic plays, similar to those identified by drilling in the southern Adriatic Italian sector, could be an interesting target for exploration, primarily in southern Albanian offshore, in the vicinity of the Sazani Zone.

Acknowledgements. The aim of this chapter is to provide a brief summary of the exploration geology of Albania. The limited amount of data released for publication results in the presentation of the issues at a very general level.

The authors would like to thank Dr. Bogdan Popescu for his continuous support and manuscript improvement, also Gerald Dixon and Theodor Felder for their constructive criticism.

References

Bakia H, Sadikaj Y, Bally A (1992) Cross-sections through central and southern Albania. Rice University, 3 pl

Curi F (1991) Oil generation and accumulation in Albanide's Ionian Basin. EAPG Conf Florence. (unpubl.)

Curi F, Stamuli Th (1990) Geochemical conditions of hydrocarbon generation, migration and accumulation in carbonate and molasse deposits. Bul Nafta dhe Gazi 1: 97–111. (in Albanian)

Dalipi H (1985) The main stages of the geologic evolution of the Outer Albanides. Bul Nafta dhe Gazi 2: 33–54. (in Albanian)

Flores G, Pieri M, Sestini G (1991) Geodynamic history and petroleum habitats of the SE Adriatic region. In: Spencer AM (ed) Generation, accumulation and production of Europe's hydrocarbons. Spec. Publ EAPG 1. Oxford University Press, Oxford, pp 389–398.

Gjoka M, Sazhadanku F, Piperi T, Mecaj D (1990) Some distribution laws of the oil beds in the Tortonian deposits of the south-eastern margin of PAD and the exploration prospects of this region. Bul Nafta dhe Gazi 1: 19–31 (in Albanian)

Lazzari A (1964) Albania. Encyclopedia del Petrolio e del Gas Naturale vol I. 190–200. Colombo, Roma

Macovei G (1938) Les gisements de petrole. Geologie Statistique, Economie. Masson et Cie, Paris 502 pp, 222 figs

Matavelli L, Novelli L, Anelli L (1991) Occurrence of hydrocarbons in the Adriatic basin. In: Spencer AM (ed) Generation, accumulation and production of Europe's hydrocarbons. Spec Publ EAPG 1. Oxford University Press, Oxford, pp. 369–380

Shehu F, Johnston D (1991) Albania has an active but difficult, drilling program. Oil and Gas Journal, Nov 18, pp 83–88

2 Geologic Structure, Petroleum Exploration Development and Hydrocarbon Potential of Bulgaria

V. VUCHEV[1], P. BOKOV[2], B. MONOV[2], A. ATANASOV[3], R. OGNYANOV[4], and D. TOCHKOV[3]

CONTENTS

[1] Geological Institute, Bulgarian Academy of Sciences, 1113 Sofia, Bulgaria
[2] Scientific Research Institute of Mineral Resources, 1505 Sofia, Bulgaria
[3] Prospecting and Production of Oil and Gas Co., 5800 Pleven, Bulgaria
[4] Committee of Geology and Mineral Resources, 1000 Sofia, Bulgaria

1 Introduction

Bulgaria covers 111,000 km^2 and is located on the Balkan Peninsula, in southeastern Europe. The search for petroleum and natural gas in Bulgaria started about 70 years ago. Locations for the first wells drilled for hydrocarbons were close to known oil and gas seeps in northeastern Bulgaria and around the Provadija dome. The first commercial discoveries were made after the World War II: oil in 1951 in the Varna Basin, at Tjulenovo, and gas in 1949 at Bliznak.

A second success came only 10 years later, in 1961, in the central part of northern Bulgaria, and marked a new era in the country's exploration for oil and gas. Several small to medium-size fields were discovered in Triassic carbonates, Lower and Middle Jurassic clastics and the Upper Jurassic to Lower Cretaceous carbonates. These fields are located both on the Moesian Platform and in the Fore-Balkan zone. Drilling was also carried out to investigate the Upper Jurassic-Lower Cretaceous turbiditic sediments in the Fore-Balkan, without commercial success but with some promising results. Finally, occasional exploration was directed toward Paleozoic sediments on the Platform in northeastern Bulgaria, which includes the large Varna coal basin.

This chapter is not intended as a detailed, comprehensive review of the geology and exploration for oil and gas in Bulgaria. For this, reference is made to Yovchev and Baluhovski (1961), Foose and Manheim (1975), Bokov et al. (1969), Bokov and Ognjanov (1990) and Bokov et al. (1993). The authors intend to give an overview of the exploration and production history and present their views, at least in part unconventional, in the hope that this may help to attract the interest of the petroleum community to unexplored and potentially hydrocarbon-rich plays in various areas of the country.

2 Morphotectonic Provinces of Bulgaria

Two branches of the Alpine-Himalaya orogenic belt and their foreland can be seen in Bulgaria: the northern branch represented by the Balkanides, consisting of two tectonic zones, the Stara Planina and the Fore-Balkan, and the southern branch represented by the Krajshtides (Fig. 1). The foreland of the Balkanides is the Moesian Platform, while the Srednogorie and Rhodope Massif "mega-anticlinoria" could be considered as representing rock series of the pre-Alpine basement of the Balkanides.

2.1 The Balkanides

Stara Planina consists of four major structural trends, from west to east: the Svoge, Berkovica and Shipka anticlinoriums and the Luda Kamchija synclinorium, all complicated by a variety of normal, reverse and wrench faults (Figs. 1 and 2),

The Fore-Balkan is also represented by several large anticlinoria, from west to east, the Belogradchik, Teteven and Preslav. The Teteven anticlinorium lies between cross-sections II and III in Figs. 1 and 2 and is the major water recharge area for the oil fields around and northwest of Pleven. In the Teteven area and generally in the Fore-Balkan, the Triassic carbonates and the Lower and Middle

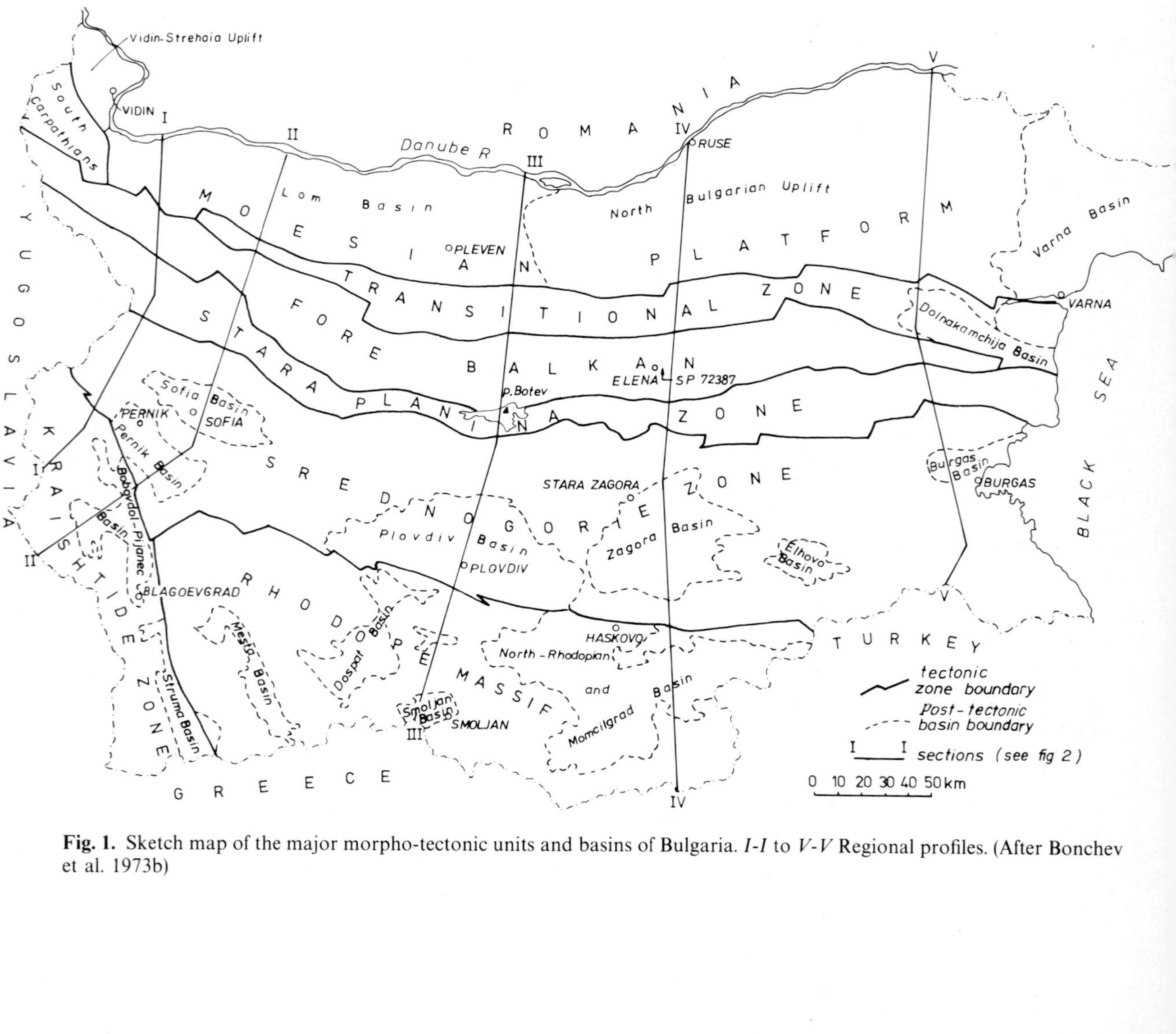

Fig. 1. Sketch map of the major morpho-tectonic units and basins of Bulgaria. *I-I* to *V-V* Regional profiles. (After Bonchev et al. 1973b)

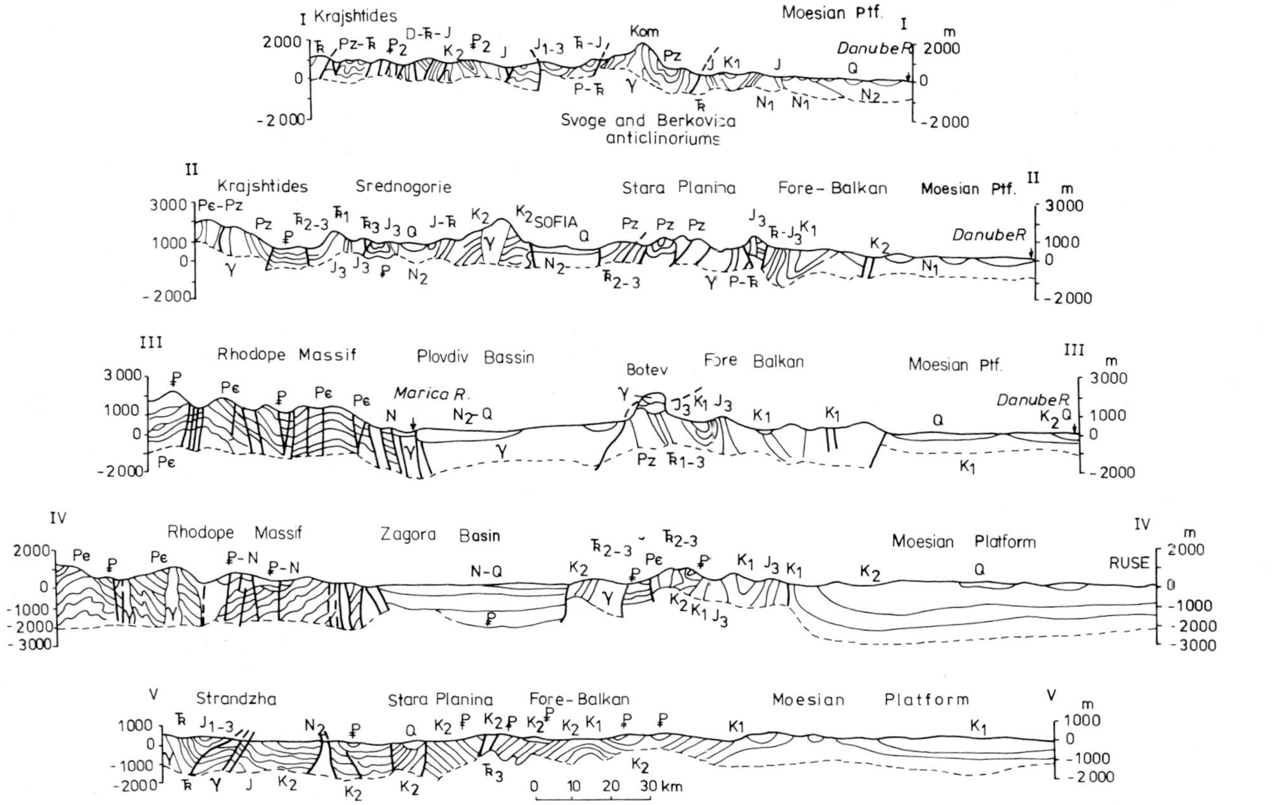

Fig. 2. Simplified regional cross-sections. (After Bonchev et al. 1973a). For location, see Fig. 1

Jurassic clastics are well exposed in outcrops in a hilly area with elevations of some 800 m; in the Moesian Platform, at depths of some 3000 m, they are the reservoirs of all known oil and gas accumulations. The Fore-Balkan anticlinoria are internally made up of second-order linear structures up to about 20 km long, some of which are still of interest for oil exploration, despite the negative results of drilling up to now.

The Southern Carpathians reach Bulgaria after a southern bending along the northwestern corner of Bulgaria (Fig. 1). This segment is made up of the Sinaia Cretaceous turbiditic sequence, folded isoclinally and striking north-south. The deeper structure of this tectonic unit and relationship with adjacent units is not fully understood. A better knowledge of the Carpathian/Balkan junction could provide valuable insight into its relationship with the adjacent Fore-Balkan and, subsequently, a better understanding of the area's hydrocarbon potential.

There are different views with respect to the tectonic relationship between the Fore-Balkan and the Moesian Platform to the north. A first interpretation considers the Transitional Zone, recently called the Periplatform, as a gently folded southern margin of Moesian Platform, as shown in Figs. 1 and 2. According to a second concept, the Fore-Balkan is overthrusted above the Moesian Platform along its entire length (Fig. 3). This mobilistic approach to the major regional tectonics of the area (Hsu et al. 1977; Bokov et al. 1993), could be a key strategic concept in exploration planning for some onshore blocks granted in the first round of bidding in 1991 (Petroconsultants FSS Petroleum Activity Map, Bulgaria, August 1991 and following).

2.2 The Krajshtides

They are the second branch of the orogen and are considered by many Bulgarian geologists as a lineament of high ranking folding, wrenching and faulting (Figs. 1, 3). An alternative view considered it as a superimposed structure over the Rhodope Massif (Yovchev et al. 1971). The Krajshtides are one of the most seismically active zones of the Balkan

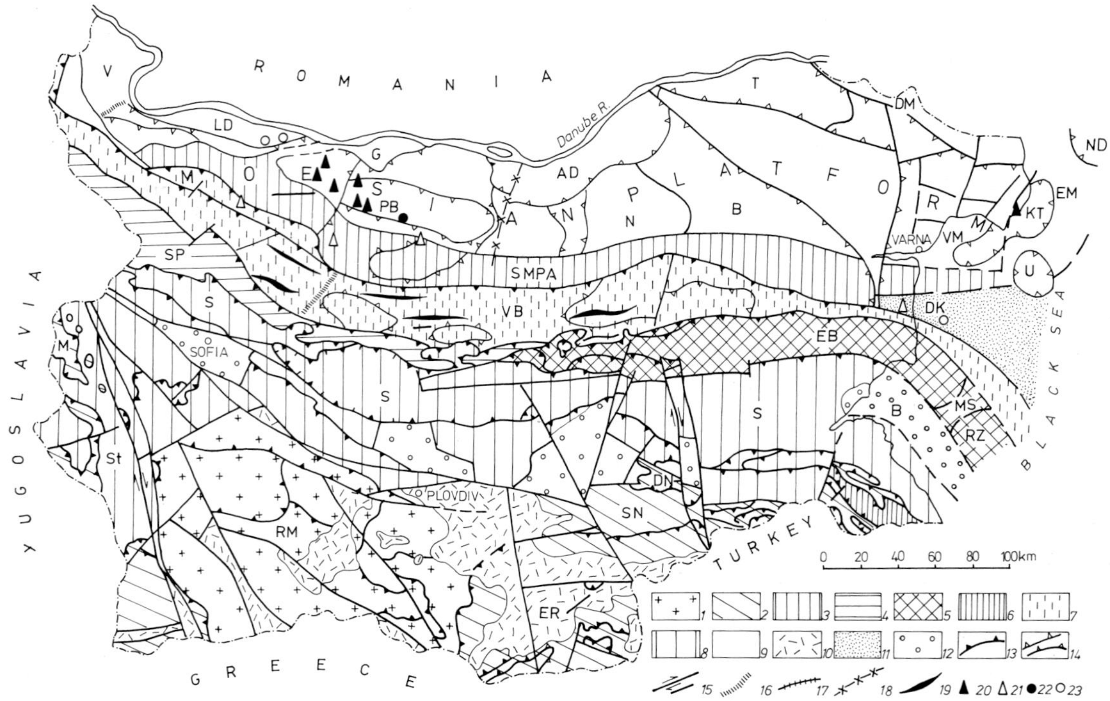

Fig. 3. Alpine tectonic maps of Bulgaria (after Bokov and Ognyanov 1990 and Bokov et al. 1993). *1* Rhodope pre-Alpine basement (RM); *2* Moravides (M) and Strandzhides (ER, SN, DN); *3* South Moesian Periplatform Zone (SMPA); *4* Stara Planina (SP); *5* East Balkan Zone (EB) including: Michurin Saddle (MS) and Rezovo Zone (RZ); *6* Zabernovo Nappe; *7* Fore Balkan (VB); *8* Srednogorie (S); *9* Moesian Platform; *10* Thracian Paleogene rift depressions; *11* Dolna Kamchija Basin (DK); *12* South Bulgarian post-collisional (Neogene-Quaternary) depressions, Burgas Basin (B); *13* Overthrust and nappe system leading lines; *14* Moesian Platform block structures (arches, saddles, uplifts, first and second order depressions); Alexandria Depression (AD); Dobrudja Massif (DM); East Moesian Monocline (EM); Gigen-Corabia Uplift (GU); Kavarna-Tjulenovo Depression (KT); North Bulgarian Uplift (NB); North Depression (ND); Pleven Block (PB); Tutracan Depression (T): Ushakov Depression (U): Lom Depression (LD); Vidin-Strehaia Uplift (V); Varna Monocline (VM); *15* Normal and strike-slip faults; *16* Cryptoruptures; *17* Reverse faults; *18* Deep seated faults; *19* Anticlinal trends in Forebalkan; *20* Oil accumulations; *21* Gas accumulations; *22* Oil discovery; *23* Gas discovery

Peninsula. A large variety of rocks, ranging in age from Precambrian to Tertiary and including Upper Jurassic to Lower Cretaceous turbidite sequences (which also cover vast areas in the Central Fore-Balkan) are found in this area. Its hydrocarbon potential is considered very low.

2.3 The Moesian Platform

This occupies the whole area of the country north of the Balkan orogen, and is believed to represent the western edge of the large Ponto-Caspian plate. The northern half of the Moesian Platform is located in Romania. Basement does not crop out in the basin, but has been encountered in boreholes in eastern Bulgaria.

The Moesian Platform actually consists of several superimposed basins (structural-sedimentary units): the Paleozoic, Early Kimmerian, the Upper Jurassic-Paleogene and the Neogene basins. The structure of these superimposed basins is complicated by swells, arches and uplifted blocks. Three larger sub-basins are distinguished on the Platform: the North Bulgarian Uplift (or Arch), the Lom Depression (also called the Moesian Basin) and the Varna Basin (Figs. 1, 3).

The westernmost part of the Moesian Platform comprises the Vidin-Strehaia Uplift, whereas a portion of the Varna Basin develops on the eastern flank of the North Bulgarian Uplift. This basin is also known as Varna Monocline (Homocline). The Varna Basin dips eastward under the Black Sea. A few small and shallow younger Paleogene basins develop along the southern border of the Moesian Platform in Bulgaria.

2.4 The Srednogorie

The Srednogorie is separated from both the Balkanides and the Rhodope Massif by Early Alpine wrench and normal faults. Srednogorie contains two large groups of rocks. The lower group consists of amphibolites, gneisses, granite gneisses, marbles and other metamorphics. The upper one consists of rhythmic volcano-clastic sequences and molasses with a total thickness of over 3000 m. Laramian granitic bodies are present.

The most important feature of the Srednogorie, however, are thought to be the older Mesozoic, mostly marine sediments overlapped by Late Cretaceous volcano-clastics and volcanic rock traps. These possibly deeply buried marine Mesozoic sediments could be a target for future oil exploration within the whole Srednogorie.

2.5 The Rhodope Massif

The Rhodopes are crucial for supporting the concept of large northward overthrusting movements. In Fig. 2 it is shown as a central, crystalline body, but recent concepts consider it to be made up of several overlapping nappes (Fig. 3). Thus, it is possible that the Rhodopes are overlying a Mesozoic sedimentary sequence, which may contain hydrocarbon accumulations.

The occurrence of a granitic body on the highest peak of the Stara Planina zone, the Botev, might be good evidence for the idea of long distance horizontal displacement (Figs. 1 and 2, section III) northward of the Rhodopes. By age, composition and structure, the above granitic body belongs to the group of southern Bulgarian granites, developed regionally within the Rhodope Massif.

A few small post-tectonic basins with thin turbiditic Tertiary sediments, containing coal beds, black shales and olistoliths of all main crystalline formations are developed over the central and western Rhodopes. Oil seepages have been found and described in many places within these sedimentary basins. Late Tertiary volcanism is also regionally developed.

3 Hydrocarbon-Producing and Prospective Basins

This section describes the main tectono-stratigraphic features of Bulgaria's hydrocarbon proven and/or prospective sedimentary provinces.

Some general characteristics were in part discussed in the previous section. The most recent issues of the Petroconsultants S.A. Petroleum Activity Map for Bulgaria should also be consulted with respect to locations of wells and oil and gas fields.

3.1 Lom Petroleum Basin

The most general features of this sub-basin (usually called "depression") of the Moesian Platform are the block fault mosaic pattern (Fig. 3) and deeply buried pre-Mesozoic rocks (Figs. 4, 5). The crystalline basement has not been reached by wells drilled as deep as 5500 m, and the oldest sedimentary rocks encountered in these wells are of Devonian age. They are overlaid by Carboniferous, Permian, Triassic, Jurassic, Cretaceous and Tertiary sediments with a total thickness of up to 9000 m. A rich variety of clastics has been encountered in the Permian, Lower and Upper Triassic, Lower and Middle Jurassic and Tertiary. The remainder of the section, i.e. Devonian, Carboniferous, Middle and part of the Upper Triassic, Upper Jurassic, Lower and Upper Cretaceous, consists of carbonates.

As a result of intensive seismic surveying, geological mapping and drilling of hundreds of exploratory wells, commercial and sub-commercial hydrocarbon accumulations have been found in reservoirs of Middle to Upper Triassic and Lower Jurassic age: Dolni and Gorni Dubnik, Bardarski Geran, Selanovci, Dolni Lukovit and Staroselci oil fields and the Butan, Dolni Dubnik, Kurpachevo and Devetaki gas pools.

Of particular interest is the Gigen field, on the southern bank of Danube River, north of Pleven, which contains heavy oil accumulated in highly porous and permeable Upper Jurassic and Lower Cretaceous carbonates, similar to the Tjulenovo oil field in the Varna basin (see also Sect. 4.3.7). Two new fields, the Bohot oil field and the Uglen gas-condensate field, are under appraisal or production testing. Oil and gas shows have been found in the entire Mesozoic section of the basin.

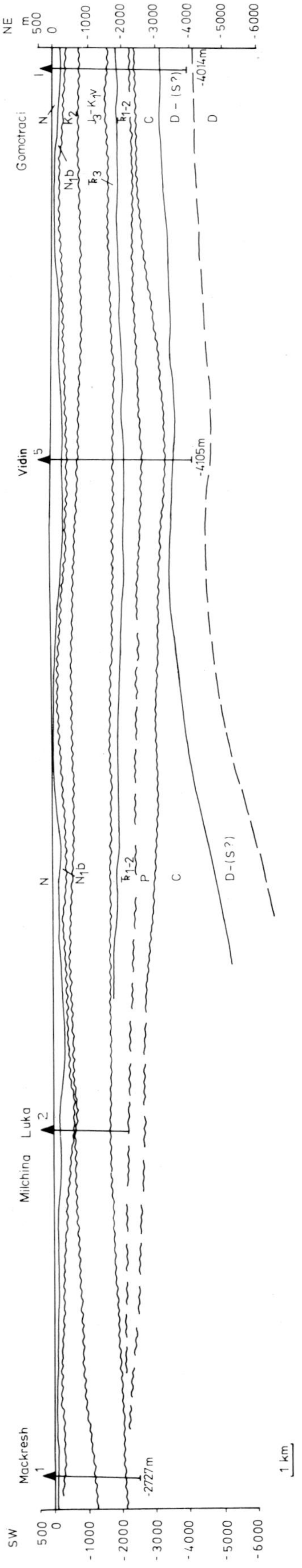

Fig. 4. Regional cross-section through the Vidin-Strehaia Uplift, north west Bulgaria. (After Bokov and Vassileva 1989 in Bokov et al. 1993)

The Paleozoic structural style is less well known because of the poor seismic results at greater depth (Figs. 4, 5). During the Triassic and Lower Jurassic, the horst and graben structural style prevailed, whereas in the Upper Jurassic, Paleogene and Neogene, mild compression prevailed, resulting in gently folded, homoclinal structure. All local positive structures are either non-faulted brachyanticlines, fault-related hemianticlines or structural noses. Their size is small to medium, varying from 5–15 km^2 to more than 100 km^2. Their amplitudes are usually 15-30 m, exceptionally up to 360 m.

The main oil and gas plays are related to the Devonian, Carboniferous, Middle Triassic and Lower Jurassic clastics and partly to the carbonates of Upper Triassic and Lower Cretaceous. The regional hiatuses – mainly Permian, Triassic, Cretaceous and Paleogene – along which all the so far discovered commercial oil and gas pools are located continues to be among the most important exploration targets in the basin.

3.2 Varna Petroleum Basin

The Varna Basin (also called Monocline, Depression or Sub-basin) is located on the eastern slope of a large positive structure known as the North Bulgarian Uplift (Figs. 1, 3). The basin dips eastwards under the Black Sea and is limited by the Dobrudja Massif to the north and by the Dolna Kamchija Basin to the south (Figs. 6, 7). Its width, from north to south, varies between 10 and 50 km and its length is 60-70 km. The thickness of the sedimentary section of the basin reaches 7000 m. Seismic surveys and intensive drilling have identified several tectono-stratigraphic stages within the basin's evolution: Epibaikalian, Hercynian, Early Kimmerian and Alpine.

The Devonian to Paleogene lithology and stratigraphy of the basin have been studied extensively. Predominant in this section are terrigenous rocks – shales, siltstones and conglomerates. Volcanics have occasionally been encountered. Carbonates are present in the Devonian, Carboniferous, Middle Triassic, Lower Cretaceous and Paleocene. Detailed studies have been carried out on Carboniferous coals and coal-bearing sediments of the Dobrudja coal depression.

The final shape and structure of the basin was achieved during the Paleogene and Neogene (Alpine) periods. The late Mesozoic and Paleogene-Neogene structural stages are the best known from drilling. A regional eastward-dipping trend with

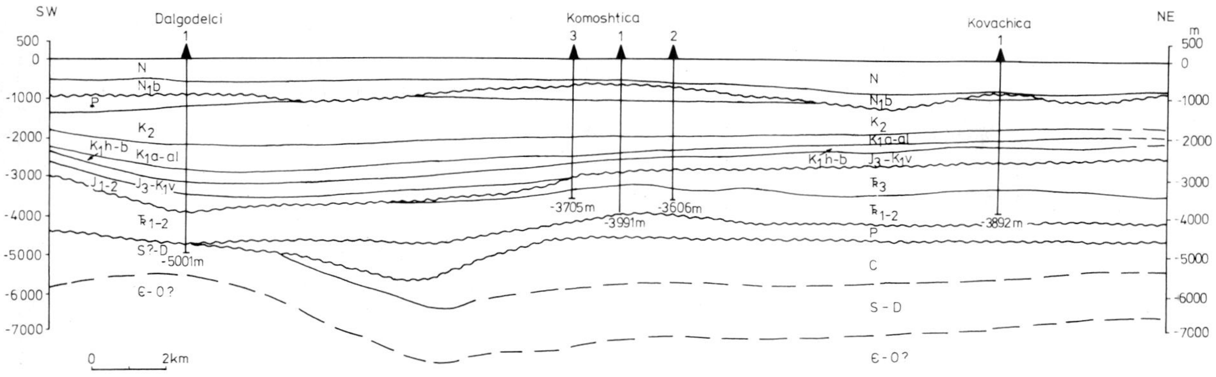

Fig. 5. Regional cross-section through the Lom Depression, North West Bulgaria. (After Bokov and Vassileva 1989, in Bokov et al. 1993)

local structures and faults has been noticed. The amplitudes of the structures vary from a few metres to 200 m and their sizes, from 8 to 150 km^2.

Tjulenovo is the only commercial field in the basin with oil and gas accumulated within karstified and fractured Upper Jurassic to Lower Cretaceous limestones and dolostones at a depth of 350–400 m.

The most promising prospects of the Varna basin are related to the Devonian dolostones, the Upper Jurassic-Lower Cretaceous limestones and dolostones and to the Eocene and Oligocene sandstones.

3.3 Fore-Balkan Petroleum Basin

This basin corresponds to the longest, about 500 km, of all Balkan orogen structural zones (Figs. 1, 3). It crosses the entire country and dips eastward under the Black Sea. The relationship to other zones was in part discussed elsewhere (Sect. 2) together with the regional tectonic framework of the country.

The sedimentary section of the basin may reach, in the east, a thickness of up to 8 km. It consists of Paleozoic, Mesozoic and Cenozoic rocks. Carbonates are predominant, but clastics represent a major portion of the Permian, Lower and Upper Triassic, Lower and Middle Jurassic and Tertiary section. As in the adjacent Lom Depression, the Middle Triassic carbonates and the Lower Jurassic terrigenous rocks form the main reservoirs in the basin.

Until now, economic small to medium-sized gas and condensate fields have been found only in the Western Fore-Balkan area: Chiren, Vurbica, Ponor, Goljamo Peshtene fields all trapped in Triassic, Jurassic and Lower Cretaceous reservoirs. However, oil and gas shows are known in many other places and are related to the whole sedimentary section.

The recent tectonic structure of this basin is Late Alpine and is made up of reworked elements of older Paleozoic, Early and Late Kimmerian structures, although some of the pre-Alpine faults reach the surface. Locally, structures appear at the surface as linear folds, but could have much more complicated, structures deeper (Figs. 8, 11). Some structures are marked by their almost vertical northern flanks, often complicated by flexures. The size of these local structures is up to 20 km in length, from 1 to 5 km in width and up to 800 m in amplitude. These sizes are remarkably larger compared to those of the structures on the Moesian Platform. Evaporites occur in the Triassic sediments of the eastern Fore-Balkan.

The whole area of the Fore-Balkan is of prime interest for oil exploration, particularly since a great number of structures have not yet been drilled. Priority should be given to the eastern part of the West Fore-Balkan, the Mikre-Sevlievo anticlinal zone of the central Fore-Balkan and to deep strata of the eastern Fore-Balkan foreland structures if a plate tectonics approach is used.

3.4 Dolna Kamchija Petroleum Basin

This basin is a small Tertiary depression, covering an area of about 350–400 km^2 between the Varna Basin and the Fore-Balkan zone (Figs. 1, 3, 6, 7). It dips eastwards under the Black Sea (Fig. 9). All boundaries of this basin are faults. The Dolna Kamchija basin is filled by thick Eocene, Oligocene and Neogene terrigenous sequences which cover

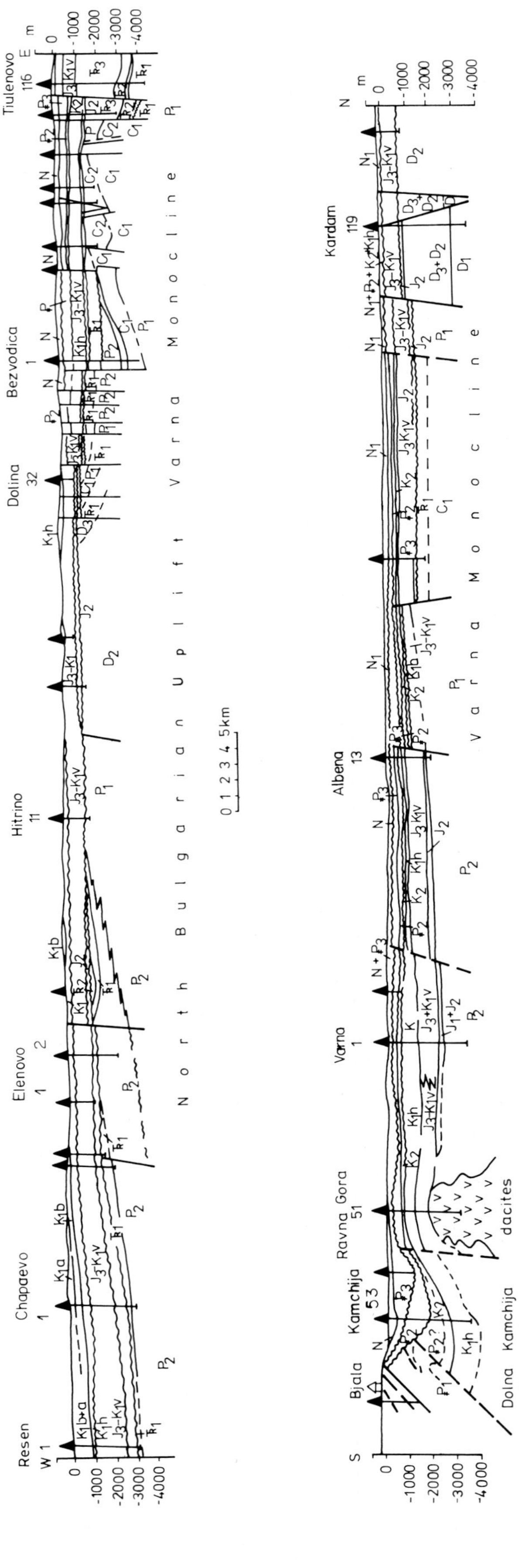

Fig. 6. Latitudinal (*top*) and longitudinal (*bottom*) cross-sections through East Moesian Platform: North Bulgarian Arch, Varna Monocline, Dolna Kamchija Depression. (After Bokov and Chemberski 1987; Dachev et al. 1988, in Bokov et al. 1993)

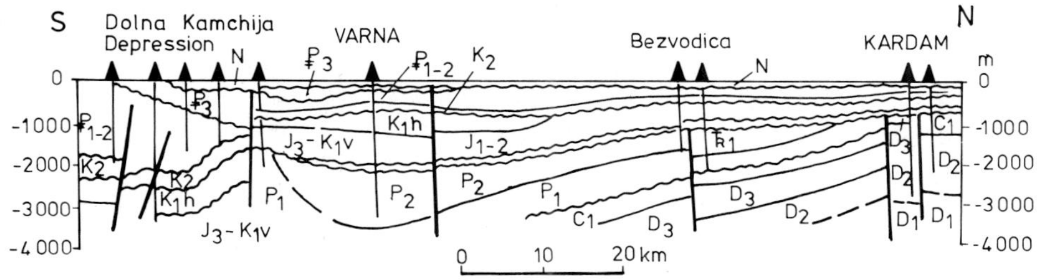

Fig. 7. Cross-section through the Varna Depression, East Moesian Platform, considered also as a part of the West Black Sea petroleum basin. (After Bokov et al., in Semenovich and Namestnikova 1981)

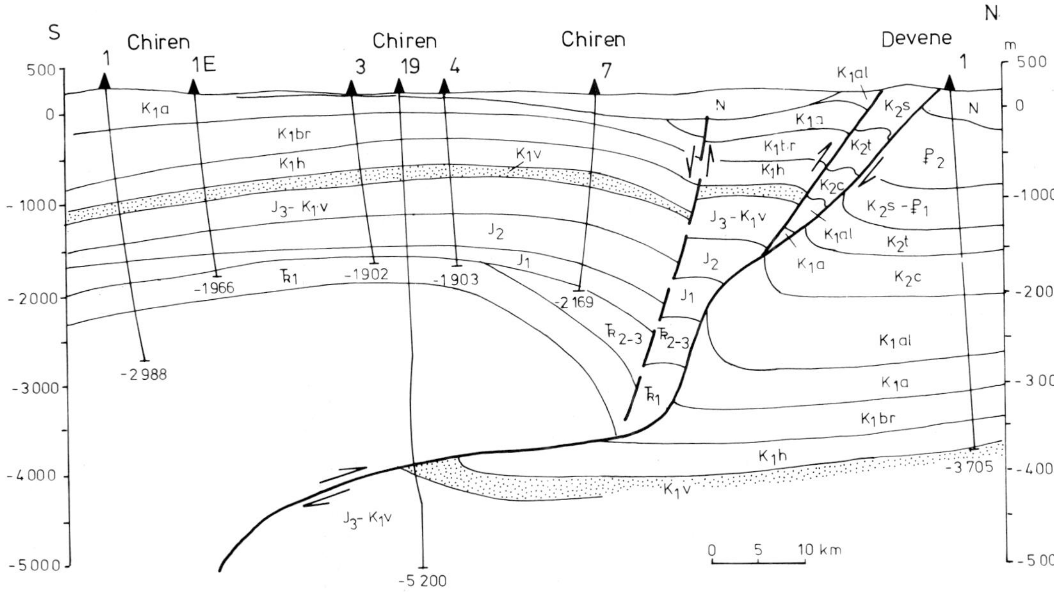

Fig. 8. Cross-section through the Chiren field in the West Fore-Balkan Devene nappe. (After Karagjuleva et al. 1982)

unconformably a Moesian Platform type of "basement".

Two structural zones can be traced within it. In the southern zone, fault-related, 2–3 km wide, overthrusted folds (e.g. Samotino, Dolni Chiflik and Rudnik anticlines) prevail. In the northern zone there are two trends of brachyanticlines buried under sediments of Eocene-Oligocene structural stage. More structures of smaller size may still be found by further exploration. East-west-striking faults of amplitude of 100–200 m are important structural elements of this basin.

Seismic surveys and drilling have resulted in the identification of several deeper structures. In one of them, the Novo Orjahovo anticline, a small gas (95% methane) and condensate field was found in 1955. The pool is in Upper Eocene sandstones. Sub-commercial gas accumulations have been found in almost all drilled structures. Therefore, the main exploration activity should be directed towards the sandstone plays of the Middle and Upper Eocene and the Oligocene. Seaward, the sandstone play become even more consistent since the basin widens and the cumulative thickness of the Paleogene and Neogene sediments reaches 5000–6000 m.

3.5 Young Post-Tectonic Basins of Southern Bulgaria

The basins of Southern Bulgaria are distributed over the Srednogorie, Krajshtides and Rhodope morphotectonic zones. The most important are shown on the map in Fig. 1 and further considerations of their hydrocarbon potential, particularly in the context of a tectonics approach, is briefly given below. Details on their stratigraphy can be found in Yovchev et al. (1971).

A number of these basins: Sofia, Pernik, Bobovdol, Struma, Smoljan, Zagora, Burgas, Elhovo and

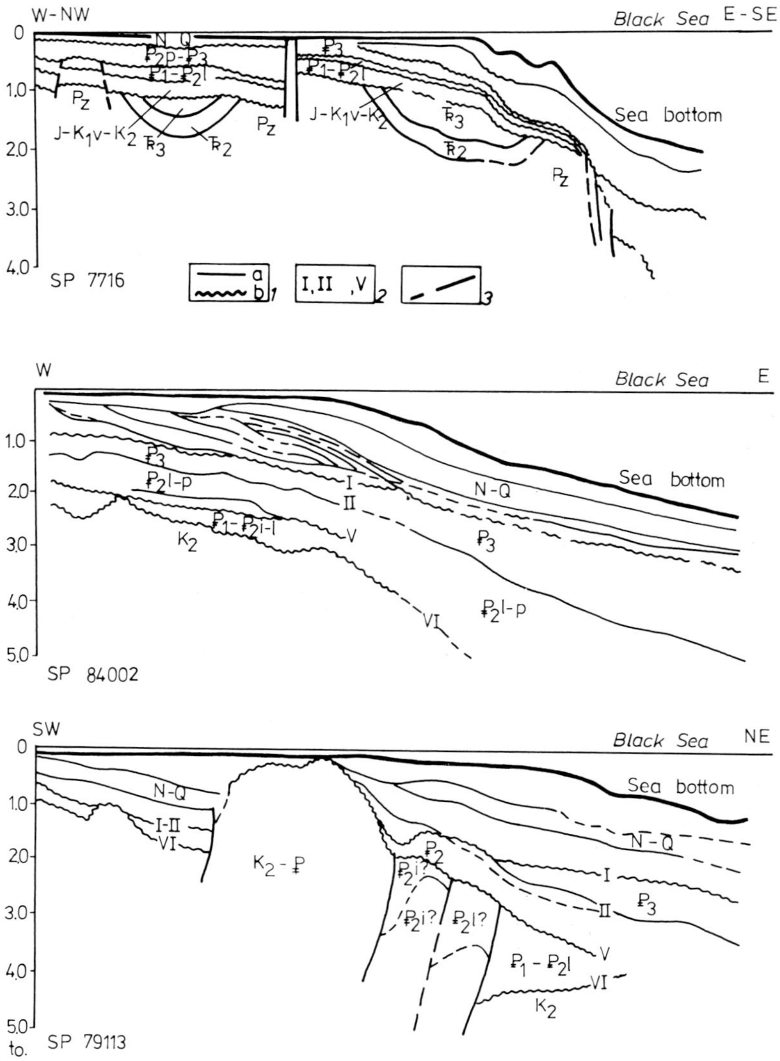

Fig. 9. Seismogeological cross-sections showing relations between Black Sea megadepression and Moesian Platform (line *SP 7716*), Dolna Kamchija depression (line *SP 84002*), and eastern Balkan (line *SP 79113*). (After Dachev et al. 1988, in Bokov et al. 1993). *1* Boundaries of seismostratigraphic complexes; *a* questionable reflector; *b* main seismic unconformity; *2* codes of seismic events; *3* faults or fault zones (see also Fig. 4)

North-Rhodopian have economic coal and oil shale deposits. Others are known to have gas and oil shows but none of them has been thoroughly explored for oil and gas. Most of the basins are immature in terms of organic matter catagenesis but have good, predominantly clastic reservoirs. Only small hydrocarbon accumulations would be discovered in these basins.

3.5.1 Pernik and Bobovdol-Pijanec Basins

These are grabens of several hundred km^2 each, filled with Paleogene mostly clastic sediments up to 1000 m thick. These sediments overlie intensively tectonized sequences of varying age and type. They have economic coal and oil shale deposits.

3.5.2 Sofia Basin

This depression, 60×20 km in size, is filled by more than 800 m of mostly Pliocene sediments which are coal-bearing and cover a tectonically complicated basement. Pleistocene and Quarternary sediments, some tens of metres thick, overlie the Pliocene unconformably. The basin area is seismically very active.

3.5.3 Struma Basin

Neogene sediments, up to 1000 m thick and tectonically deformed, are deposited in a fault-dependent basin over a complex basement. The sedimentary sequence, mostly clastic, has also coal and oil shale deposits, tuffs and tuffo-breccias. It is believed that older Tertiary sediments have been deposited, thus increasing the potential thickness of the sedimentary section to more than 1500 m.

3.5.4 Mesta Basin

Up to 1000 m of normal or volcano-sedimentary sequences of a Paleogene age fill a narrow graben developed over a crystalline complex. The graben had a very active period of subsidence during the Tertiary.

3.5.5 Smoljan Basin

This is a remarkable although not very large basin. Lacustrine turbiditic sediments of up to some hundreds of metres were deposited over a wide variety of basement structures and additionally complicated by slumping and some peculiar volcanic activity. The age of the turbiditic series is Paleogene. They are partly coal-bearing. Many oil seepages have been reported.

The Rhodope crystalline massif is regionally polymetallic ore-bearing and the ore bodies are related to the above-mentioned Tertiary volcanism and related hydrothermal activity. A few smaller basins of the same isolated character are developed nearby.

3.5.6 North Rhodope and Momchilgrad Basins

An area of some hundreds of km^2 is covered by mixed Paleogene sediments, up to 2500 m thick, partly tectonized but generally gently folded. Marine and lacustrine sediments are predominant, with some volcanic intercalations. Tephroidal turbiditic sequences, containing oil-bearing shales, are also developed.

3.5.7 Plovdiv Basin

This graben, around the second largest city in the country, is up to 60 × 25 km in size, bounded by faults and filled with up to 1000 m of Tertiary sediments. In the central part the uplifted Plovdiv Horst is made up of Cretaceous to Paleocene syenites.

3.5.8 Zagora Basin

The total subsided area of this basin is over 1000 km^2 and the thickness of Tertiary sediments can be more than 1300 m. It is known as the largest lignite basin in Bulgaria with open mining and thermal electric power station production.

Interestingly enough, the deepest well drilled so far is located between the Plovdiv and the Zagora grabens. At a total depth of about 3000 m, it was still in Tertiary volcano-clastic sequences.

3.5.9 Burgas Basin

Onshore, this basin covers an area of 35 × 20 km of gently eastwards-dipping Paleogene sediments, containing a commercial coal field. The thickness of Tertiary sediments is not known yet, but it could reach some hundreds of metres slightly east of the coast line. The basin developed over Upper Cretaceous volcano-clastic sequences of the Srednogorie zone. It is speculated that this basin may have a moderate to good oil potential.

3.6 Basin Correlations

Some of the regional correlations and results of previous exploration for hydrocarbons have been mentioned in previous sections. Here, an attempt is made to provide an alternative interpretation or to illustrate better the structural details and facies relationship of some areas that have recently been under scrutinity by Bulgarian geoscientists. New concepts detailed below could be helpful for further petroleum potential assessment.

3.6.1 Correlations of Tectonic Styles of Major Units

Several interpretations are available on the structural style of Southern Bulgaria and, in particular, of Strandzhides, the major part of which is located in the western Thracian area of Turkey and of eastern Srednogorie in Bulgaria (Fig. 10). This unit is probably an overthrusted melange over folded

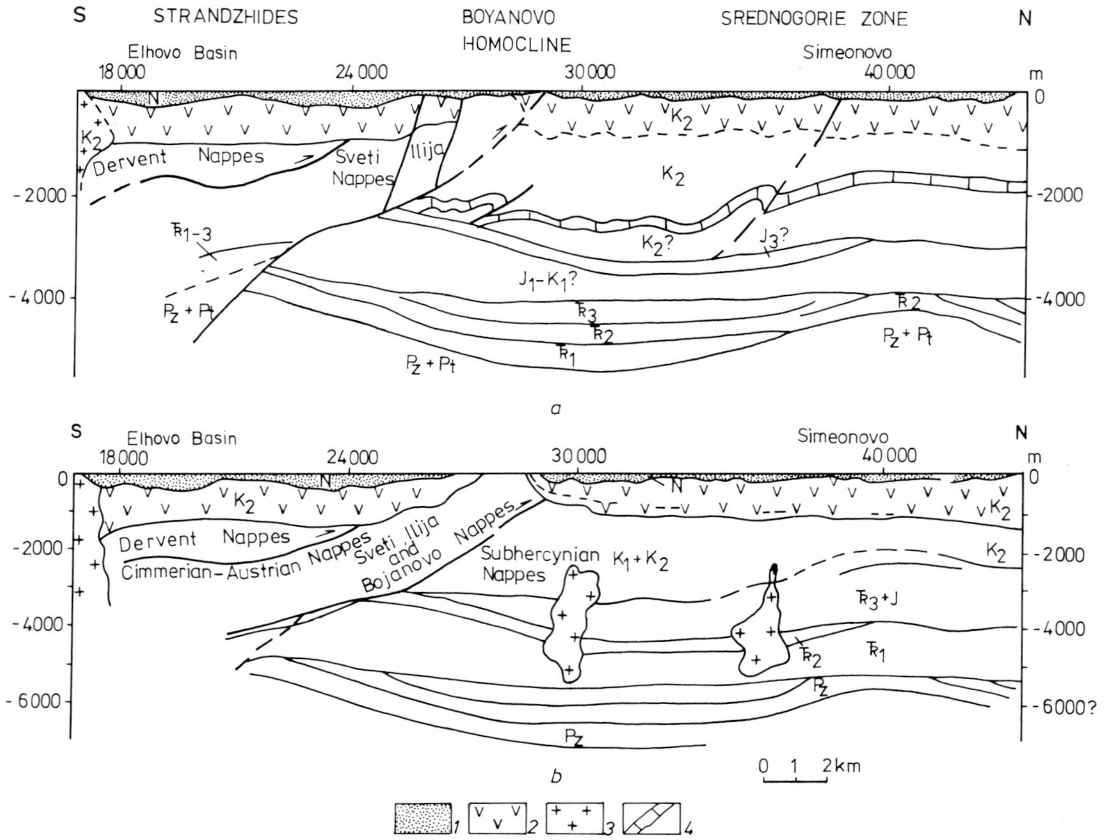

Fig. 10. Two alternative interpretations of seismic cross-sections from the southern margin of the East Srednogorie zone. (After Gochev in Bokov et al. 1993). *1* Neogene and Pilo-Pleistocene depressions; *2* Upper Cretaceous Srednogorie complexes; *3* Upper Cretaceous igneous rocks; *4* Upper Cretaceous limestone

marine Mesozoic sediments. According to some authors' opinions, the underlying marine, slightly deformed Mesozoic should be placed under active hydrocarbon exploration.

To the north, the higher tectonic style of Balkanides is met and comparatively simple plastic deformations of Mesozoic sequences in the Fore-Balkan are significantly complicated by normal and thrust faults (Fig. 11). The relationship between the Moesian Platform and the Fore-Balkan, considered in the light of plate/foreland coupling, is not accepted unanimously by Bulgarian geologists. Classical concepts, considering this relationship as a progressive, slightly disturbed, passage from platform to a flexure, then a foredeep, were introduced and developed by the leading tectonicians Bonchev (1940, 1986) and Jaranov (1960). Recently, the Moesian Platform-Fore-Balkan relationship was considered in a plate tectonics approach by Gochev (1986, 1991).

The essence of this model, as stated earlier, is northward-overthrusting, as a result of which normal marine Mesozoic sediments are deeply buried in Southern Bulgaria, and Balkanides folded belts overlie the southern, sub-thrusted, margin of the Moesian Platform.

The Moesian Platform still had a relatively simple structure in Middle Cretaceous (Fig. 12) compared with the intense post-Paleogene structuring (Fig. 13). The active Triassic paleorelief and its subsequent complex paleogeographic development during the Jurassic over the whole of Northern Bulgaria was described by Sapunov and Tchoumatchenco (1987) and Sapunov et al. (1991).

3.6.2 Lithologic and Facies Correlations

Based on drilling information from the whole Moesian Platform in both Romania and Bulgaria and some locations in the Fore-Balkan and the SW slope of Dobrudja Massif, a few representative regional lithostratigraphic correlations are given in figs. 14 to 17. They show the consistent, although

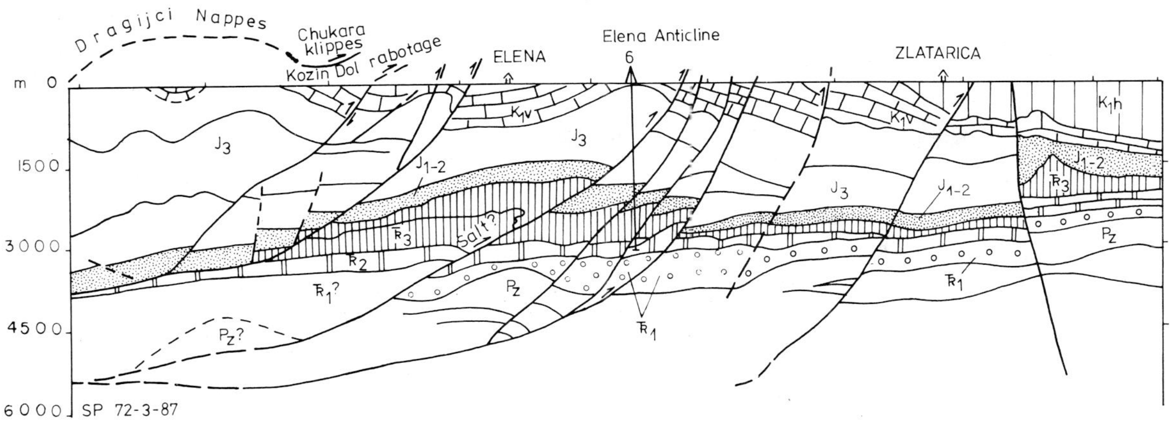

Fig. 11. Geologic interpretation of the seismic line SP 72-3-87 in the Central Fore-Balkan. (After Gochev in Bokov et al. 1993)

locally varying, paleogeological development of the whole Moesian Platform.

The correlations between lithostratigraphic units of the Triassic in northeast Bulgaria are reliable enough, but the thickness and the composition of sediments vary widely (e.g. Fig. 18) The Lower Cretaceous is the thickest sedimentary sequence of North Bulgaria. Few wells have penetrated the Lower Carboniferous, but at all locations a large hiatus between the Carboniferous and overlying sediments has been identified. The thickness of the Permian on the North Bulgarian Uplift is remarkable. There is also a large regional hiatus between Paleozoic and Mesozoic. The most diverse in thickness is the Jurassic.

The continuity in time and the paleogeographic unity of the Moesian Platform sedimentary province is also demonstrated by the isopach maps of selected most characteristic, sedimentary units (Fig. 19). The same is partly true for the Varna and Dolna Kamchija basins, which are marked by a very active development in the Tertiary particularly during Neogene (Figs. 20, 21, 22).

3.7 Exploration and Production History

The interest in petroleum exploration can be traced back to the post-World War I period when, in 1925, British companies were granted four blocks (Owen 1975). A well drilled near the Varna railway station in 1927 encountered salt water and gas shows, while a 460-m-deep well drilled on the Provadija salt dome (Moesian periplatform area) penetrated Tertiary oil shales and tested salt water.

During World War II, the newly organized Geological Survey invited German companies to run seismic and drill wells in the same general area. After World War II, the exploration was carried out with Soviet assistance. The Bliznak gas pool was discovered in the Dolna Kamchija Basin in 1949 and the Tjulenovo oil and gas field was found in the Varna Basin in 1951. These two commercial discoveries gave an impetus to the activity of the young local exploration companies. During the next 40 years, the bulk of discoveries was made in the Moesian Platform under the supervision of the State Department of Geological Prospecting through three local branches in Varna, Pleven and Mihaijlovgrad.

First offshore drilling was performed in 1985. Some 17,600 MCFD of gas was tested from Tertiary clastics in the Samotino wildcat. The other three wells were abandoned dry or with minor shows. Offshore activity was abandoned due to lack of technical facilities and funding.

In 1989, foreign companies were invited to explore on- and offshore Bulgaria. Of 16 blocks offered, six offshore and three onshore concessions were granted in 1991.

3.7.1 *Surface Exploration*

Geological studies were initiated by the Bulgarian government after the liberation from Turkey at the end of the last century. The first modern studies and maps were made by foreign geologists at the beginning of this century.

Regional geophysical surveys and deep seismic profiling, together with a national field geological

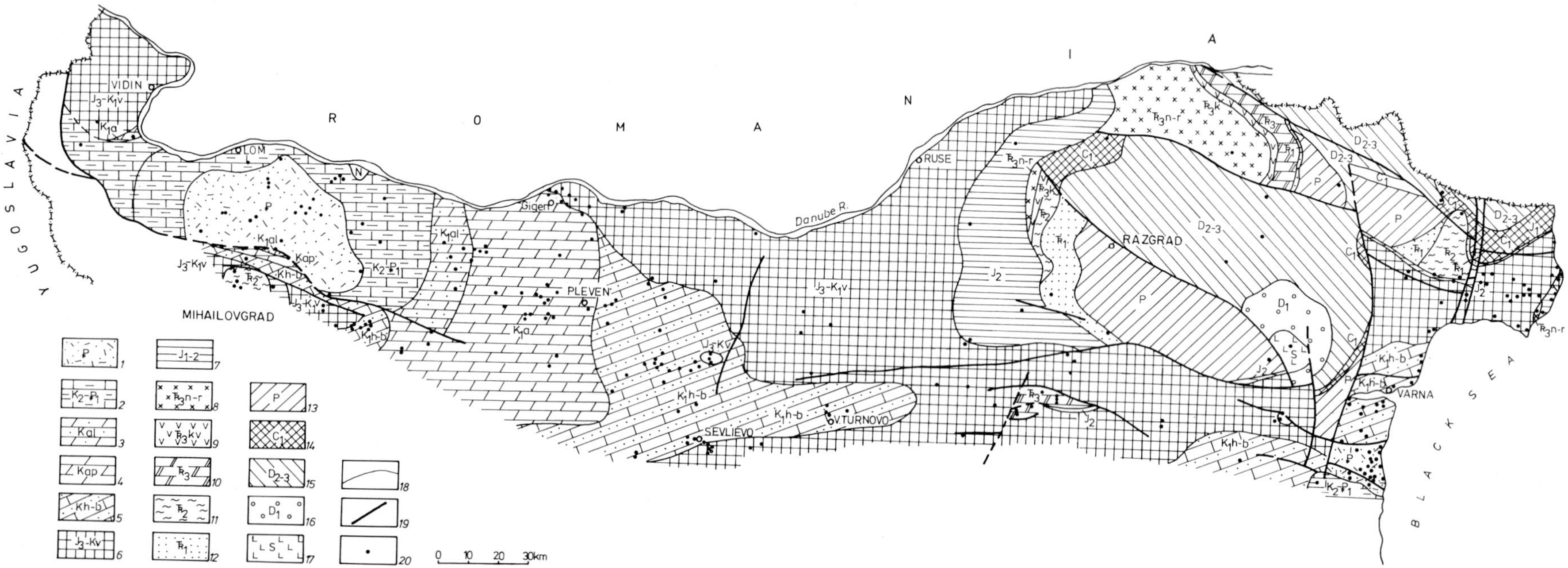

Fig. 12. Map of the top surface of the Upper Jurassic-Lower Cretaceous (Valanginian) carbonate complex. *1* Paleocene; *2* Upper Cretaceous-Lower Paleocene; *3* Albian; *4* Aptian; *5* Lower Cretaceous-Hauterivian-Barremain; *6* Upper Jurassic-Valanginian; *7* Lower-Middle Jurassic; *8* Norian-Rheatian; *9* Carnian; *10* Upper Triassic; *11* Middle Triassic; *12* Lower Triassic; *13* Permian; *14* Lower Carboniferous (?); *15* Middle-Upper Devonian; *16* Lower Devonian; *17* Silurian; *18* Stratigraphic boundary; *19* fault; *20* well location

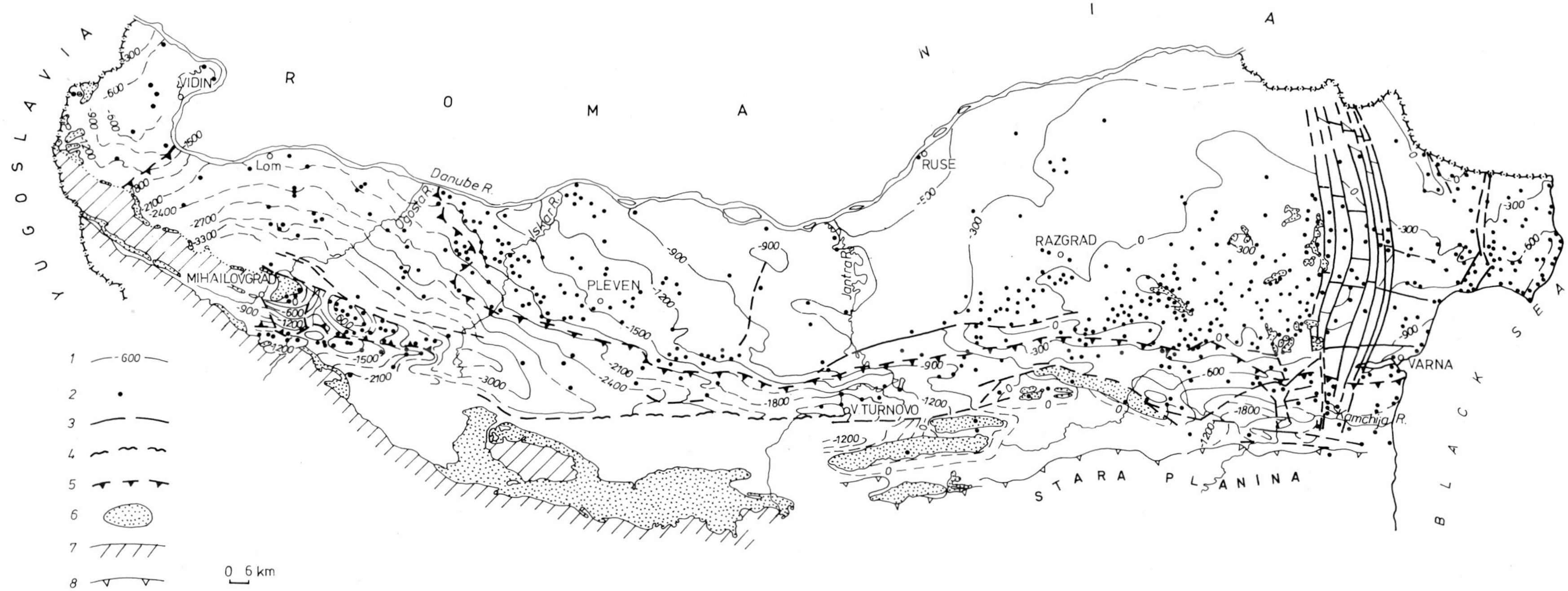

Fig. 13. Geologic map of North Bulgaria at 1000 m deep horizontal section. (After Bokov et al. 1978). *1* Isohypse contour; *2* well location; *3* fault; *4* northern edge of the Tithonian-Berriasian flysch sediments; *5* southern edge of the carbonate paleoshelf; *6* outcrop in folded areas; *7* outcrop of folded rocks older than Upper Jurassic; *8* Stara Planina line

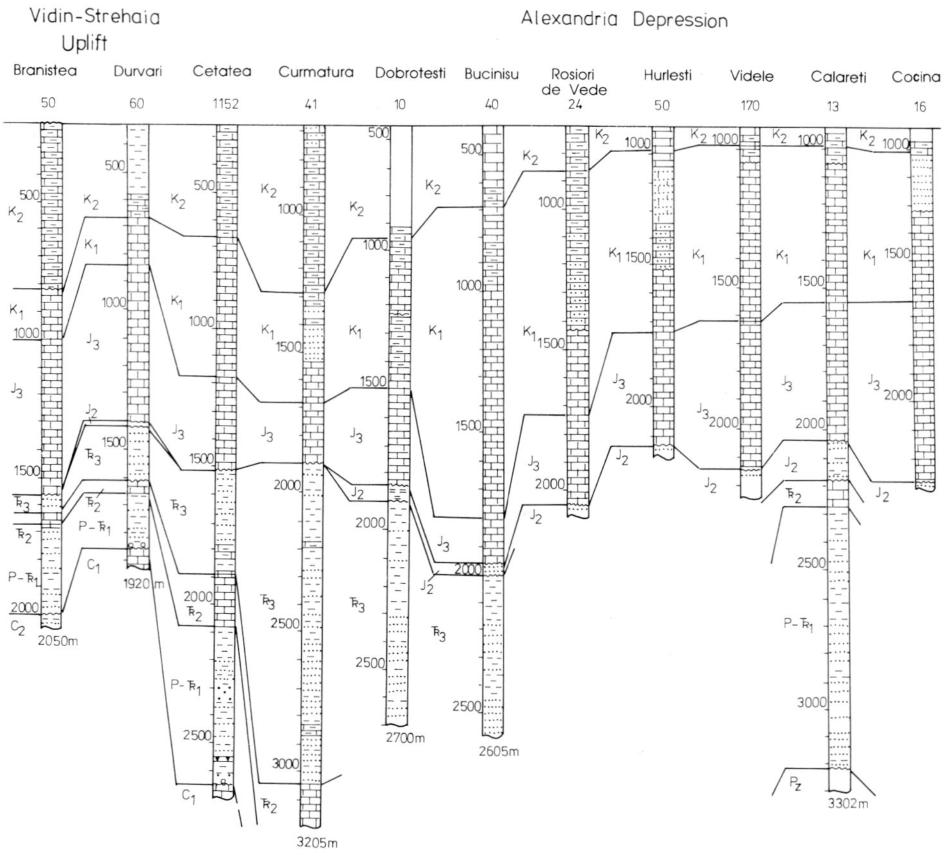

Fig. 14. Well correlation chart along line I-I in Romanian sector of the Moesian Platform. (After Strelcova in Semenovich and Namestnikova 1981) For the location of line I-I, see Fig. 17

mapping programme at a scale of 1:25,000, have been completed during the past 40 years. Results were used for the editing of several versions of geological and tectonic maps at scales of 1:1,000,000 and 1:500,000 (see map reference list). They are based on data from maps at 1:200,000 published in the late 1970s. A recent programme of editing maps at a scale of 1:100,000 is in progress. This last map programme is entirely based on international lithostratigraphical standards and includes a nation-wide accepted lithostratigraphic unit definition.

First geophysical work was performed in Bulgaria in 1936 and delineated the Provadija salt dome in the Varna area. In 1940, the newly organized Geological Prospecting Division invited the German company Seismos to carry out a gravimetric campaign around Varna.

After World War II, the geophysical work was centrally managed in Sofia by the Geophysical Enterprise. They used only one crew until 1953, increasing to three in 1955. During 1949-1953 the annual production was 200-250 line-km, and 600 line-km since 1955, i.e. some 7550 line-km cumulative in this period. From 1960 to the present the average annual seismic was 2000 line-km, bringing the country's cumulative total to 67,550 line-km. Most of the nation's seismic effort was directed to in the Moesian Platform.

Recently, Jakimov (1991) published a review of the seismic studies carried out during all periods of petroleum exploration activities in northeastern Bulgaria, since 1949 (Figs. 23 to 27). The illustrations to this paper are highly informative for the activity as a whole, covering one of the best seismically studied area of this country. The same area is covered completely by the on-shore blocks of the first round of open bidding in 1990-1991 in Bulgaria.

All known methodologies and techniques have been used, and some good results have been obtained. For example, the discovery of Dolni Dubnik oil and gas field was made by drilling based on seismic surveys carried out more than 30 years ago.

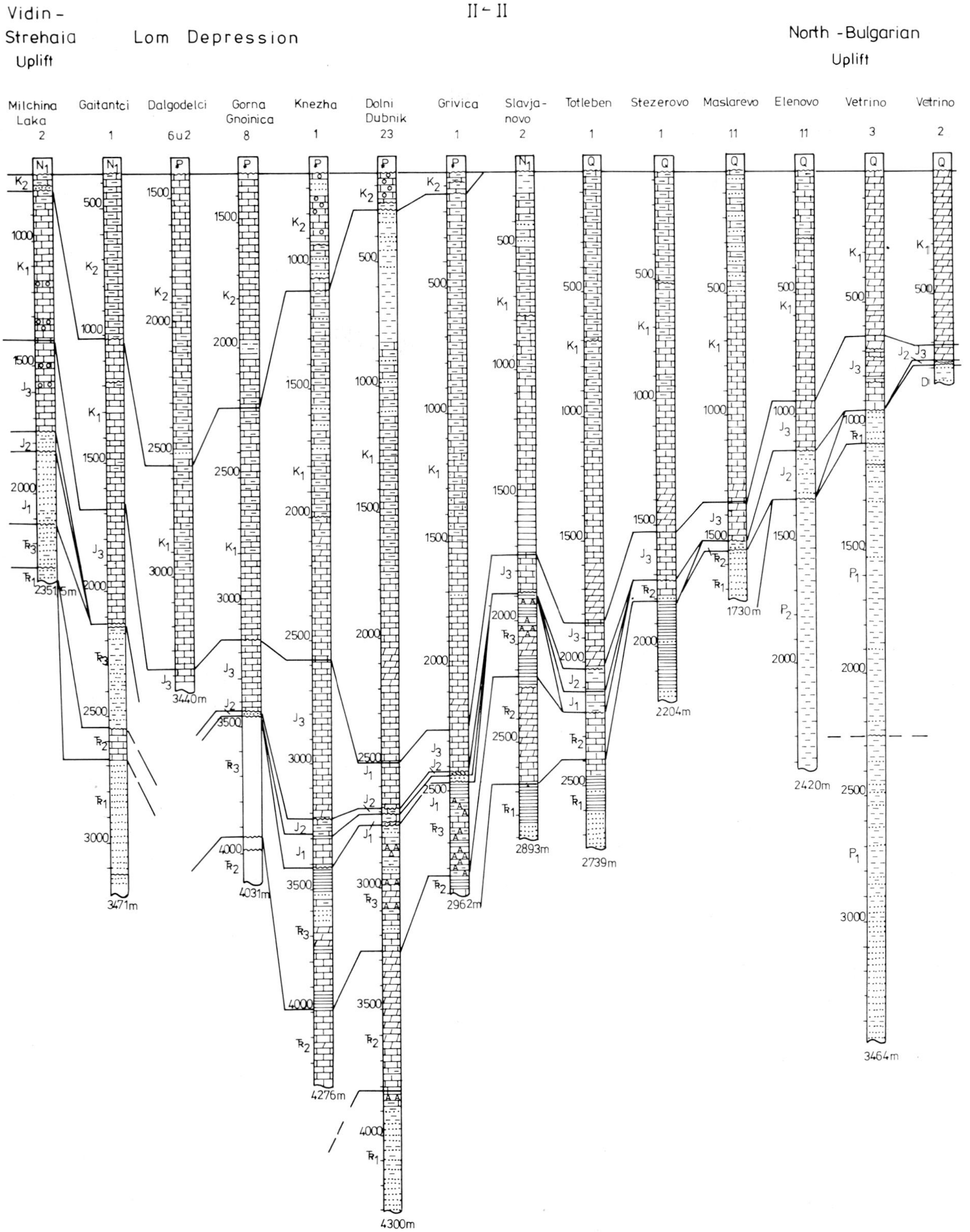

Fig. 15. Well correlation chart along line II-II in Bulgarian sector of Moesian Platform. (After Strelcova in Semenovich and Namestnikova 1981) For location of line II-II, see Fig. 17

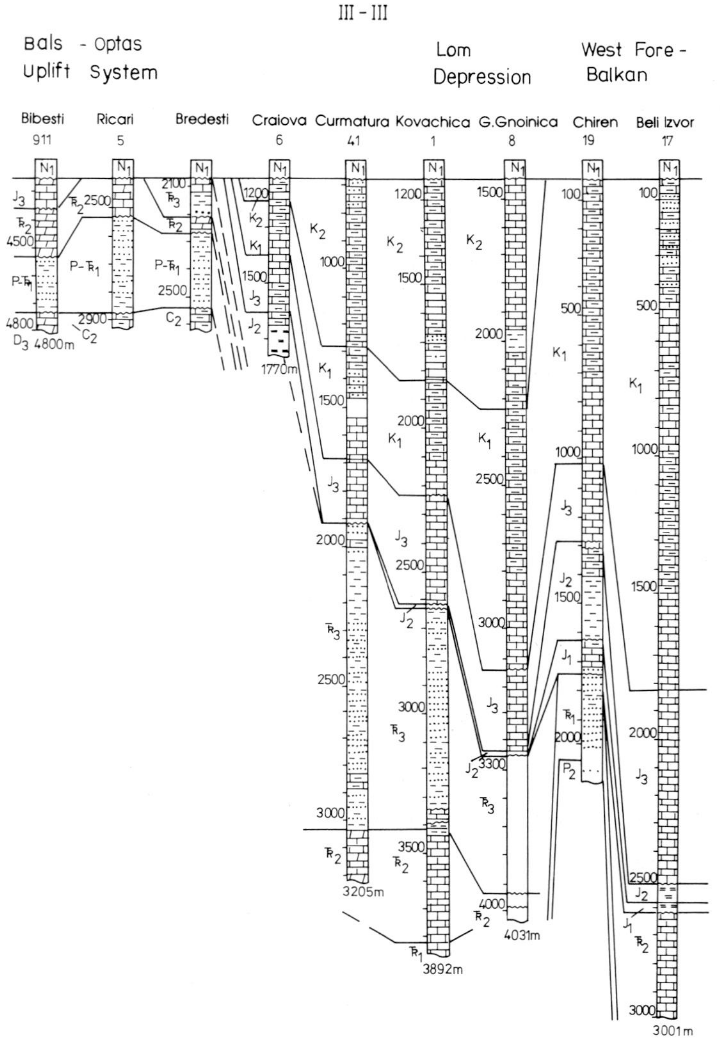

Fig. 16. Well correlation chart along line III-III crossing both Romanian and Bulgarian sectors of the Moesian Platform and the Western Fore-Balkan. (After Strelcova in Semenovich and Namestnikova 1981) For location of line III-III, see Fig. 17

Some syntheses and models on the Bulgarian deep structure were made using seismic data acquired during the past 40 years. These are mainly theoretical and a series of them was published by Dachev (1988). They all need to be clarified and tested by drilling in order to be successfully used for the goals of petroleum exploration and potential estimation.

3.7.2 Geothermal Studies

Based on the data of the thermometric measurements in 148 wells drilled during the last 30 years (Petrov et al. 1991), the geothermal field and the thermal regime in the country are described by temperature distribution maps at depths −500, −1000, −2000, −3000 and −5000 m. Mean temperature versus depth curves were computed for the major morphostructural zones. The heat flow results are in good correlation to these zones (Figs. 28 to 32). The highest temperatures of 115–120 °C were measured at 3 km depth in the Lom Basin. They decrease to 70–80 °C at the same depth in the Varna Basin.

There is evidence of a slight overheating of the structures in a closed hydrodynamic regime. A significant cooling effect is related to North Bulgaria Upper Jurassic-Lower Cretaceous carbonates, while overheating in South Bulgaria is related to the deep-seated large crystalline rocks. The temperature curves are almost linear-dependent on

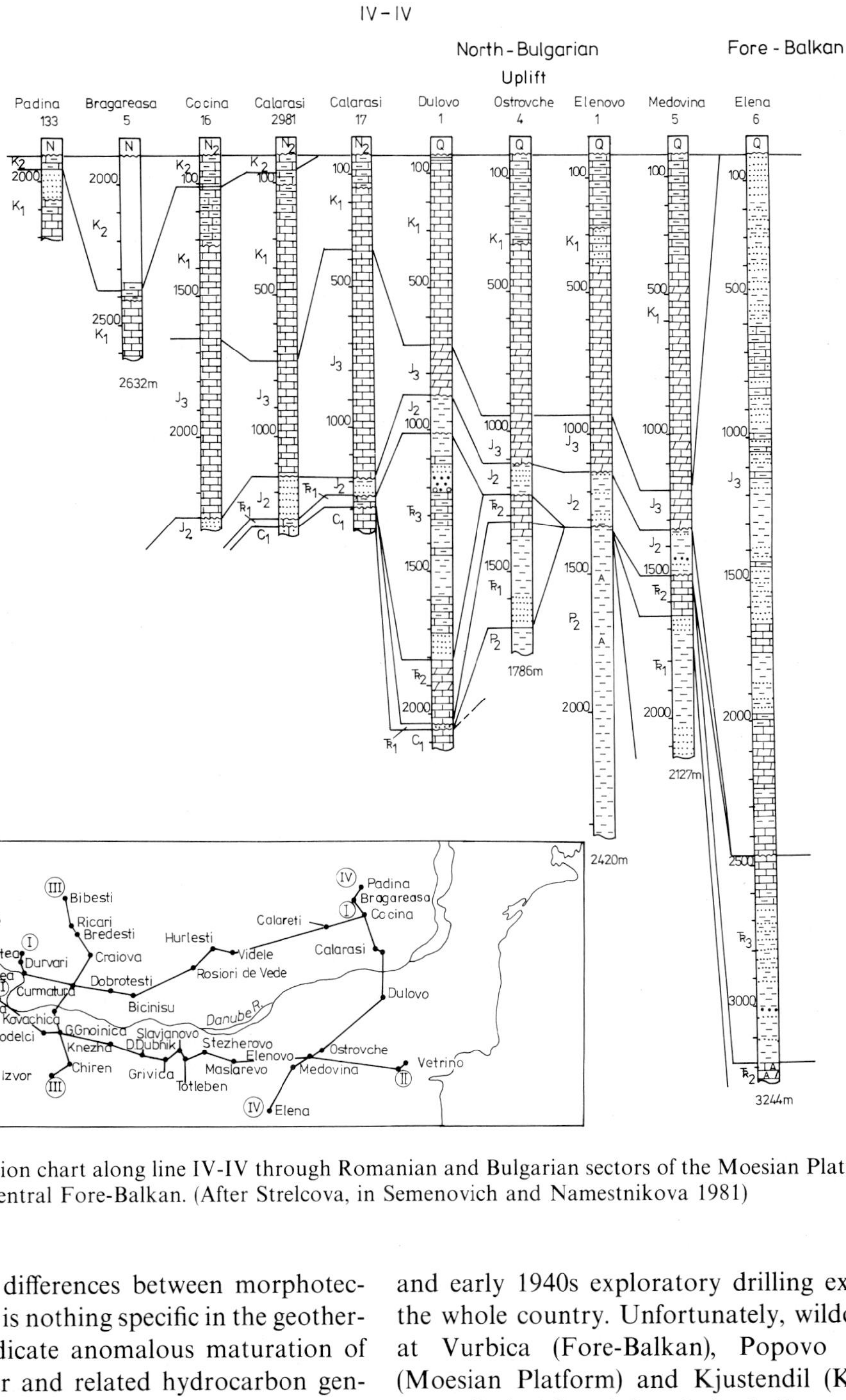

Fig. 17. Well correlation chart along line IV-IV through Romanian and Bulgarian sectors of the Moesian Platform and the eastern part of the Central Fore-Balkan. (After Strelcova, in Semenovich and Namestnikova 1981)

depth with slight differences between morphotectonic zones. There is nothing specific in the geothermal regime to indicate anomalous maturation of the organic matter and related hydrocarbon generation. Studies show good maturation of DOM in oil sediments from Paleozoic to Tertiary (Bokov and Chemberski 1987).

3.7.3 Drilling

Drilling for hydrocarbons was first performed after World War I in the Varna area. In the late 1930s and early 1940s exploratory drilling expanded to the whole country. Unfortunately, wildcats drilled at Vurbica (Fore-Balkan), Popovo and Lom (Moesian Platform) and Kjustendil (Krajshtides) were all abandoned dry. Only shallow wells drilled around the Provadija salt dome encountered shows and a deep, 1840 m, hole near Konstantinovo (Varna area) tested salt water from the Valanginian and gas from the Oligocene. In total, a few hundreds shallow holes were drilled between the two World Wars.

The first commercial discoveries made in northwest Bulgaria, after World War II, with Soviet

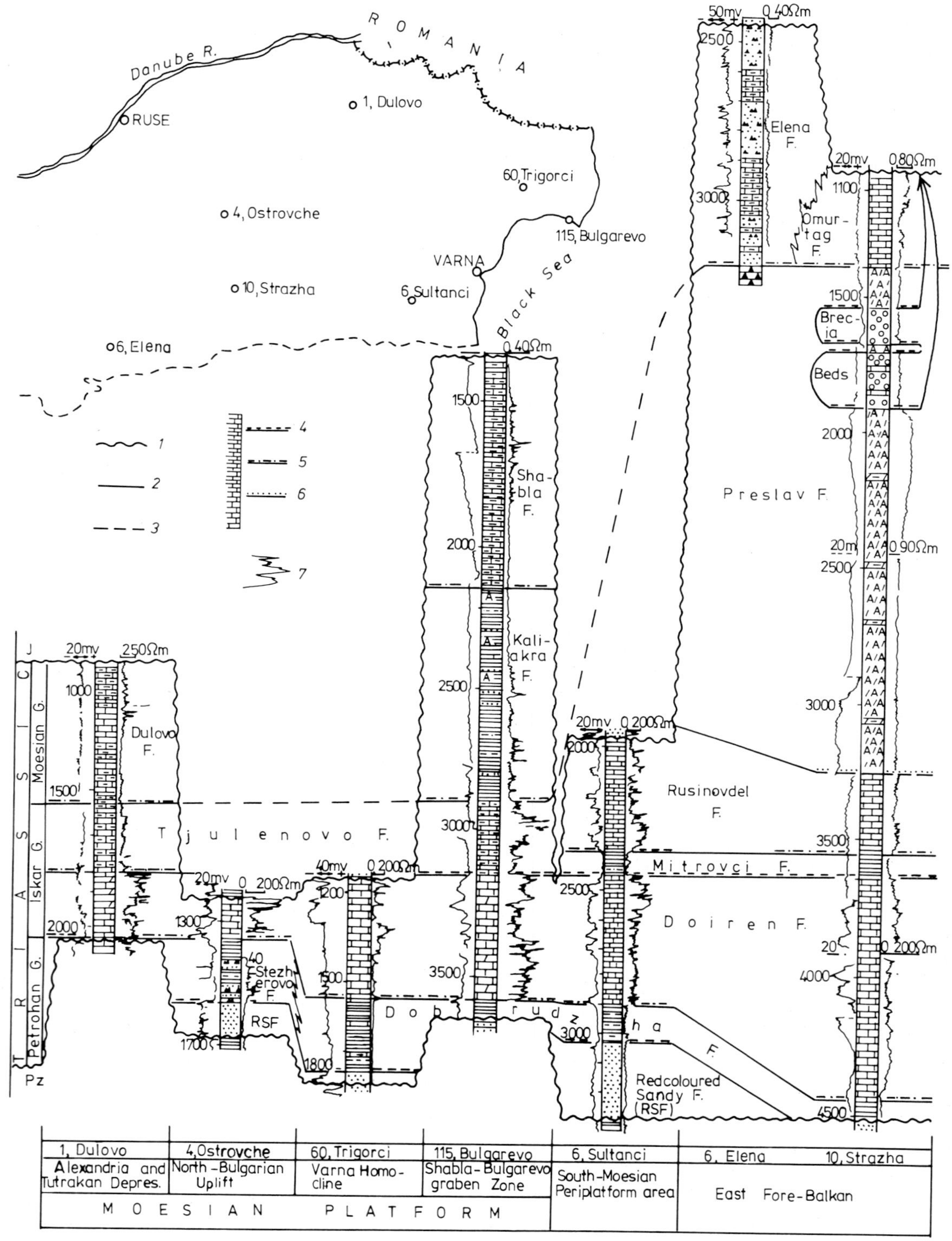

Fig. 18. Well correlation chart of the Triassic lithostratigraphic units in north-eastern Bulgaria. (After Chemberski and Doskova, in Bokov and Chemberski 1987)

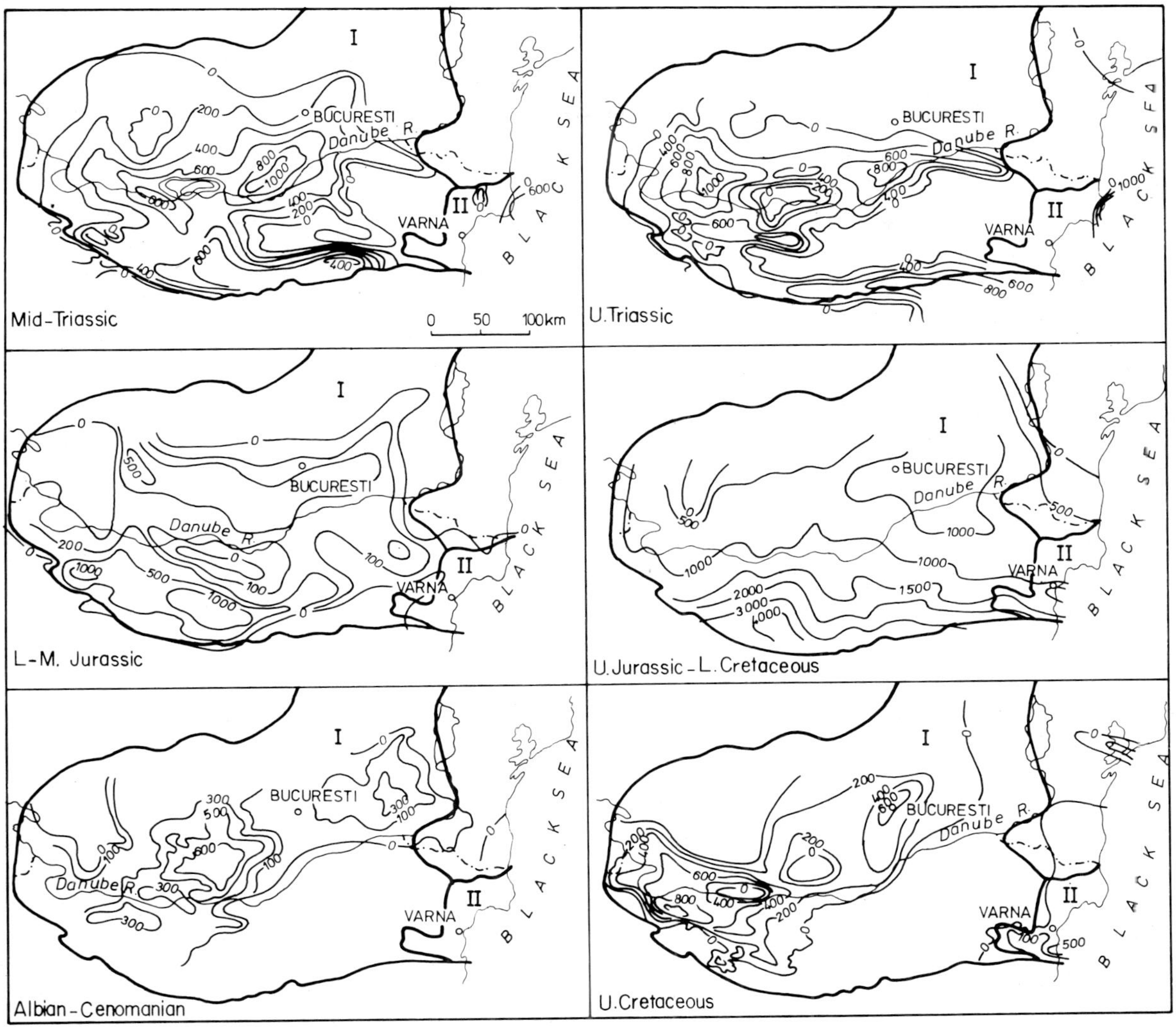

Fig. 19. Contour maps of the Mesozoic sediments thickness in the Carpathian and Balkan foreland (*I*) and west Black Sea (*II*) areas. (After Strelcova, in Semenovich and Namestnikova 1981)

assistance, stimulated a systematic exploratory effort of the Varna and Dolna Kamchija basins. Wells were generally shallow (up to 1000 m) and discovered, with the exception of Tjulenovo, only non-commercial hydrocarbon accumulations. During the period 1949–1954 some 90,000 m of stratigraphic core hole and 27,350 m of deeper exploratory and development hole, with an average depth of 600 m, were drilled in the northeast of Bulgaria. In the 1955–1959 period, about 135,000 m of stratigraphic core hole and some 55,000 m of exploratory and development wells with an average depth of 1400 m were drilled.

After the discovery in the 1950s of sizeable hydrocarbon fields in the Romanian sector of the Moesian Platform, drilling was focussed on the Bulgarian sector of the basin, with the first commercial discoveries occurring in the 1960s at Dolni Dubnik and Chiren in Triassic carbonates.

During the following 15 years, exploration was centred on this carbonate play, with the only commercial find at Devetaki. Introduction of CDP seismic recording led to the discovery of Lower Jurassic sandstone oil reservoirs in the Dolni Lukovit and Bardarski Geran stratigraphic traps.

Since the mid-1970s, all exploratory drilling has been directed towards the search for stratigraphic traps in the Moesian Platform and has resulted in the discovery of four more fields. Recently, some stratigraphic drilling was conducted in the onshore Central and Eastern Fore-Balkan, Burgas Basin, the Stara Planina and offshore Black Sea (Table 1).

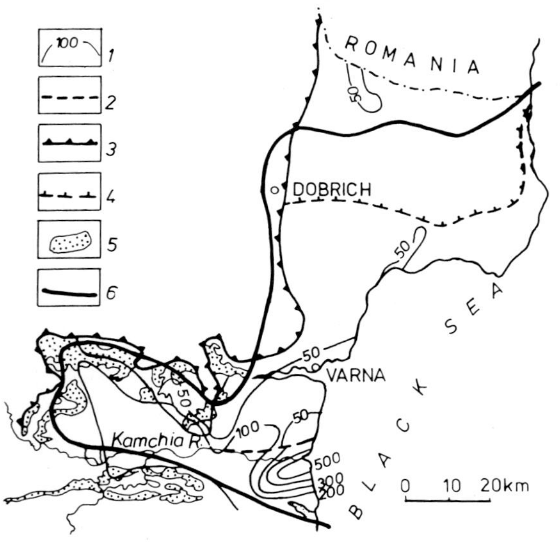

Fig. 20. Map of the Upper Cretaceous in west Black Sea basins. (After Bokov et al. 1978, in Semenovich and Namestnikova 1981). *1* Isopach contours (m); *2* Bliznak fault; *3* extension of Upeer Cretaceous sediments; *4* extension of Maastrichtian sediments; *5* outcrop of Upper Cretaceous sediments; *6* petroleum basin boundary

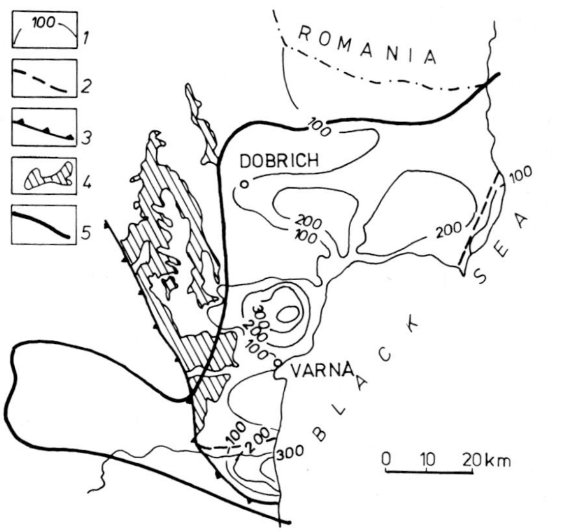

Fig. 22. Map of Neogene in west Black Sea basins. (After Bokov et al. 1978 in Semenovich and Namestnikova (1981). *1* Isopach contours (m); *2* Bliznak fault; *3* extension of Neogene sediments; *4* outcrop of older sediments; *5* petroleum basin boundary

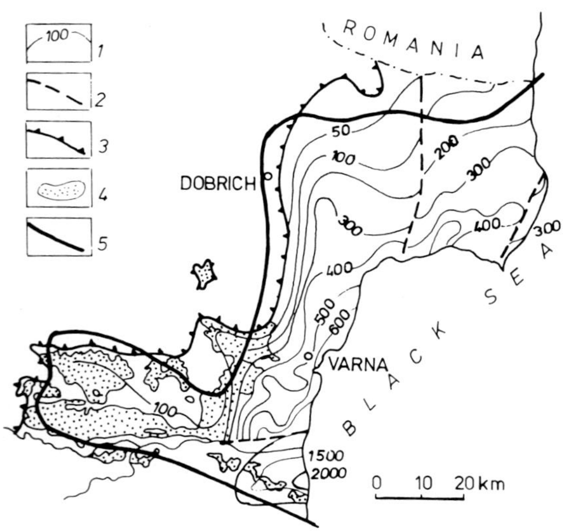

Fig. 21. Map of the Paleogene in west Black Sea basins. (After Bokov et al. 1978 in Semenovich and Namestnikova (1981). *1* Isopach contours (m); *2* Bliznak fault; *3* extension of Paleogene sediments; *4* outcrop of the Paleogene sediments; *5* petroleum basin boundary

3.7.4 Production and Reserves

The first 4700 tonnes of heavy oil were produced from the Tjulenovo field in 1951. Production increased continuously until 1968 when it reached the all time record high of 432,330 tonnes (3.3 million bbl) of oil, 505.1 million cu m (17.8 billion cu ft) of gas and 15,774 tonnes (110,000 bbl) of condensate.

After 1968 the decrease was almost continuous, with a short jump in the late seventies and 1980. In 1989 production was 72,800 tonnes (509,600 bbl) of oil and 9.28 million cu m (327.6 million cu ft) of gas.

Production comes from some 25 oil and gas fields discovered since 1949, most of which are depleted or at the end of their natural production life (Table 2). Recently, new fields, Uglen and Butan, were under active appraisal. The first is producing and the second will be put on stream soon.

4 Petroleum Geology

4.1 Organic Matter in Sediments and Source Rocks

Geochemical analyses have been carried out on a great number of rocks and recent sediments samples, crude oils and natural gases from both boreholes and outcrops. Sediments of all types and ages contain disseminated organic matter (DOM) of various source types, amount and degree of maturation.

The analytical data have been broadly published and are available from many sources (Vuchev et al. 1965a, 1976; corresponding chapters in: Kalinko 1976; Vuchev 1973; Atanasov 1980; Mandev and

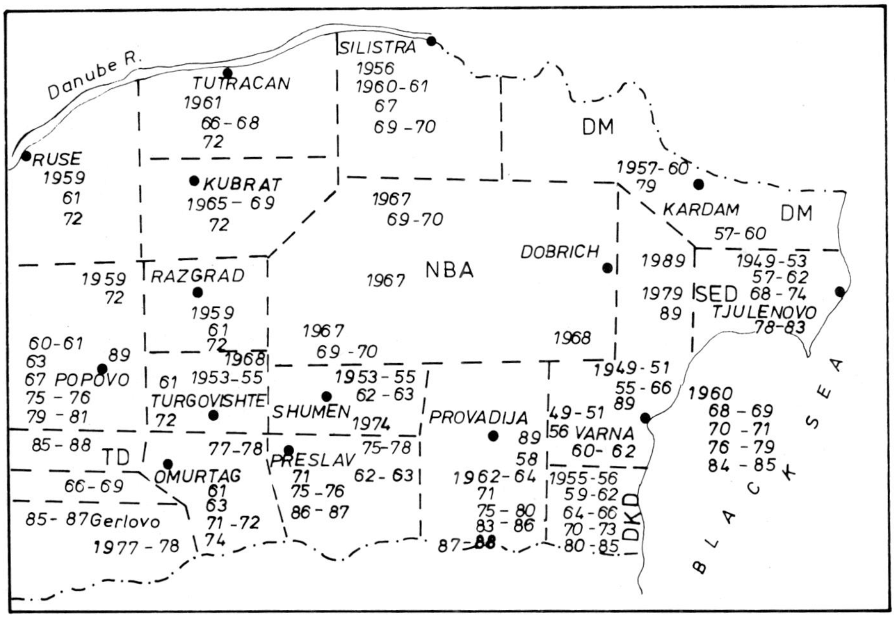

Fig. 23. Sketch map showing the seismic surveys in north-eastern Bulgaria during 1949-1989. (After Jakimov 1991) *NBA* North Bulgarian Arch (uplift); *DKD* Dolna Kamchija Depression; *SED* South Eastern Dobrudja; *DM* Dobrudja Massif; *TD* Turnovo Depression

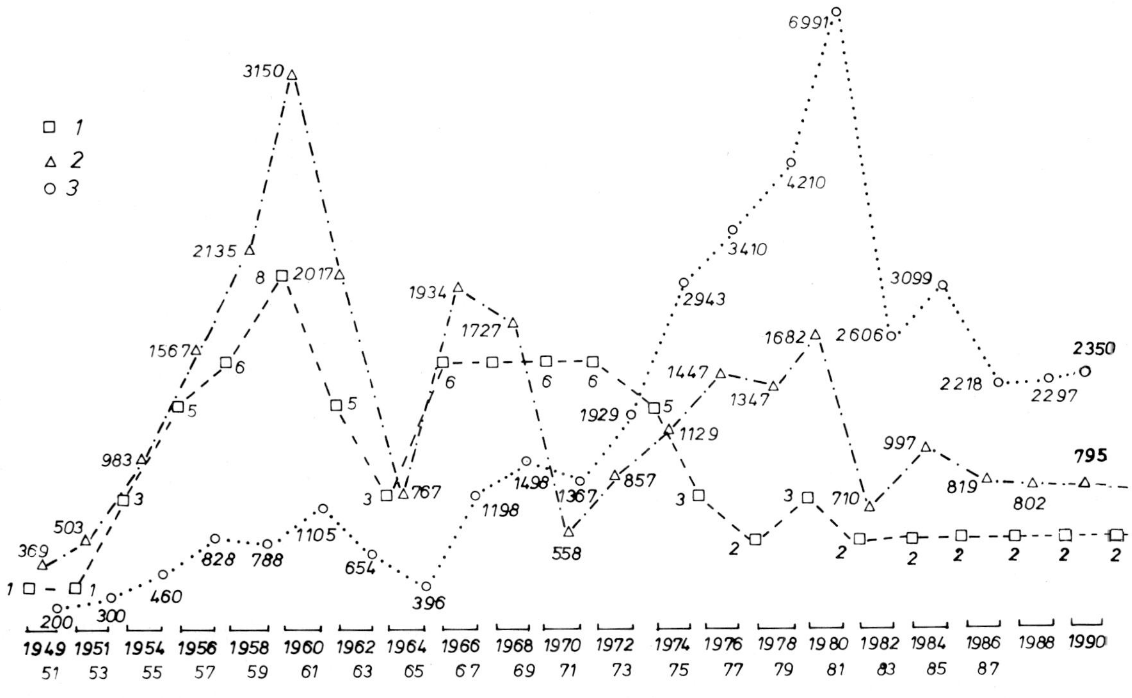

Fig. 24. Evolution of seismic activity in North-Eastern Bulgaria during 1949–1990. (After Jakimov 1991). *1* Number of seismic crews; *2* line-km; *3* Cost (thousands of levas)

Nachev 1981; Atanasov and Bokov 1983; Bokov and Chemberski 1987).

The oldest studied Ordovician and Silurian clayey and calcareous-clayey rocks from cores of mainly central and northeastern Bulgaria contain 0.40–3.40 % (% wt.) of DOM and the Lower Devonian calcareous-clayey and silty rocks are less rich, with DOM of 0.13–0.44 % and about 36 ppm hydrocarbon content (HC). The sediments are marine and the DOM is mostly sapropelic or mixed

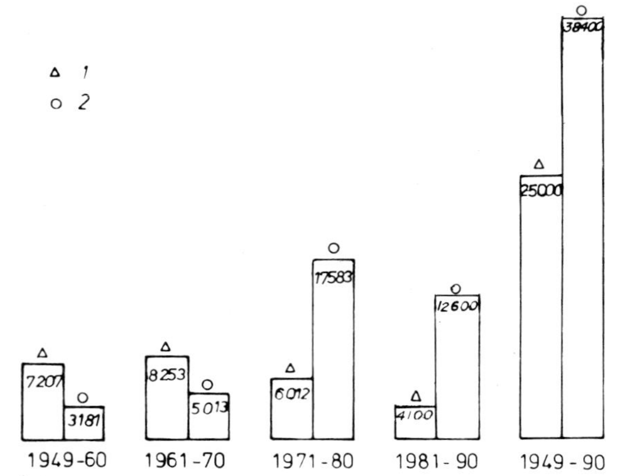

Fig. 25. Seismic activity summary by decade in northeastern Bulgaria. (After Jakimov 1991). *1* Line-km; *2* Cost (thousands of levas)

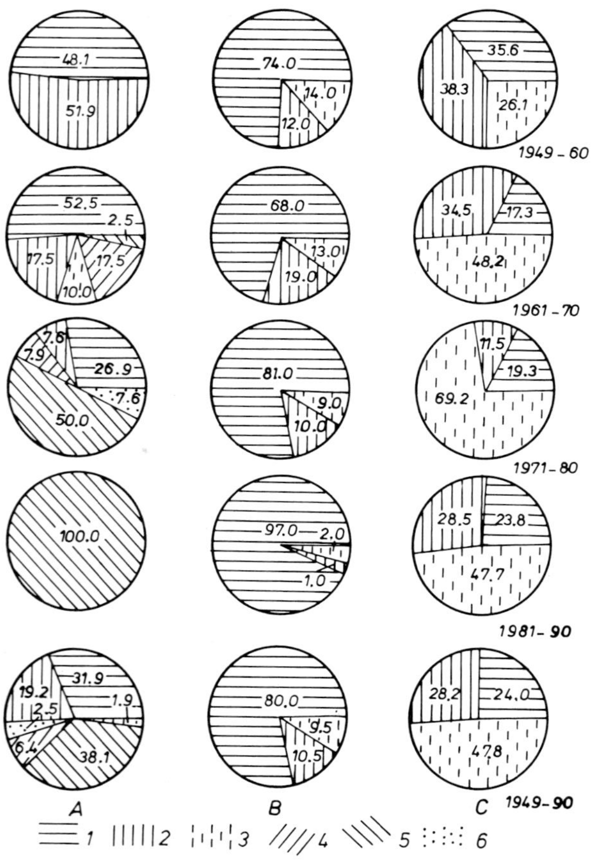

Fig. 26A–C. Method, type and scale of the seismic surveys in northeastern Bulgaria by decades. (After Jakimov 1991). **A** Methods: *1* reflection survey; *2* refraction survey; *3* deep seismic profiling; *4* sonic survey; *5* CDP; *6* transmitting exchange waves. **B** Type: *1* routine industrial; *2* experimental-methodological; *3* experimental-industrial; **C** scale: *1* regional and reconnaissance; *2* semi-detailed, *3* detailed

sapropelic-humic. Both have reached high stages of maturation and are considered as good source rocks.

The sediments of the Middle Devonian to Lower Carboniferous, 550 to 2500 m thick and highly tectonized, also in the same parts of Northern Bulgaria, correspond to two marine and lagoonal complexes: sulphate-carbonate and dolomitic. Their rock units contain the following amounts of DOM: calcareous-clayey rocks of the sulphate-carbonate complex – 0.9–3.0 %; similar rocks of the dolomitic complex – 0.65–1.0 %; carbonate rock types of both complexes – 0.05–0.50 %; marls – up to 0.3%. The amount of syngenetic HC varies between 50 and 200 ppm, but there is evidence for allochthonous accumulations where in some samples the amount of HC is up to 3000 ppm. There are source rocks for both oil and gas in the above complexes.

The Visean to Upper Carboniferous carbonate-terrigenous and coal bearing complexes, marine, lagoonal and terrestrial, covering – with a huge hiatus – older Paleozoic formations on the Moesian Platform and in the folded belts have a thickness ranging from 200 to 2500 m; the maximum corresponds to the Varna basin. The Visean sandstones and siltstones contain DOM of 0.20–1.50 %, claystones 0.10–0.70 % and limestones up to 0.20 %. On the other hand, the Westphalian sandstones and siltstones are richer in DOM – 0.7–2.0 % and coally rocks are richest of all with 3.5–9.8 % of DOM. The HC never exceed 150 ppm. Source rocks may exist but are not yet identified within this complex.

The terrestrial terrigenous and effusive volcanogenic Lower Permian, regionally distributed, but lacking on many local positive structures in northern Bulgaria, is highly unconformable, both stratigraphically and angularly, on the older formations and is, therefore, from 0 to 3000 m thick. The maximum thickness is measured in the western Fore-Balkan. It contains from 0.3 to 0.9 % humic DOM and is not known to include source rocks.

The Upper Permian evaporitic-terrigenous complex, also regionally present in northern Bulgaria, again unconformable on older formations, is from 300 to 2000 m thick and contains DOM of 0.2–0.6 % in claystones and marls, 0.4–0.5 % in clayey siltstones of the upper part of the complex and up to 4.0 % in argillites and marls of its lower part. The DOM is of both the sapropelic and humic type and the presence of source rocks within the complex is possible.

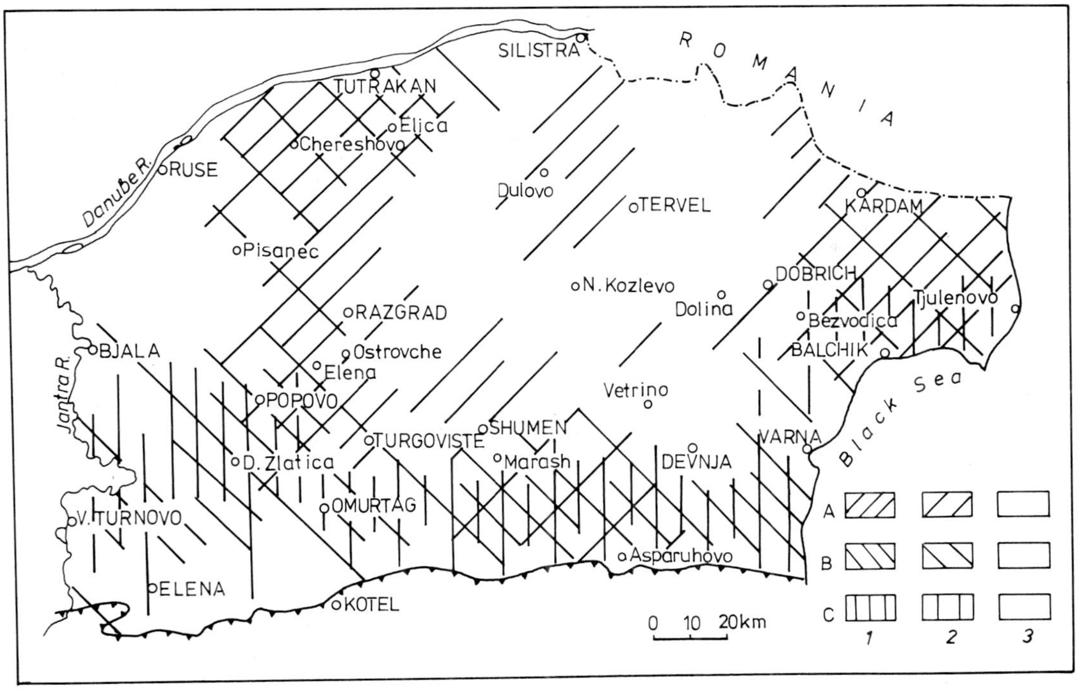

Fig. 27. Map showing the quality of seismic methods used in northeastern Bulgaria during 1949–1988. (After Jakimov 1991). *1* good; *2* medium; *3* poor. Methods applied: *A* refraction, deep seismic profiling, transmitting exchange waves; *B* reflection, sonogramming; *C* CDP

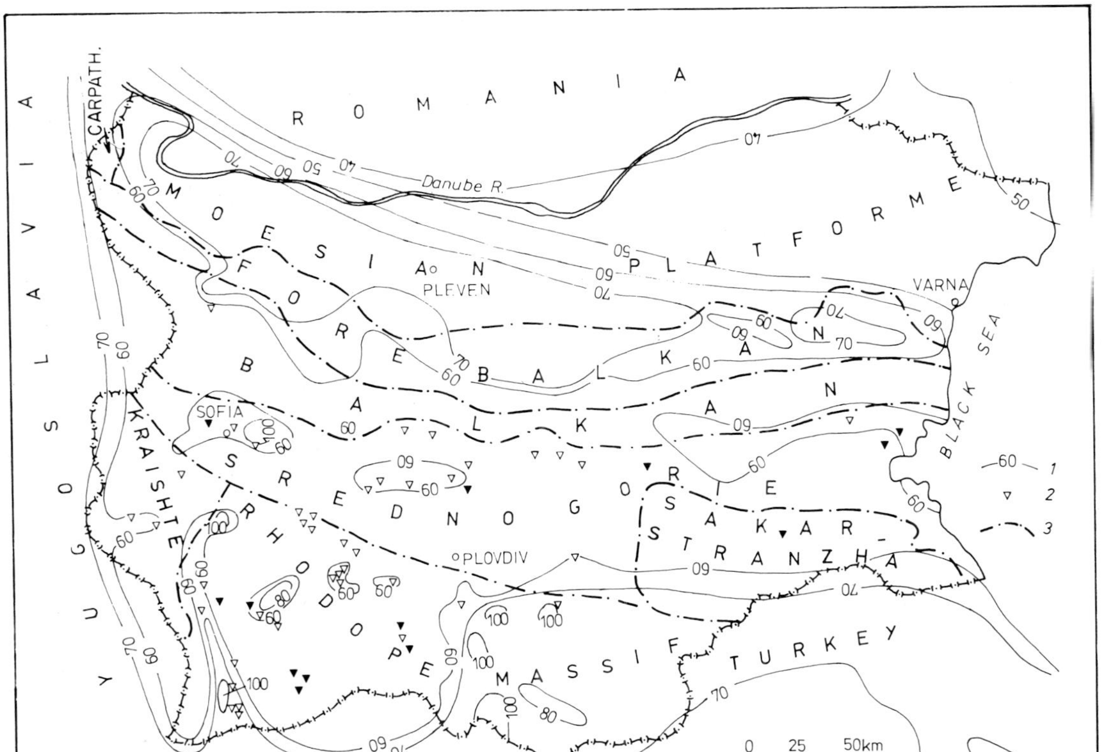

Fig. 28. Heat flow map of Bulgaria. (After Petrov et al. 1991). *1* heat flow contour line (mW/m^2); *2* heat source; *3* morphotectonic unit boundary

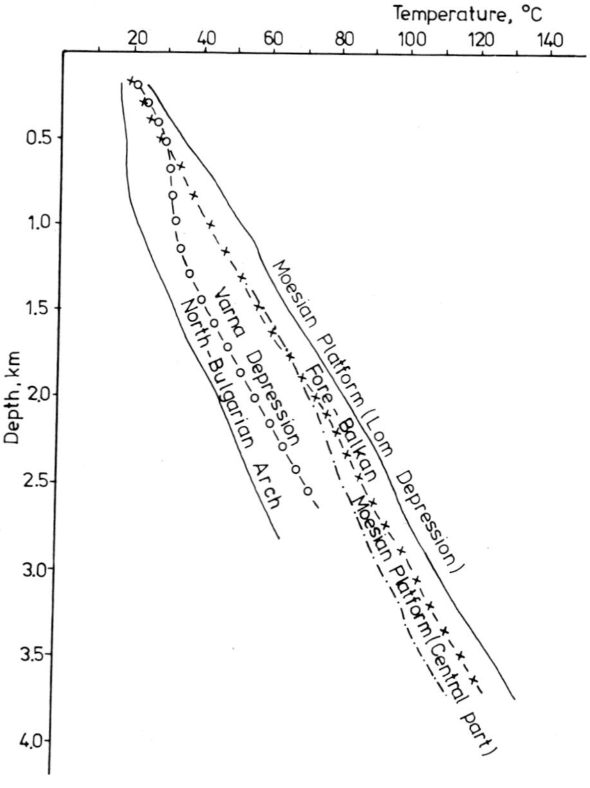

Fig. 29. Generalized temperature curves, north Bulgaria. (After Petrov et al. 1991)

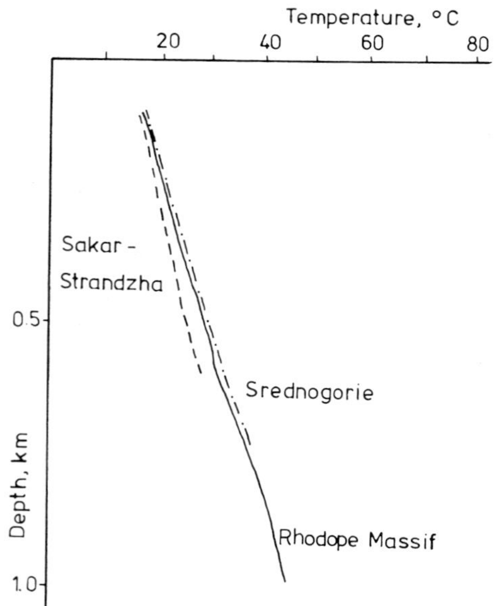

Fig. 30. Generalized temperature curves, south Bulgaria. (After Petrov et al. 1991)

The Lower Triassic carbonate-terrigenous complex is regionally distributed, penetrated by many wells in the Moesian Platform and in the Fore-Balkan, and crops out southwards in the folded belts. It is up to 600 m thick stratigraphically unconformable or lithologically transitional from the Permian and dominated by a variety of red or variegated sandstones. Calcareous-clayey sediments are related to the upper part of the complex. The average content of DOM is 0.7 % and the HC content 110 ppm independently of rock type. There is evidence of migration, since HC in some sandstones reaches 2600 ppm. The organic solvent extractable materials (EOM) or bitumens of this formation are asphaltene-resin rich, and naphtenic-aromatic HC are predominant over alkanes in oil fractions.

The Triassic carbonate complex (Middle and part of the Upper Triassic) which has a regional wide distribution, both on the Moesian Platform and in the folded areas, is the main hydrocarbon-containing, -producing and prospective unit of the whole Phanerozoic in Bulgaria, and was subjected to the most representative organic geochemical studies. The complex most likely includes its own source rocks, although they are not definitively identified. It is transitional from the previous one and has a thickness of up to 2000 m. The total volumetrically calculated amount of DOM in the whole complex over northern Bulgaria is about 2.7 billion tonnes. The content of predominantly sapropelic DOM in the complex varies between 0.06 and 1.3 %, and is even higher in the coally argillites. The means based on hundreds of analytical measurements are: 0.2 % for carbonates, 0.25–0.35 % for calcareous clays, 0.5–0.6 % for terrigenous rocks and 0.6–0.8 % for the matrix of the micritic clayey limestones. The EOM is 0.002–0.15 % and the syngeneic EOM on average do not exceed 0.01–0.015 %. However, allochtonous bitumens are present almost everywhere within the complex, since there are many rock samples containing HC variations between 40 and 3000 ppm. The elemental composition of the EOM is as follows: C: 69–85 %; H: 8–12 % and H/C (atomic): 1.3–1.8.

Less promising is the Upper Triassic carbonate-terrigenous complex transitionally overlying the previous one or placed on its eroded surface and having a thickness of up to 1000 m. Calcareous clays, siltstones and sandstones have a content of DOM between 0.03 and 0.8 % (average 0.2 %) and the EOM averages 0.02 % with HC ranging from 50 to 300 ppm.

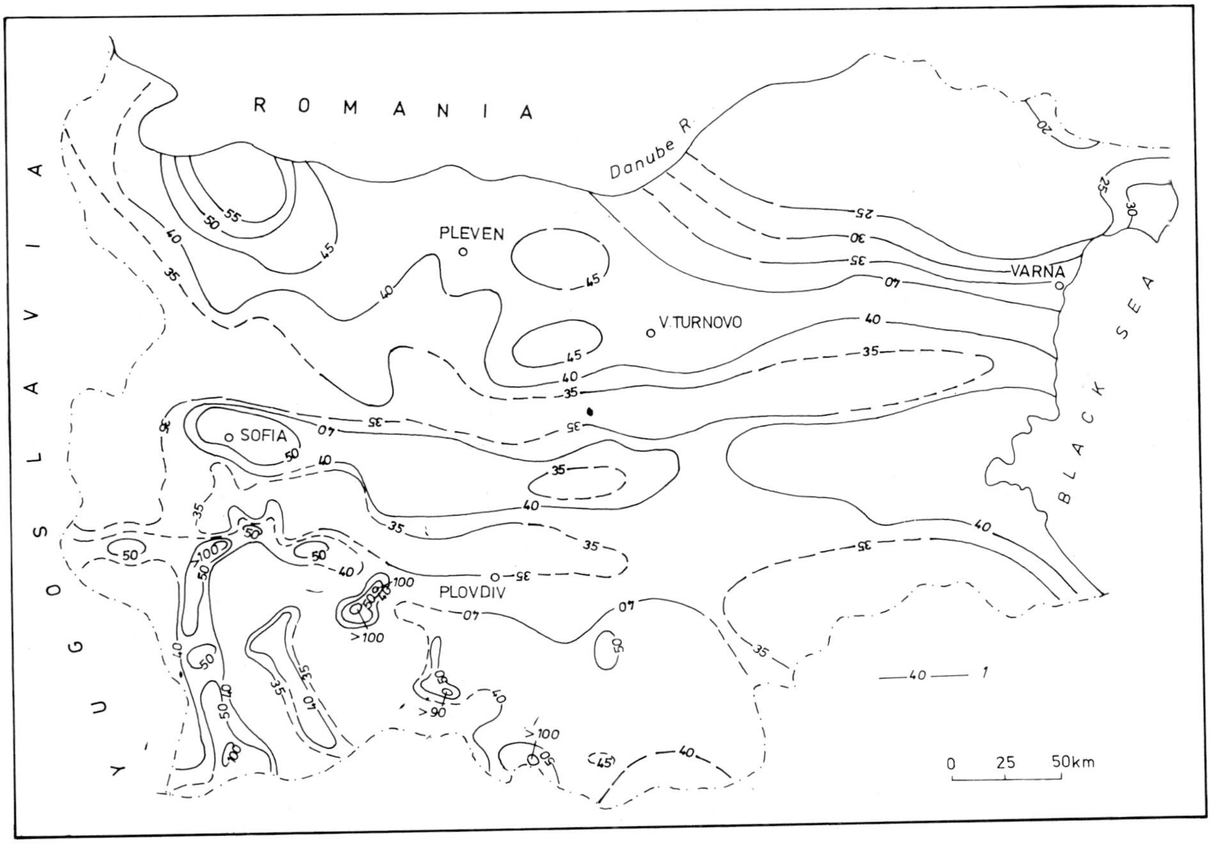

Fig. 31. Contour temperature map of Bulgaria at the depth of 1000 m. Compare with Fig. 13. (After Petrov et al. 1991). *1* isotherm, (centigrades)

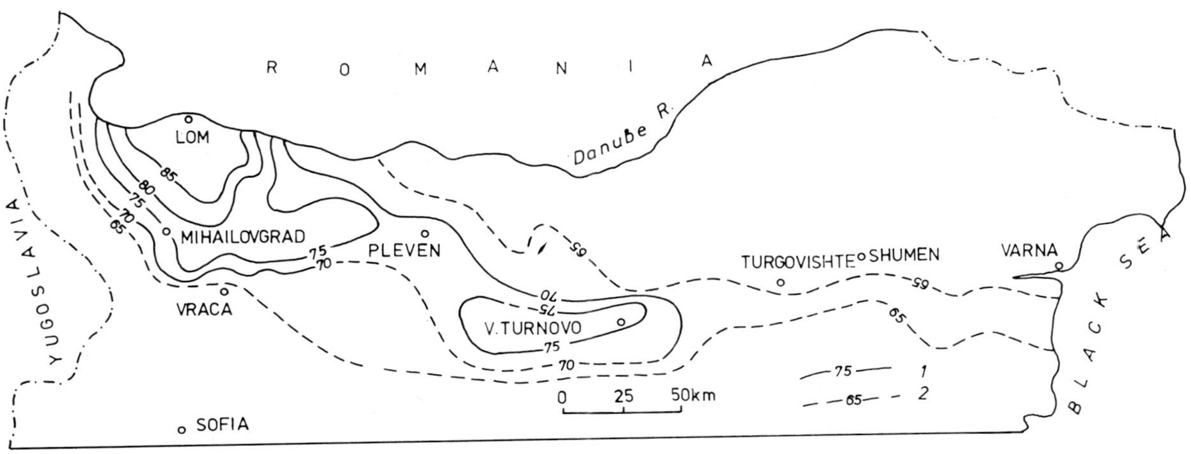

Fig. 32. Contour temperature map of North Bulgaria at the depth of 3000 m. (After Petrov et al. 1991). Isotherm, (centigrades): *1* measured; *2* estimated

The Lower Jurassic, up to 600-m-thick clastics, has DOM between 0.04 and 3.2 %. Slightly enriched are the silty marls (0.5–0.8 %) and the richest are the clayey rocks (2.7–3.2 %), but all the rocks are poor in HC, not exceeding 150 ppm. By the same volumetric calculations, around 90 million tonnes of disseminated HC are still present in this complex in northern Bulgaria.

The Middle Jurassic terrigenous and calcareous-terrigenous complex has a broad regional setting over northern Bulgaria, reaching 450 m in thickness and being richer in DOM compared to the

Table 1. Cumulative drilled meterage during the 1960 to 1992 period

Year	Wildcat	Appraisal	Stratigraphic	Structural	Development
1960	72,943	—	—	—	—
1970	448,291	265,780	101,840	423,250	65,000
1980	754,733	265,780	101,840	423,250	265,000
1989	1,200,550	441,207	202,972	423,250	637,000
1990	1,215,497	459,659	206,556	423,250	640,621
1991	1,227,356	472,549	207,823	423,250	640,621
1992	1,236,633	475,228	208,466	423,250	640,621

Table 2. Cumulative production and reserves as of January 1, 1993 for some of the oil and gas fields in Bulgaria

Dolni Dubnik oil field (Lom Basin)	
Oil	
Initial in place	9,400,000 t
Initial recoverable	3,930,000 t
Cumulative production	3,200,000 t
Gorni Dubnik oil field (Lom Basin)	
Oil	
Initial in place	1,600,000 t
Initial recoverable	482,000 t
Cumulative production	497,000 t
Dolni Lukovit oil and gas field (Lom Basin)	
Oil	
Initial in place	6,450,000 t
Initial recoverable	1,950,000 t
Cumulative production	1,440,000 t
Gas	
Dissolved in place	1,750,000,000 m^3
Free in place	380,000,000 m^3
Initial recoverable	844,000,000 m^3
Staroselci oil field (Lom Basin)	
Oil	
Initial in place	2,100,000 t
Initial recoverable	320,000 t
Cumulative production	34,000 t
Bardarski Geran oil field (Lom Basin)	
Oil	
Initial in place	2,000,000 t
Initial recoverable	300,000 t
Devetaki gas field (Lom Basin)	
Gas	
Initial in place	1,028,000,000 m^3
Initial recoverable	777,000,000 m^3
Cumulative production	317,000,000 m^3
Uglen gas field (Lom Basin)	
Gas	
Initial in place	1,100,000,000 m^3
Initial recoverable	700,000,000 m^3
Gomotraci gas field (Lom Basin)	
Gas	
Initial in place	17,000,000 m^3
Cumulative production	10,000,000 m^3

Table 2. (continued)

Tjulenovo oil and gas field (Varna Basin)	
Oil pool (Lower Cretaceous)	
Initial in place	5,980,000 t
Initial recoverable	3,090,000 t
Cumulative production	3,105,000 t
Bulgarevo gas field (Varna Basin)	
Gas	
Initial in place	176,000,000 m^3
Initial recoverable	160,000,000 m^3
Cumulative production	30,000,000 m^3
Krapec–Blatnica gas field (Varna Basin)	
Gas	
Initial in place	46,000,000 m^3
Cumulative production	7,500,000 m^3
Chiren gas-condensate field (Fore-Balkan Basin)	
Gas	
Initial in place	3,526,000,000 m^3
Initial recoverable	2,935,000,000 m^3
Condensate	
Initial recoverable	153,000 t
Vurbica gas field (Fore-Blakan Basin)	
Gas	
Initial in place	23,500,000 m^3
Cumulative production	5,000,000 m^3
Goljamo Peshtene gas field (Fore-Balkan Basin)	
Gas	
Initial in place	22,000,000 m^3
Cumulative production	13,600,000 m^3

previous complex. The richest are the shales or argillites and siltstones of the Etropole Formation (Aalenian and Bajocian) – the DOM is up to 2.4–6.0 %. The carbonates contain 0.3 % of DOM. However, the sediments are not rich in disseminated HC, which does not exceed 300 ppm. The DOM is sapropelic and humic. These sediments are considered as possible source rocks but it has not yet been proven.

The Upper Jurassic to Lower Cretaceous sediments are regionally distributed over all sedimentary morphotectonic zones and belong to two different complexes: the carbonate complex, up to 1400 m thick and covering the whole of the Moesian Platform, parts of the Fore-Balkan, Strandzha and Krajshtides, and the turbiditic complex located within the central and southwestern parts of the folded belts including central Fore-Balkan, Stara Planina and Krajshtides. The white and light-coloured carbonates are very poor in both DOM – 0.01–0.08% and HC–24–40 ppm but the transitional clayey and silty limestones and marls are richer with 0.1–0.5% of DOM and up to 100 ppm of HC. On the other hand, the turbiditic rocks are enriched in DOM 0.8–2.0%, and HC 180–200 ppm. There are indications of allochthonous HC in these sediments exceeding 400 ppm with strong predominance of paraffinic-naphthenic over naphthenic-aromatic structures. The total present amount of disseminated HC in both complexes is estimated volumetrically at 2.0 billion kilolitres, but no specific source rocks are nominated and proven.

The Lower Cretaceous (Hauterivian to Albian) terrigenous-carbonate complex has a total thickness of up to 4500 m measured in hundreds of boreholes and in the outcrops. The DOM is not abundant: 0.08–2% in silty limestones and 0.25–0.5% in marls (both are shallow marine sediments), 0.6–1.4% in deeper facies types. The same is true for HC distribution: 30–100 ppm for shallow marine facies and 190–230 ppm for deeper sediments. In some analyzed EOM, the asphaltenes and resins are predominant over oil fractions. Source rocks in this complex are likely but not clearly identified.

The Upper Cretaceous terrigenous-carbonate complex is distributed in both basins of the Moesian Platform: Lom (Moesian) and Varna basins, in the Transitional Zone and the Fore-Balkan including the Dolna Kamchija Basin. The greatest thickness (up to 1300 m) of this complex is in the Moesian Basin, the least (up to 120 m) in the Varna Basin and up to 600 m in the Transitional zone. The sediments are among the poorest of all, containing 0.1–0.2% of DOM and less than 80 ppm of HC. The petroleum potential of this complex is very limited.

Special studies were done on oil shales from young Tertiary basins, to determine the quality and amount of organic substance in the organic-rich Tertiary black shales in Bulgaria (Vuchev et al. 1977). Production of oil from these shales has been considered for some time.

The Paleogene calcareous-terrigenous complex is restricted to the Moesian and Varna basins (up to 550 m thick), and the western and eastern Fore-Balkan (up to 1000 m) and reaches its maximal thickness of 3000 m in the Dolna Kamchija Basin. The distribution of DOM is as follows: 0.05–0.08% in pure limestones, 0.10–0.16% in silty limestones, 0.2–0.5% in marls, 1% in calcareous siltstones and 2.3–2.9% in deep marine silty sediments. The same trend is seen in variations of EOM: 0.015–0.04% and HC: 65–230 ppm. In the group composition of EOM, resin fractions are up to 70%, oils 24–26% and asphaltenes 3–7%. The paraphinic-naphthenic to naphthenic-aromatic ratio of the oil fractions of sapropelic matter is 6:10.

In conclusion, the Middle Jurassic shales are richest in DOM and are considered as one of the most probable sources for generation of the major part of Bulgarian oil and gas. It should be mentioned that the DOM of these shales bears considerable differences (Vuchev et al. 1972) in molecular composition in comparison with that from the Triassic carbonates. A sufficient amount of high quality DOM is also found in Paleozoic sediments (Kovacheva et al. 1974). As stated earlier, there are only a few proven source rocks within sedimentary successions in Bulgaria. Instead, DOM is widely found in all sediments and all can contribute to the forming of oil accumulations where other conditions are favourable.

4.2 Reservoirs

Among Phanerozoic rocks, a great variety of reservoirs were distinguished based on extensive sedimentological and special laboratory studies. Many thousands of physical measurements on cores and samples from outcrops led to the recognition of reservoirs and cap rocks in the whole sedimentary section. They were published by Vuchev et al. (1965b, 1969), Shnirev et al. (1974), Kalinko (1976), Atanasov (1980), Mandev and Nachev (1981), Atanasov and Bokov (1983), Bokov and Chemberski (1987).

Good reservoirs on the Moesian Platform and in the Fore-Balkan are found within Devonian and Carboniferous, Middle and Upper Triassic, Lower and Upper Jurassic and Tertiary sediments. They belong to all known types: porous (Carboniferous, Upper Triassic, Lower Jurassic and Tertiary), fractured and mixed porous-fractured and karstified (Devonian, Middle and Upper Triassic, Upper Jurassic, Lower Cretaceous).

Predominant are the reservoirs of low porosity and permeability among both clastic and carbonate rocks. Samples are rare in which the porosity reaches 20% and the permeability 500 md. Highly porous and permeable are the regionally scattered in Northern Bulgaria, Upper Jurassic-Lower Cretaceous limestones and dolostones, bearing clear similarities, both lithological and physical, to the known excellent reservoirs of the Arabian Upper Jurassic carbonates (Powers 1962). Some Lower Jurassic sandstones, in particular the unconformable Kostina Formation, appearing in erosional wedges and lenses, as well as beds of other clastic formations, are also good reservoirs. With respect to all carbonates, predominant in the sedimentary section of the Phanerozoic, karstification and intense fracturing should also be considered as improving reservoir properties.

Buried karstified horizons were described within Paleozoic, Middle and Upper Triassic, Upper Jurassic, Lower Cretaceous, including biothermal Urgonian type limestones, Upper Cretaceous and Tertiary carbonates. Systematic fracturing, especially studied, developed cathetally in all sediments and rocks, greatly improves both the capacity and the permeability of all reservoirs. Additionally, it influences the interbed relations and contributes to the favourable geometry of reservoirs made by all kinds of pores, caves, stylolites, interbed surfaces and fractures (Vuchev et al. 1965b, Vuchev 1985).

The sedimentary rocks sequence of all described complexes contains good combinations of reservoir and impervious cap rock everywhere. They could provide traps of different kinds, which should be explored for hydrocarbons as illustrated by the selected fields in the next section.

4.3 Hydrocarbon Fields

Altogether 25 oil and gas fields were found in Bulgaria up to the end of 1991: 13 oil and oil and gas fields, including the two pools of heavy oil in Upper Jurassic-Lower Cretaceous carbonates of Tjulenovo and Gigen fields and 12 gas and gas-condensate fields. Light oil and methane gas are predominant in all corresponding cases (for field locations see Petroconsultants 1991). The fields are small to medium in size.

Most of the nation's fields were partly described in publications by Yovchev and Baluhovski (1961); Atanasov et al. (1965); Kalinko (1976); Velev (1978); Atanasov (1983); Atanasov and Monov (1984). Described and illustrated below are only a few significant fields.

4.3.1 Chiren Gas-Condensate Field

It is located in the west Fore-Balkan and corresponds to a linear NW-SE oriented anticline, 17 km long, 7 km wide and 700 m in amplitude (Fig. 33). The hydrocarbon pool is located in the Middle Triassic to Lower Jurassic carbonates and clastics and is capped by Middle Jurassic shales of the Etropole Formation. The upper part of the structure is thrusted over its northern flank (Fig. 8). The field is a combination structural and stratigraphic trap buried at 1600–2000 m. The initial production of some wells used to be 1 million m^3/d of gas. Exploitation started in 1965 and after 3 billion m^3 of gas were produced, the field was converted into the first, still operating, natural sub-surface gas storage in Bulgaria.

4.3.2 Dolni and Gorni Dubnik Oil and Gas Fields

These fields are located in the southern part of the Lom (Moesian) Depression and are structurally related to the Pleven group of uplifts. The two separated, but geographically very close, pools within Middle Triassic carbonates at a depth of 3300–3500 m are in combined structural and stratigraphic traps (Fig. 34). The Dolni Dubnik pool itself is 6×3.5 km and has a structural closure of 350 m. The Gorni Dubnik pool is much smaller. The reservoir of the Dolni Dubnik field is capped by the dolomites, also Middle Triassic. The small gas and condensate pool in the Upper Triassic carbonates on top of the uplift is capped by the evaporites of the same age.

There is an angular and stratigraphic unconformity between Triassic and Jurassic sediments and the whole uplift is normally and intensively faulted. These structures are not expressed in younger sediments.

This field was discovered in 1962 with initial production of the wells between 75 and 130 m^3/d of light oil.

4.3.3 Dolni Lukovit Oil and Gas Field

This belongs to the same group of uplifts, located within the same basin. The stratigraphic gas caped oil pool, 5×4 km, with total column of 100 m, has been accumulated within sandstones of a Lower Jurassic age (Kostina Formation), at depth of

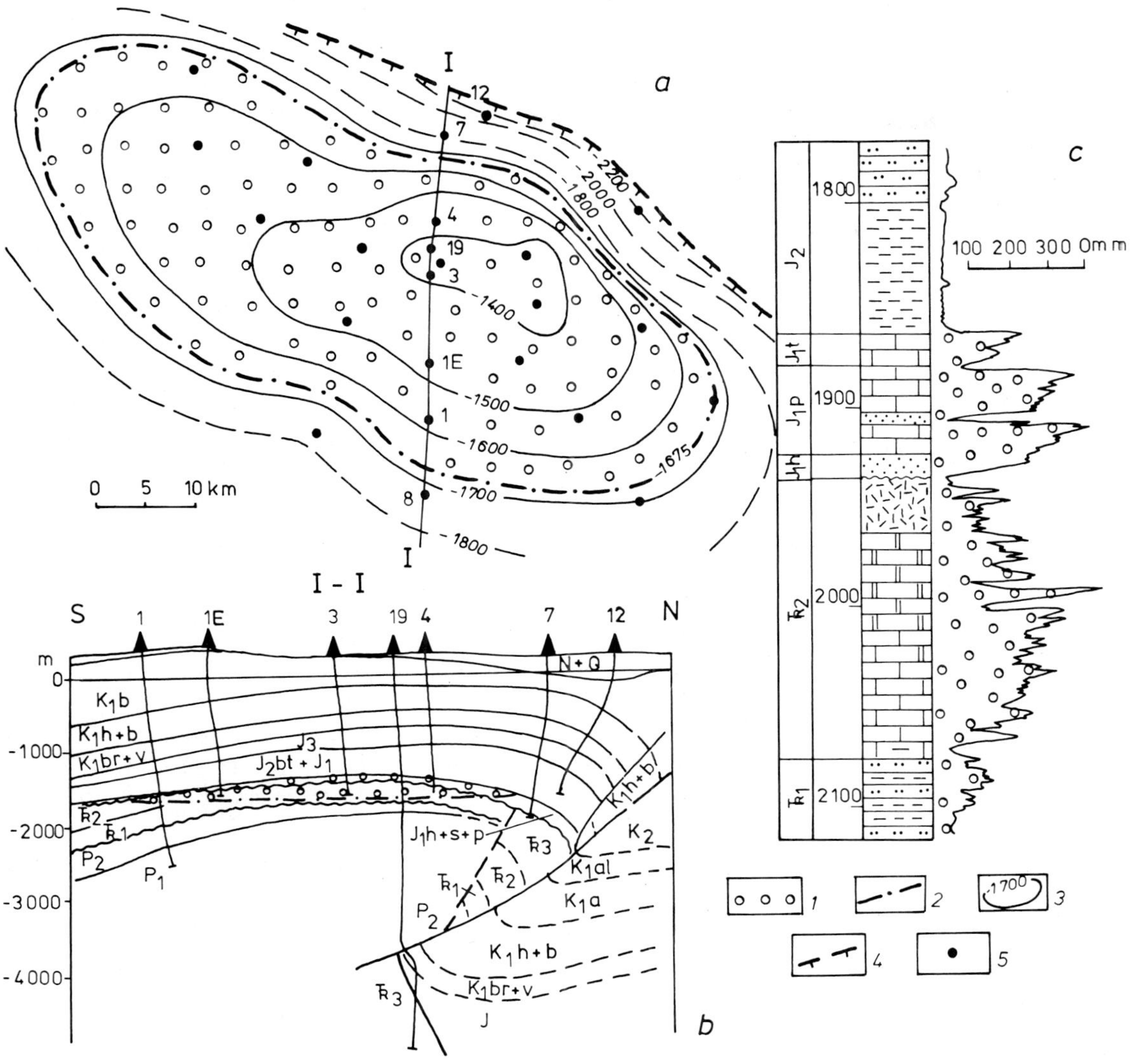

Fig. 33a–c. Chiren gas-condensate field in west Fore-Balkan. (After Vladov et al. in Kalinko 1976). **a** Structural map at the top of the Pliensbachian (Lower Jurassic) carbonates. **b** Cross-section along line I–I. **c** Lithology and log response of the reservoirs and seal. *1* gas-condensate pool; *2* initial gas-water contact; *3* contour line of the top of the Pliensbachian (m); *4* overthrust line; *5* well location

3100–3300 m (Fig. 35). The initial production of wells was between 16 and 76 m^3/d of oil.

4.3.4 Uglen Gas Field

This is located in the southern part of the Lom (Moesian) Depression on an uplift stratigraphically controlled by the relating Triassic and Jurassic sediments (Fig. 36). The pool is in the Middle Triassic carbonates at a depth of 5100–5300 m and is 8 × 2.2 km with gas column of 275 m. The initial productions of wells are 90,000–270,000 m^3/d of gas with some gas-condensate. The condensate factor is 9 cm^3/m^3 of gas. Production started in May, 1992.

4.3.5 Devetaki Gas-Condensate Field

Located in the most southern part of the Lom (Moesian) Depression, the structural and stratigraphic pool is accumulated in the Middle Triassic carbonates of a faulted block structure at depth of 3800–4000 m (Figs. 37 and 38). The size of the pool is 5 × 2 km and the gas column is 200 m. The pool is capped by the Lower Jurassic shales and limestones overlying unconformably older Triassic rocks. The

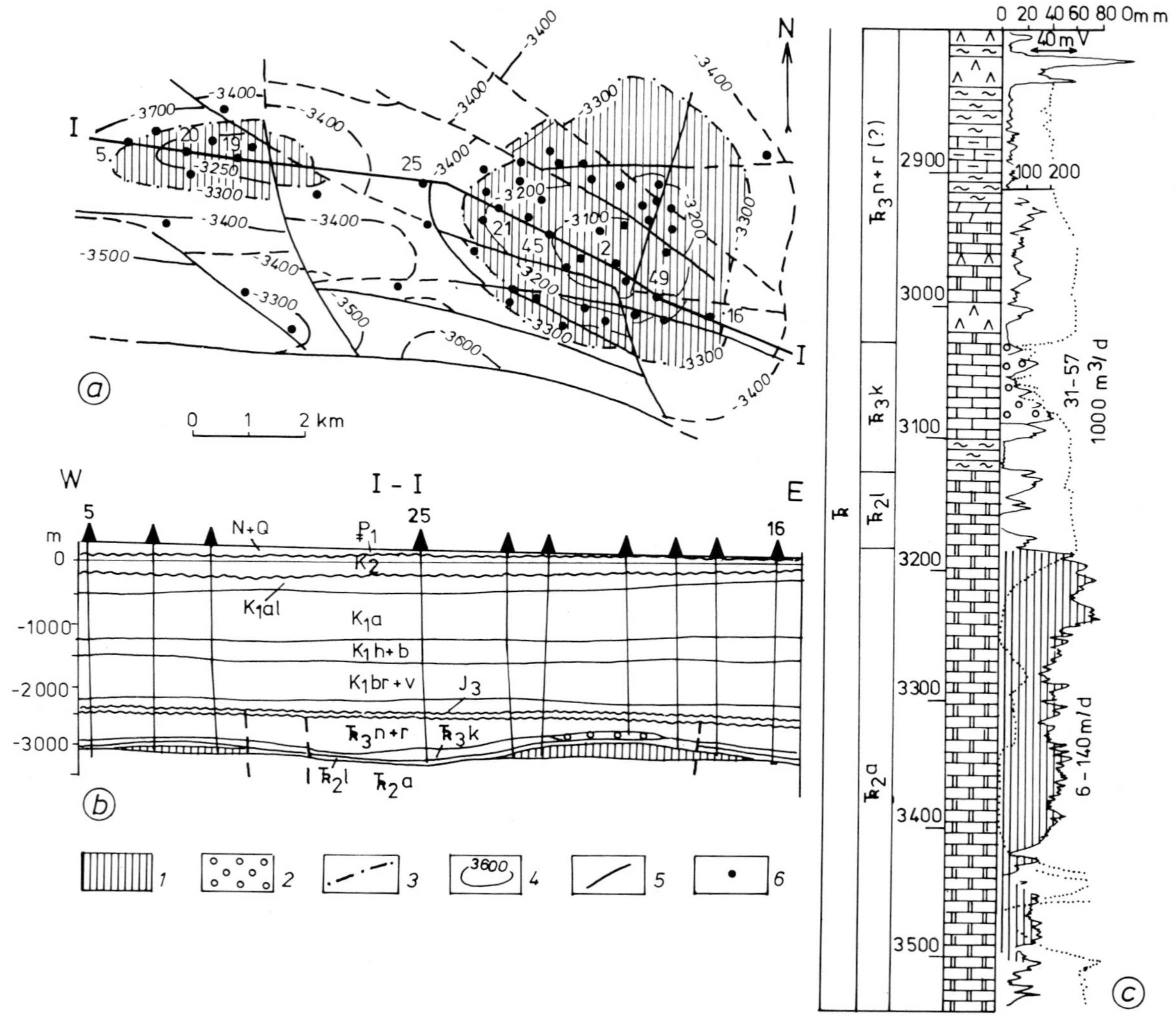

Fig. 34a–c. Dolni and Gorni Dubnik oil and gas fields in Central North Bulgaria (Moesian Platform). (After Doncheva et al., in Kalinko 1976). **a** Structural map at the top of the Anisian (Middle Triassic). **b** Cross-section along line I-I. **c** Lithology and logs of the reservoirs and cap rocks. *1* oil; *2* gas and condensate; *3* initial oil-water contact; *4* contour lines of the top of the Anisian (m); *5* fault; *6* well location

initial maximal production of some wells was 240,000 m^3/d of gas with 224 m^3/d of condensate.

4.3.6 Tjulenovo Oil and Gas Field

This is located at the Black Sea coast on the northern slope of the Varna basin and is related to the block faulted structure striking NE-SW along the coast line (Fig. 39). The field was discovered in 1951 and is still producing, largely exceeding the initially calculated recoverable reserves. The normal faults have displacements from 10 to 400 m and they cause the partition of the pools. The full size of the structure is 25 × 8 km, but there are several separated small pools within it.

The heavy oil pools (0.938 g/cm^3) of the Upper Jurassic and Lower Cretaceous carbonates are at a depth of 350–400 m and have an oil column of up to 38 m capped by a gas column of 22 m. The reservoir characteristics are very similar to the well-known Upper Jurassic Zone of the Persian Gulf (Powers 1962, Vuchev et al 1969). A small free gas pool in the Oligocene sandstones at 140–170 m has a column of 15–20 m. The initial production of the wells used to vary between 4 and 300 m^3/d as the factor of dissolved gas in oil is 10 m^3/t. The estimated recoverable reserves were initially slightly over

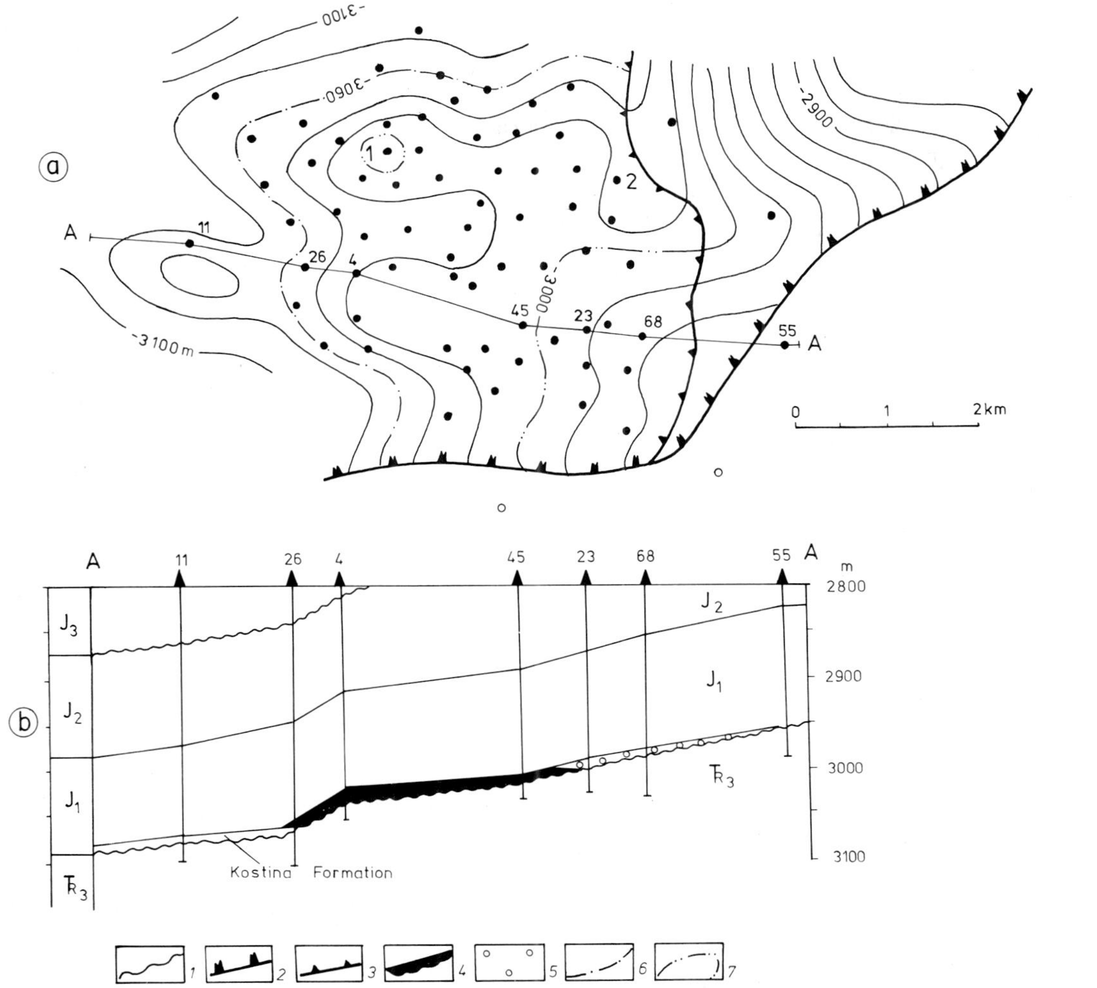

Fig. 35a, b. Dolni Lukovit oil and gas field in Central North Bulgaria (Moesian Platform). **a** Structural map at the top of the Kostina Formation (Lower Jurassic) **b** Cross-section along A-A line. *1* Unconformity; *2* extension of the Kostina Formation; *3* outline of reservoirs; *4* oil; *5* gas; *6* oil-water contact; *7* gas-oil contact

2 million metric tonnes, compared to a present cumulative production of 3 million metric tonnes.

A special study of the field using a newly created, 500-well database is being made, and a detailed structural and paleogeological model is expected to be available soon.

4.3.7 Gigen Oil Field

As mentioned earlier, this heavy oil field is located in the Lom (Moesian) Depression, on the southern bank of the Danube River, at a depth of 950–1000 m. The pool accumulated in a small positive erosional form of the Upper Jurassic-Lower Cretaceous carbonates, 2.5 × 1.5 km, with an oil column of 35 m and restricted reserves. It is much like the Tjulenovo field.

4.3.8 Staroselci Oil Field

It belongs to the same group of fields of the Lom (Moesian) Depression. The reservoirs are Upper Triassic clastics forming a structural and stratigraphic wedging trap at depth of 3100–3200 m, 4 × 1.5 km, 60 m of oil column and restricted reserves.

4.3.9 Pisarovo Gas-Condensate Field

Another field of the above group, related to the Middle Triassic carbonates, having a stratigraphic trap at a depth of 3700–3800 m, size 2.5 × 1.5 km, 80 m total column of gas condensate and oil and limited reserves.

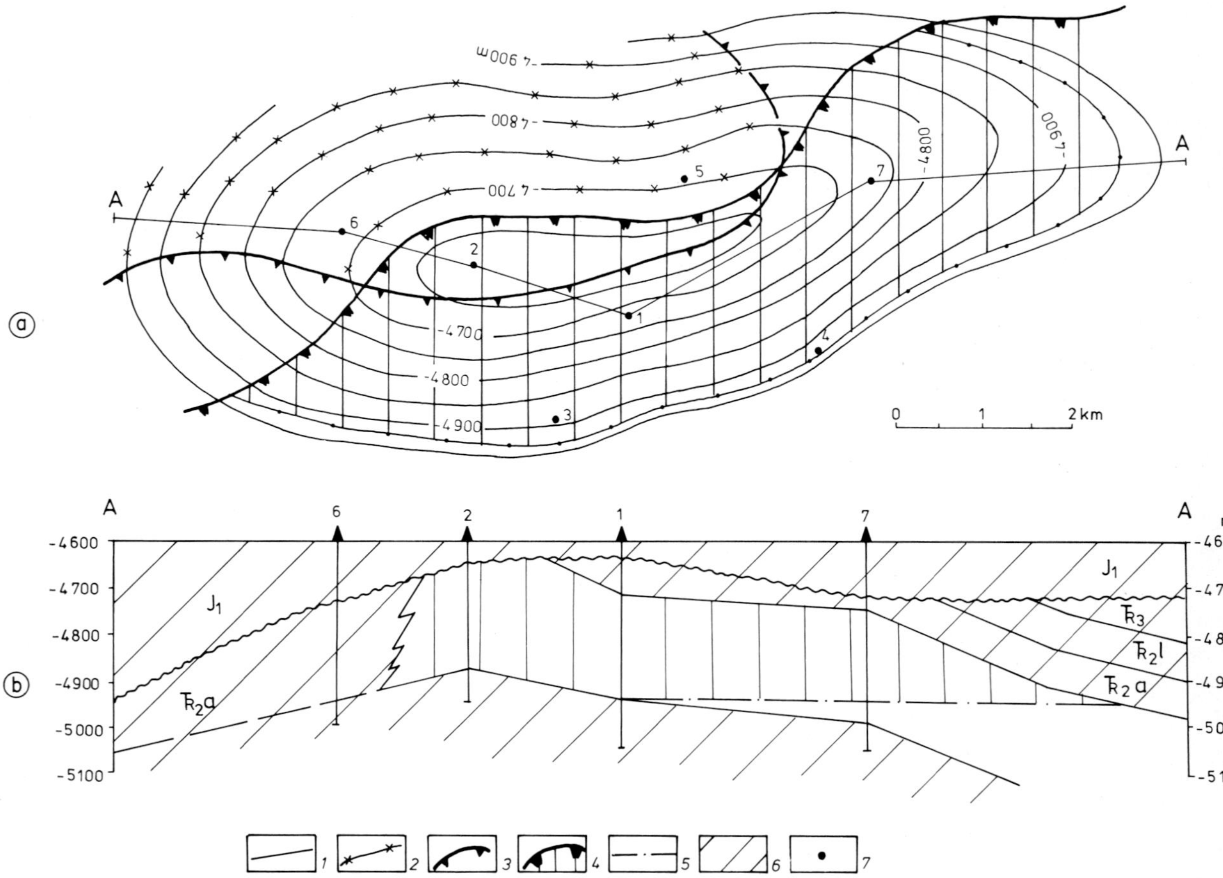

Fig. 36a, b. Uglen gas field in Central North Bulgaria (southern part of Moesian Platform). **a** Structural map at the top of the reservoir and the erosional Triassic-Jurassic boundary. **b** cross-section along A-A line. *1* contour lines at the top of the reservoir (m); *2* contour lines of the erosional Triassic-Jurassic boundary (m); *3* extension of the eroded, in part saturated, reservoirs; *4* boundary of the maximum reservoir extension; *5* gas-water contact; *6* cap rocks and other impervious sediments; *7* well location

4.3.10 Gomotraci Gas Field

This is a field of different type, located within the Vidin-Strehaia Uplift to the Western edge of the Moesian Platform (Fig. 4).

The small (2.8 × 1.8 km and closure of 18–20 m) brachyanticline holds several separate pools, in up to 2-m-thick Upper Meotian siltstone beds, buried at a depth of 50–80 m. The initial production of wells used to be 20,000–40,000 m^3/d of gas. The limited reserves of this field are used locally.

4.3.11 Koshava Gas Field

Similar and closely located to the previous, this field is again a brachyanticline holding several small traps (1 × 0.6 km and closure of up to 15 m) among Lower Meotian sandstones at depth 180–200 m. The initial production of the wells was 20,000–53,000 m^3/d of gas and the limited reserves are also used locally.

4.3.12 Other Fields

In the northern Bulgarian basins, a number of other small oil and gas pools have been discovered. Some gas pools have estimated reserves of up to 700,000 million m^3, while those of oil are not yet calculated, being considered non-commercial. The gas fields are: Krapec-Blatnica, Bulgarevo, Priselci and Dolna Kamchija in the Varna and Dolna Kamchija basins; Turnak, Hajredin, Dobrolevo, Butan, Ponora, Goljamo Peshtene, Vurbica, Glavaci in the Lom (Moesian) Depression and

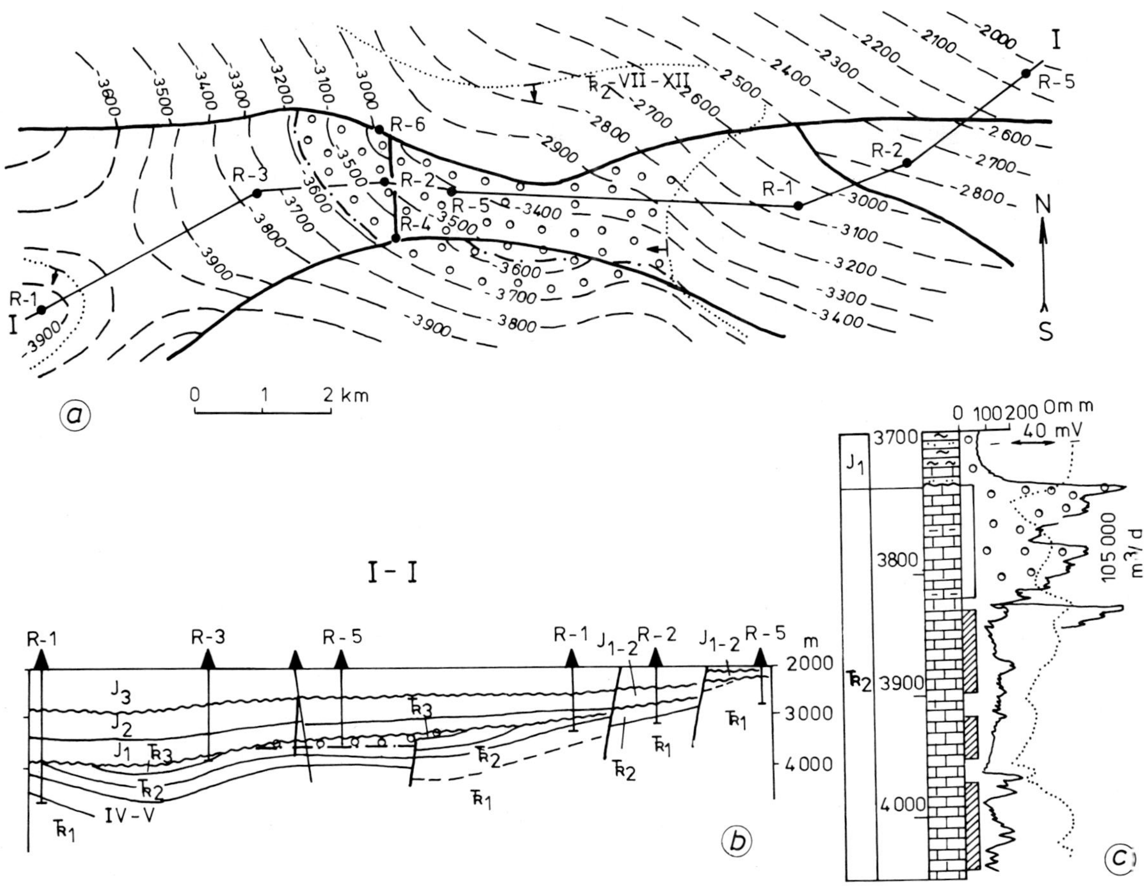

Fig. 37a–c. Devetaki gas-condensate field in Central North Bulgaria (southern part of Moesian Platform). (After Golovackij et al. in Kalinko 1976). **a** Structural map contoured at the top eroded surface of the Triassic sediments within the Umarevci, Devetaki and Karpachevo area. **b** Cross-section along line I-I. **c** Lithology and log responses of the reservoir. Legend as in Fig. 34. *Dotted lines* in **a** are the distribution boundaries line of the Middle Triassic (units VII-XII). Shown in **b** are also the Triassic units IV-V.

Fore-Balkan. The oil accumulations are: Bohot, Gradina, Burdarski Geran, Marinov Geran, Selanovci, Novo Orjahovo, Kozloduj in the Lom (Moesian) depressions and Beli Izvor in the Fore-Balkan.

4.4 Classification of Traps

All fields described and mentioned are entered into a simple classification scheme (Fig. 40, Table 3) which can be referenced during further exploration in the country.

Structural traps of all varieties are found in the Fore-Balkan (Fig. 40, class I, types 1–3) The oil fields are related to tilted blocks and anticlines, having massive or bedded reservoirs. In the Lom (Moesian) Depression structural traps were discovered in Triassic carbonates, homoclines overlapped by younger sediments (Dolni Dubnik), or in fault-controlled traps (Devetaki, Ponor, Bohot) (see also Figs. 33–39).

In both the Fore-Balkan and the Moesian Platform stratigraphic traps (Fig. 40, class II, types 1–3) were tested and can generally be classified as follows: lithologically controlled (Dolni Lukovit) or lithologically trapped (Staroselci); stratigraphically trapped, i.e. erosional cutout of the Triassic reservoirs unconformable overlapped by Lower and Middle Jurassic sediments (Pisarovo); and paleorelief traps (Gigen).

During the last few years, special attention has been given to the reef build-ups which have been found in the Middle Triassic (e.g. Uglen field) and Lower Cretaceous carbonate sequences (Bokov and Petrov 1981; Bokov 1989).

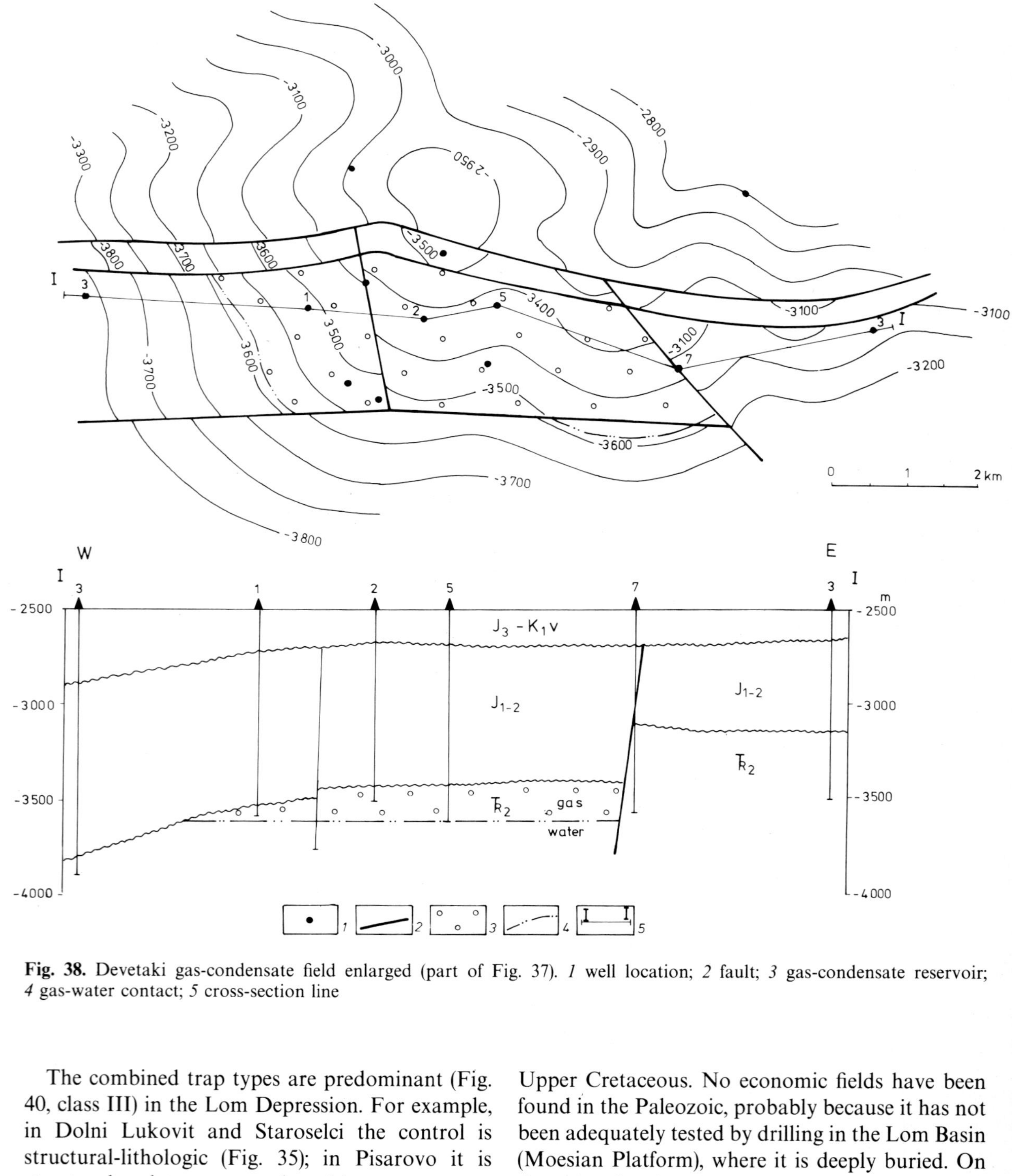

Fig. 38. Devetaki gas-condensate field enlarged (part of Fig. 37). *1* well location; *2* fault; *3* gas-condensate reservoir; *4* gas-water contact; *5* cross-section line

The combined trap types are predominant (Fig. 40, class III) in the Lom Depression. For example, in Dolni Lukovit and Staroselci the control is structural-lithologic (Fig. 35); in Pisarovo it is structural and tectono-stratigraphic.

4.5 Age Distribution of Fields and Shows

The age spectrum of all hydrocarbon manifestations in Northern Bulgaria covers the entire Phanerozoic (Fig. 41) with the exception of the Upper Cretaceous. No economic fields have been found in the Paleozoic, probably because it has not been adequately tested by drilling in the Lom Basin (Moesian Platform), where it is deeply buried. On the other hand, Middle and Upper Triassic and Lower Jurassic sediments are the best oil and gas reservoirs in the same basin. In the Varna Basin, the carbonates of Upper Jurassic and Lower Cretaceous have been prolific shallow reservoirs. Free gas pools have been discovered in the Oligocene of the basin. The Dolna Kamchija Basin has gas pools and shows in the Tertiary clastics only. Small gas

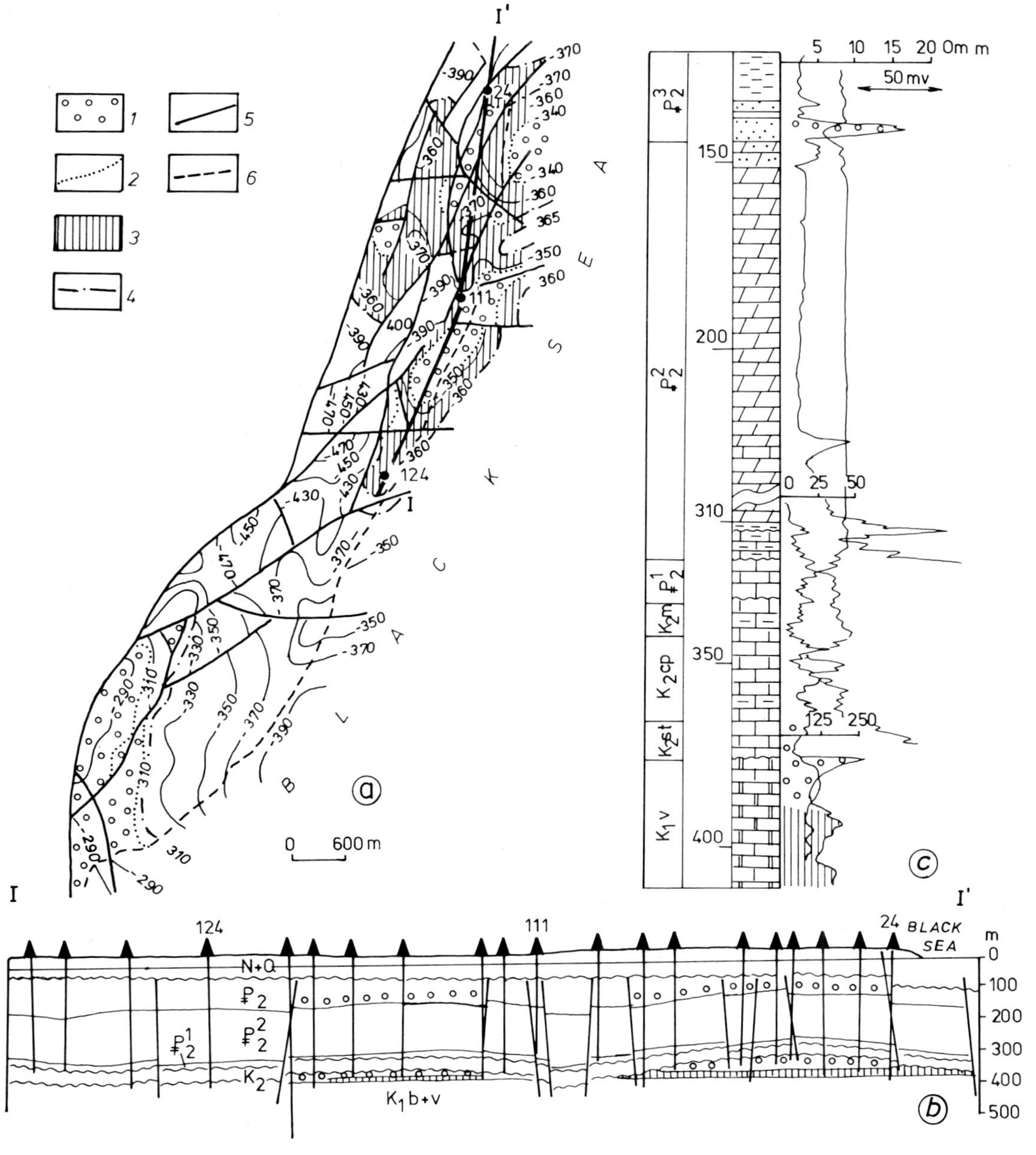

Fig. 39a–c. Tjulenovo oil and gas field in east-northern Bulgaria (eastern part of Moesian Platform). (After Golovackij et al., in Kalinko 1976). **a** Structural map at the top of the Berriasian-Valanginian carbonate complex (m) **b** Cross-section along line I-I. **c** Lithology and log response of the producing intervals. *1* Gas pool and free-gas cap above oil pool; *2* initial extension of the gas reservoir; *3* oil pool; *4* initial extension of the oil pool; *5* fault; *6* sea shore line

pools in Neogene clastics have been found in the western part of the Lom Basin.

No oil fields has yet been discovered in the Fore-Balkan, but gas shows are known from the whole sedimentary section and the oil shows are located within Jurassic and Lower Cretaceous formations only.

5 Conclusion

The geologically rich and well-studied area of Bulgaria is of definite interest for oil and gas exploration. Although not very many and only small to medium-sized hydrocarbon fields have been found to date, further studies should consider all

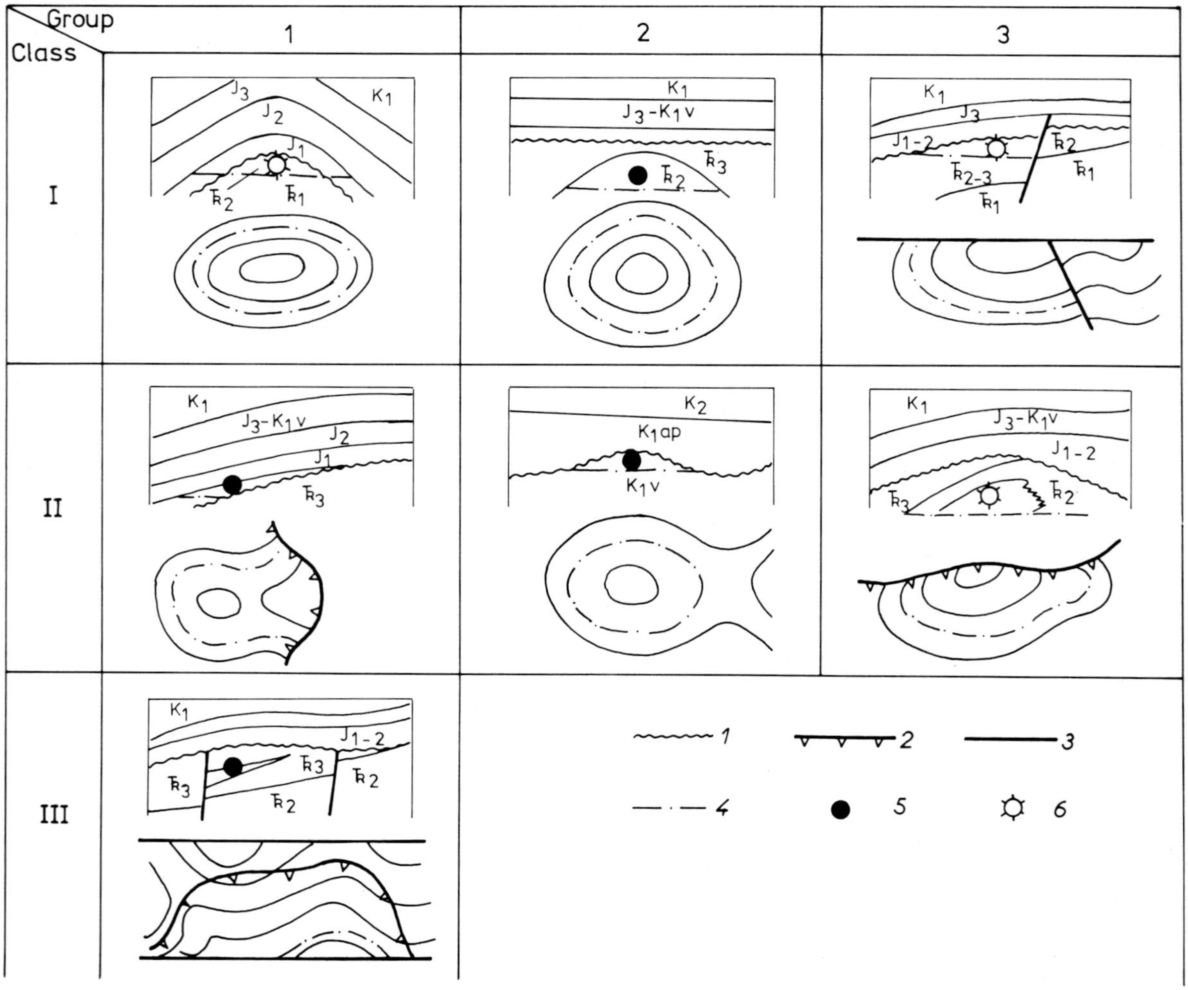

Fig. 40. Chart showing main types of traps and pools of hydrocarbons in Bulgaria (for details see Table 3). *1* Unconformity; *2* lithologic boundary; *3* fault; *4* gas or oil-water contact; *5* oil; *6* gas

Table 3. The classification of main trap of hydrocarbon accumulations in Bulgaria

Class	Group	Type	Basin	Example
I Structural	1. Anticlinal and structural swell, simple	Arch	Fore-Balkan	Chiren
	2. Anticlinal and swell, different structural plans	Arch	Moesian	Dolni Dubnik
	3. Anticlinal and monoclinal and faults	Structural	Moesian Fore-Balkan	Devetaki Ponor
II Stratigraphic	1. Lithologic wedged out and pinched out	Lithologic screened	Moesian	Dolni Lukovit
		Lithologic restricted	Moesian	Staroselci
	2. Stratigraphic unconformity	Stratigraphic		
	a) anticlines and monoclines		Moesian	Pisarovo
	b) uplifts of palerorelief		Moesian	Gigen
	3. Bioherms	Reefs	Moesian	Uglen
III Combined	1. Structural Stratigraphic	Stratigraphic and tectonic	Moesian Moesian	Staroselci Pisarovo

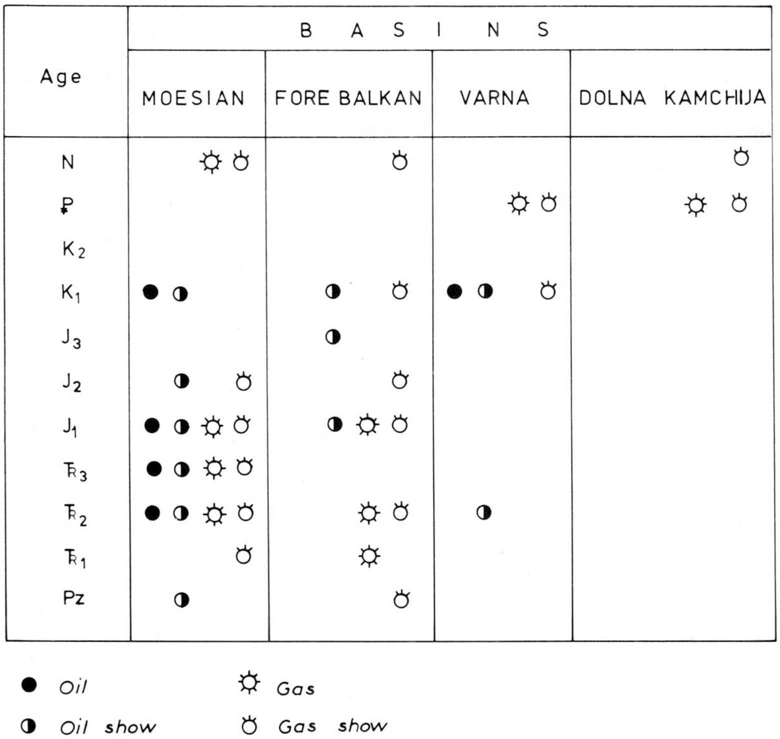

Fig. 41. Stratigraphic distribution of hydrocarbon fields and shows in Bulgaria's major structural zones

favourable factors for large-scale generation of oil and gas and their accumulation in various kinds of reservoirs and traps.

On-shore exploration should continue in the hope that unconventional thinking will lead to the discovery of new and larger fields. Considering the plate tectonics approach to the zones in South Bulgaria, Rhodope Massif and Srednogorie, the overthrusting of the Balkanides on the southern margin of the Moesian Platform and the possible nappe structures in all folded zones, exploration strategy in this country should take some new directions. At least some of the younger post-tectonic basins should not be excluded from this strategic planning.

Special attention should be paid to the hiatuses and unconformities, which are dramatic and probably quite factorial in the geologic evolution of the country. Among these the Triassic and Jurassic, as well as Paleozoic and Mesozoic boundaries and events are most important. They fit best the Hedberg (1964) concept of geotiming as a leading factor in petroleum generation and accumulation.

New expectations are placed in the Bulgarian Black Sea shelf and the adjacent areas, where foreign companies acquired new seismic data using the state-of-the-art technology.

Acknowledgements. This chapter was originally written for the Symposium *Hydrocarbon Exploration Opportunities in Central-Eastern Europe and the USSR*, held October 22–24, 1990, in Geneva, Switzerland, organized by Petroconsultants S.A. The data was collected and compiled by the authors themselves, but originally produced by many colleagues and friends working for the geological enterprises of the Committee of Geology and Mineral Resources and for the research institutions, including the Bulgarian Academy of Sciences.

All are cordially thanked for their efforts and accomplishments. Support was provided by the above-mentioned institutions. The authors of the research papers mentioned herein are sincerely thanked also for the use of their figures in this work.

The final version of the chapter was written after helpful discussions with all participants at the Symposium during the last week of October 1990, including our hosts from Petroconsultants. For this we express our cordial gratitude to all of them. Special thanks are due to the scientific editor of this Volume, Dr. Bogdan Popescu, for this help and valuable suggestions.

References

Atanasov A (ed) (1980) Oil and gas bearing of the Fore-Balkan. Technika, Sofia, 207 pp (in Bulgarian)

Atanasov A (1983) Nonanticlinal type of traps of oil and gas in the Triassic and Lower Jurassic sediments on the territory of Central Northern Bulgaria. Ann Comm Geol 24: 71–76 (in Bulgarian, English summary)

Atanasov A, Bokov P (eds) (1983) Geology and oil and gas prospectivity of the Moesian Platform in Central Northern Bulgaria. Technika, Sofia, 287 pp (in Bulgarian)

Atanasov A, Monov B (1984) The Triassic-Lower Jurassic natural reservoir in Lom Depression and Central North Bulgaria and possible types of oil and gas traps. Rev Bulg Geol Soc 45 (2): 213–220 (in Bulgarian, English summary)

Atanasov A, Monahov I, Shimanov JU (1965) Dolni Dubnik oil and gas field. CBGA 7th Congr Sofia Bulgaria Rep vol 4: 7–12 (in Russian)

Bokov P (1989) Deficient sedimentation zones. Science and Arts, Sofia, 382 pp (in Bulgarian, English summary)

Bokov P, Chemberski CH (eds) (1987) Geological factors for oil and gas bearing in northeastern Bulgaria. Technika, Sofia, 332 pp (in Bulgarian)

Bokov P. Ognyanov R (1990) Structural-tectonic and lithologic-stratigraphic features of the territory of Bulgaria and its petroleum prospects. Proc Symp Hydrocarbon exploration opportunities in Central Eastern Europe and the USSR. Petroconsultants, Geneva, 25 pp

Bokov P, Petrov JV (1981) Forms of pre-Jurassic relief in North Knezha region (Moesian Platform). CR Bulg Acad Sci 34 (8): 1131–1134 (in Russian)

Bokov P, Vuchev V, Dimov G, Mandev P, Monahov I, Troshanov V (1969) Distribution of economic oil and gas pools and of their shows in Bulgaria. Rev Bulg Geol Soc 30 (3): 321–330 (in Russian, English summary)

Bokov P, Vitanova N, Monanov I (1978). Geological maps of northern Bulgaria by horizontal slices. Geol. Balc 8 (4): 79–86 (in Russian, English summary)

Bokov P, Gochev P, Ognyanov R (1993) Tectonic position, hydrocarbon exploration and future potential of Bulgaria. Geol Balc 23 (3): 3–30

Bonchev E (1940) Alpidic tectonic manifestations in Bulgaria. Rev Bulg Geol Soc 12 (3): 155–247 (in Bulgarian, German summary)

Bonchev E (1986) The Balkanides: geotectonic position and development. Bulg Ac Sci, Sofia, 273 pp (in Bulgarian, Russian and English summaries)

Dachev C (1988) Structure of the Earth's crust in Bulgaria. Technika, Sofia, 334 pp (in Bulgarian, English summary)

Foose R, Manheim F (1975) Geology of Bulgaria: a review. Am Assoc Petrol Geol Bull 59 (2): 303–335

Gochev P (1986) Modèle pour une nouvelle synthese tectonique de la Bulgarie. Dari de Seama Inst Geol Geofiz Bucuresti 70–71 (5): 97–107

Gochev P (1991) The Alpine orogen in the Balkans—a polyphase collisional structure. Geotect Tectonophys Geodynam Geol Inst Bulg Ac Sci 22: 3–44 (in Bulgarian, Russian and English summaries)

Hedberg H (1964) Geologic aspect of origin of petroleum. Am Assoc Petrol Geol Bull 48 (11): 1755–1803

Hsu KJ, Nachev I, Vuchev V (1977) Geologic evolution of Bulgaria in light of plate tectonics. Tectonophysics 40: 245–256

Jakimov J (1991) Seismic studies in Northeast Bulgaria—state and problems. Rev Bulg Geol Soc 52 (2): 67–74 (in Bulgarian, English summary)

Jaranov D (1960) Tectonics of Bulgaria. Technika, Sofia, 282 pp (in Bulgarian, French summary)

Kalinko MK (ed) (1976) Geology and oil bearing of Northern Bulgaria. Proc VNIGNI 165: 243 pp (in Russian)

Karagjuleva J, Gochev P, Pironkov P (1982) Types and features of Alpine nappes in Bulgaria. Alpine Struct Elem Carpath Balkan Caucasus Pamir Orog zones, Bratislava pp 57–93

Kovacheva J, Vuchev V, Kerakova E, Kovachev G (1974) Hydrocarbons in Devonian sediments of Kardam-Krapec uplift in Northeastern Bulgaria. CR Bulg Ac Sci Sofia 27 (3): 375–378 (in Russian)

Mandev P, Nachev I (eds) (1981) Geology and oil and gas bearing of Northeastern Bulgaria. Technika, Sofia, 134 pp (in Bulgarian, Russian and English summaries)

Owen EW (1975) Trek of the oil finders: a history of exploration for petroleum (Bulgaria). APPG Mem 8: 1530–1531

Petroconsultants SA (1991–1992) Bulgaria FSS reports and maps

Petrov P, Bojadzhieva K, Gasharov S, Velinov, T (1991) Thermal field and geothermal regime in Bulgaria. Rev Bulg Geol Soc 52 (1): 60–64

Powers R (1962) Arabian Upper Jurassic carbonate reservoir rocks. Symp. Classification Carbonate Rocks. Am Assoc Petrol Geol Mem. 1 (Tulsa): 122–192

Sapunov I, Tchoumatchenco P (1987) Geological development of Northeast Bulgaria during the Jurassic. Paleontol Stratigr Lithol Geol Inst Bulg Ac Sci Sofia 24; 3–59 (in Bulgarian, English and Russian summaries)

Sapunov I, Tchoumatchenco P, Atanasov A, Marinkov A (1991) Central North Bulgaria during the Jurassic. Geol Balc 21: 3–68 (in Russian, English summary and captions)

Semenovich V, Namestnikova J (eds) (1981) Petroleum and gas-bearing basins of European socialist countries and Cuba. Comecon, Moscow, 400 pp (in Russian)

Shnirev V, Veneva R, Balinov V (1974) On the natural reservoirs of oil and gas in the Mesozoic sediments in Northern Bulgaria. Rev Bulg Geol Soc 35 (2): 191–202 (in Russian, English summary)

Velev V (1978) Types of hydrocarbon accumulations in Triassic-Lower Jurassic sediments of Northern Bulgaria. Geol Balc 8 (3): 59–84 (in Russian, English summary)

Vuchev V (1973) Studies of organic matter in the Triassic carbonates of Northern Bulgaria. Adv Org Geochem 1973. Edit Technip, Paris, pp 489–504

Vuchev V (1985) Tectonic fracturing of sedimentary rocks in the Fore-Balkan. DSci Thesis Geol Inst Bulg Ac Sci, Sofia vols 1 & 2, 260 pp + 104 pp (in Bulgarian, English captions)

Vuchev V, Mateeva O, Nikolov T (1965a) Geochemical characteristics of disseminated organic matter in Mesozoic sediments of Fore-Balkan. CBGA 7th Congr Sofia, Bulgaria Rep Vol 4: 19–27 (in Russian)

Vuchev V, Najdenova A, Radev G (1965b) Basic results of studies of fracturing and reservoir properties of Mesozoic rocks of the Fore-Balkan. CBGA 7th Congr Sofia Bulgaria Rep Vol 4: 29–34

Vuchev V, Radev G, Balinov V (1969) Lithologic and reservoir properties of the producing horizon of the Tjulenovo oil field. Rev Geol Inst Bulg Ac Sci Ser Petrol Coal Geol Sofia 18: 63–76 (in Bulgarian, English summary)

Vuchev V, Howels G, Burlingame A (1972) The presence and geochemical significance of organic matter extractable from Jurassic and Triassic sediments of Northern Bulgaria. Adv Org Geochem 1971. Pergamon Press, Braunschweig, pp 365–386

Vuchev V, Kovachev G, Petrova R, Stojanova G, Tsakov, K (1976) Organic matter in the Triassic sediments of Bulgaria. II. Disseminated organic matter in the Middle Triassic carbonates from the holes near the city of Knezha (Northern Bulgaria). Petrol Coal Geol Bulg Ac Sci 4: 3–22 (in Russian, English summary)

Vuchev V, Vucheva A, Kabakchieva M, Petrov L (1977) Geochemical characteristics of Bulgarian bitumoliths II. Comparative analysis of bitumoliths from Krassava, Mandra, Iskra, Gorkovo and Borov Dol. Petrol Coal Geol Geol Inst Bulg Ac Sci 7: 37–45 (in Russian, English summary)

Vuchev V, Monov B, Atanasov A, Tochkov D (1990) Hydrocarbon exploration, geology and prospectivity of Bulgaria. Proc Symp Hydrocarbon exploration opportunities in Central Eastern Europe and the USSR. Petroconsultants, Geneva, 27 pp

Yovchev Y, Baluhovski N (1961) Mineral resources of Bulgaria. Petroleum and gas. Technika, Sofia, 120 pp (in Bulgarian, English summary)

Yovchev Y, Atanasov A, Bogdanov S, Bojadzhiev S, Bojanov I, Jordanov M, Kanchev I, Savov S, Cheshitev G (eds) (1971) Tectonic structure of Bulgaria (The explanatory notes to the tectonic map of Bulgaria, scales 1:200,000 and 1:500,000). Technika, Sofia, 558 pp (in Bulgarian, English summary)

Maps

Atanasov A, Bojadzhiev S, Cheshitev G (1976) Tectonic map of Bulgaria, scale 1:500,000, and profiles. Central Acm Geodezy Cartogr, Sofia (Bulgarian and English legend)

Bonchev E, Karagjuleva J, Kozhucharov D, Kostadinov V, Savov S (1973a) Geologic map of Bulgaria in the scale of 1:1,000,000 and geological profiles. National Atlas of Bulgaria. Geodezy and Cartogr, Sofia, pp 24–25 (in Bulgarian)

Bonchev E, Bojanov I, Gochev P, Dabovski H, Karagjuleva J, Kostadinov V, Savov S, Hajdutov I, Tsankov TS (1973b) Tectonic map of Bulgaria of the scale 1:1,000,000. National Atlas of Bulgaria. Geodezy and Cartogr, Sofia, pp 28–29 (in Bulgarian)

Cheshitev G, Kanchev I, Vulkov V, Marinova R, Shiljafova J, Russeva M, Illiev K (1989) Geologic map of Bulgaria, scale 1:500,000. Committee of Geology, sofia (Bulgarian and English legend)

Petroconsultants SA (1991) Petroleum activity map, Foreign Scouting Service Aug 1991 Bulgaria, scale 1:500,000. Geneva

3 Exploration History, Geology and Hydrocarbon Potential in the Czech Republic and Slovakia

MILAN BLIZKOVSKY[1], AUGUSTIN KOCAK[1], MILAN MORKOVSKY[1], ANTONIN NOVOTNY[1], BEDRICH GAZA[2], PETR KOSTELNICEK[2], VLADIMIR HLAVATY[3], STANISLAV LUNGA[3], DIONEYZ VASS[4], JURAJ FRANCU[5] and PAVEL MULLER[5]

1 Generalities

1.1 Geological Structure of the Czech Republic and Slovakia

Two major geologic units are known in the Czech Republic and Slovakia: the Bohemian Massif in the west and the West Carpathians in the east (Fig. 1).

[1]Geofyzika, Jecna 29a, 61246 Brno, Czech Republic
[2]Moravian Oil Company, Uprkova 6, 69530 Hodonin, Czech Republic
[3] Nafta, 90845 Gbely, Slovakia
[4]Geological Institute of Dionyz Stur, Mlynska dolina 2, 81704 Bratislava, Slovakia
[5]Geological Survey, Leitnerova 22, 60200 Brno, Czech Republic

The Bohemian Massif consists of a metamorphosed Precambrian-Early Paleozoic basement, unconformably covered by Paleozoic clastics and carbonates in a shallow water facies. During the Hercynian orogeny the Bohemian Massif was uplifted and became the source for sediment supply in the Tethyian basin. Southeastwards the Bohemian Massif plunges beneath the Alpine-Carpathian system.

The tectonic units of the West Carpathians are: the Carpathian Foredeep, the Flysch and Klippen Belts, the Central and Inner Carpathians, the interior Paleogene and Neogene depressions and the neovolcanites of a post-Cretaceous age.

The Carpathian Foredeep developed on the slopes of the Bohemian Massif, the foreland of the Alpine-Carpathian arc. It is filled by a Neogene molasse, approximately 1000 m in thickness.

The Flysch Belt is folded and overthrust and received its final configuration in the Miocene. The Flysch zone overthrusts the Carpathian Foredeep and in places, the Bohemian Massif. It consists mainly of Paleogene flysch-type deposits. In the south, the thickness of Flysch Belt sediments reaches up to 10,000 m.

The Upper Cretaceous-Paleogene Klippen Belt is located between the Flysch Belt arc and the Central Carpathians. It has an extremely complex structure.

The Central Carpathians are made up of crystalline cores, overthrust (up to 50 km) by nappes with Triassic to Lower Cretaceous pelitic-carbonate components. According to geophysical data, the Klippen Belt, the Flysch Belt, and the Bohemian Massif were overthrust by the crystalline complexes of the Central Carpathians. The Inner Carpathians form a Paleozoic anticlinorium covered by Mesozoic nappes.

The Central Carpathian Paleogene and the Neogene basins developed on the Alpine-type folded West Carpathian units: the Vienna Basin, the Danube Basin, the South Slovakian Basin and the East Slovakian Basin (or Transcarpathian depression).

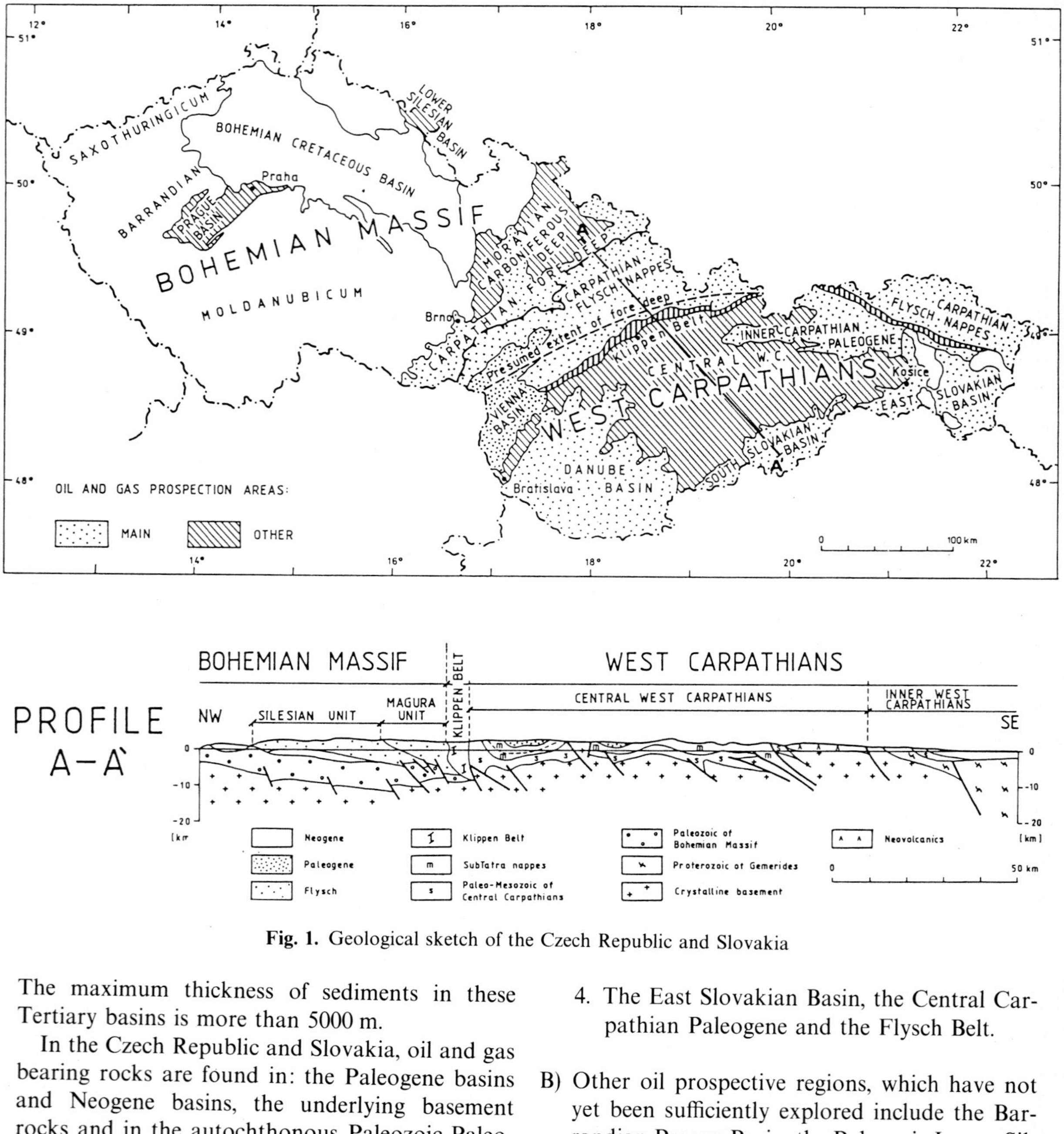

Fig. 1. Geological sketch of the Czech Republic and Slovakia

The maximum thickness of sediments in these Tertiary basins is more than 5000 m.

In the Czech Republic and Slovakia, oil and gas bearing rocks are found in: the Paleogene basins and Neogene basins, the underlying basement rocks and in the autochthonous Paleozoic-Paleogene formations of the Bohemian Massif. The Bohemian Massif and the Carpathian Foredeep are overthrust by the nappes of the Flysch Belt.

In the Czech Republic and Slovakia, four proven prospective regions, A (Fig. 2), and four potentially prospective regions, B (Fig. 46), are recognized:

A) 1. The Southeastern Slopes of the Bohemian Massif.
2. The Vienna Basin.
3. The Danube Basin and the South Slovakian Basin.
4. The East Slovakian Basin, the Central Carpathian Paleogene and the Flysch Belt.

B) Other oil prospective regions, which have not yet been sufficiently explored include the Barrandian Prague Basin, the Paleozoic Lower Silesian Basin, the Lower Carboniferous (Culm) in Moravia, and the Central Carpathians

1.2 Exploration and Production

During the last few years, some 26 drilling rigs drilled an average of some 80,000 m/year. Three seismic crews have been active in oil and gas prospection in the region, one of them gathering 3-D data. Annual CDP output has averaged some

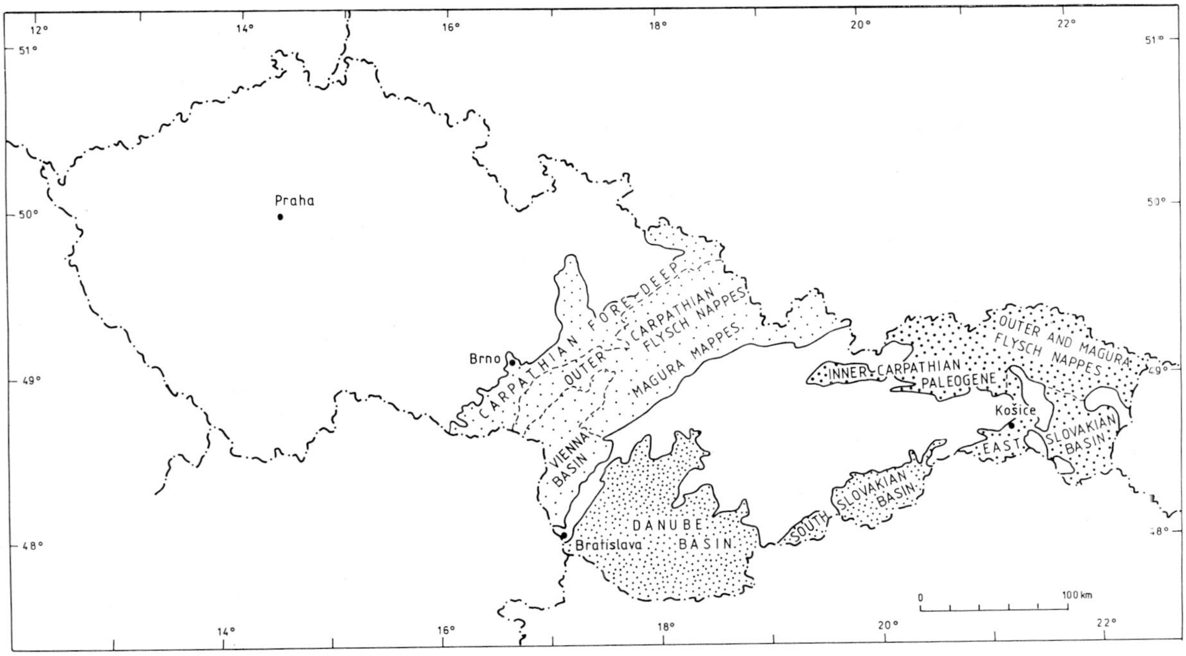

Fig. 2. Sketch of main hydrocarbon exploration and production areas

1200 km; 3-D seismic surveys were recorded in two areas. Production is slightly over 1 MM bbl/year.

2 Sedimentary Provinces with Proven Hydrocarbon Potential

2.1 Exploration History

2.1.1 Geophysics

Geophysical prospection has been carried out extensively in the main oil, gas and coal regions of the Czech Republic and Slovakia. The main geophysical methods used are seismic, gravimetry, magnetics and geoelectric. In the past few years satellite imagery interpretation has been increasingly used.

Geophysical measurements conducted in the Vienna Basin were among the first in the world applied to oil prospection. In the years 1915 and 1916, Lorand Eötvos used his torsion balance for gravity measurements in the area of Gbely (Fig. 3). Here, he revealed a structural high at the place where in 1914 drilling encountered an accumulation of oil at a depth of 164 m, and the same place where a farmer, J. Medlen, exploited natural gas seeps for domestic use.

In the 1940s, large-scale gravity, geomagnetic, geoelectric and seismic (fan shooting) surveys were carried out by the German company Gesellschaft fur Praktische Lagerstattenforschung, and the Austrian Austrogasko in the Carpathian Foredeep, the Vienna Basin, the Danube Basin and the East Slovakian Basin. The 1950s saw a rapid development of applied geophysics, with prospection commencing in the Carpathian Flysch and in the Central Carpathian Paleogene.

Gravimetry. All Czech and Slovakian territory has been explored at a density of 1 point per 3–5 km^2. Maps at the scale of 1:200,000, with 2×10^{-5} ms^{-2} intervals have been issued by the Geological Survey in Prague in 1964. Gravity mapping at the scale of 1:25,000 has been carried out at a density of 4–6 points per km^2. By 1990, 80% of the territory had been surveyed (Fig. 4). Gravity meters Worden, Scintrex and Sodin with an accuracy of 3×10^{-7} ms^{-2} were used. Thirty-four sheets of gravity maps at the scale of 1:100,000, with 0.5 $\times 10^{-5}$ ms^{-2} intervals have been issued by Geofyzika Brno between 1979 and 1989.

Geomagnetic and Air-Borne Geophysics. A ground magnetic survey (delta Z) for oil prospection was accomplished in 1967. Torsion balance with an

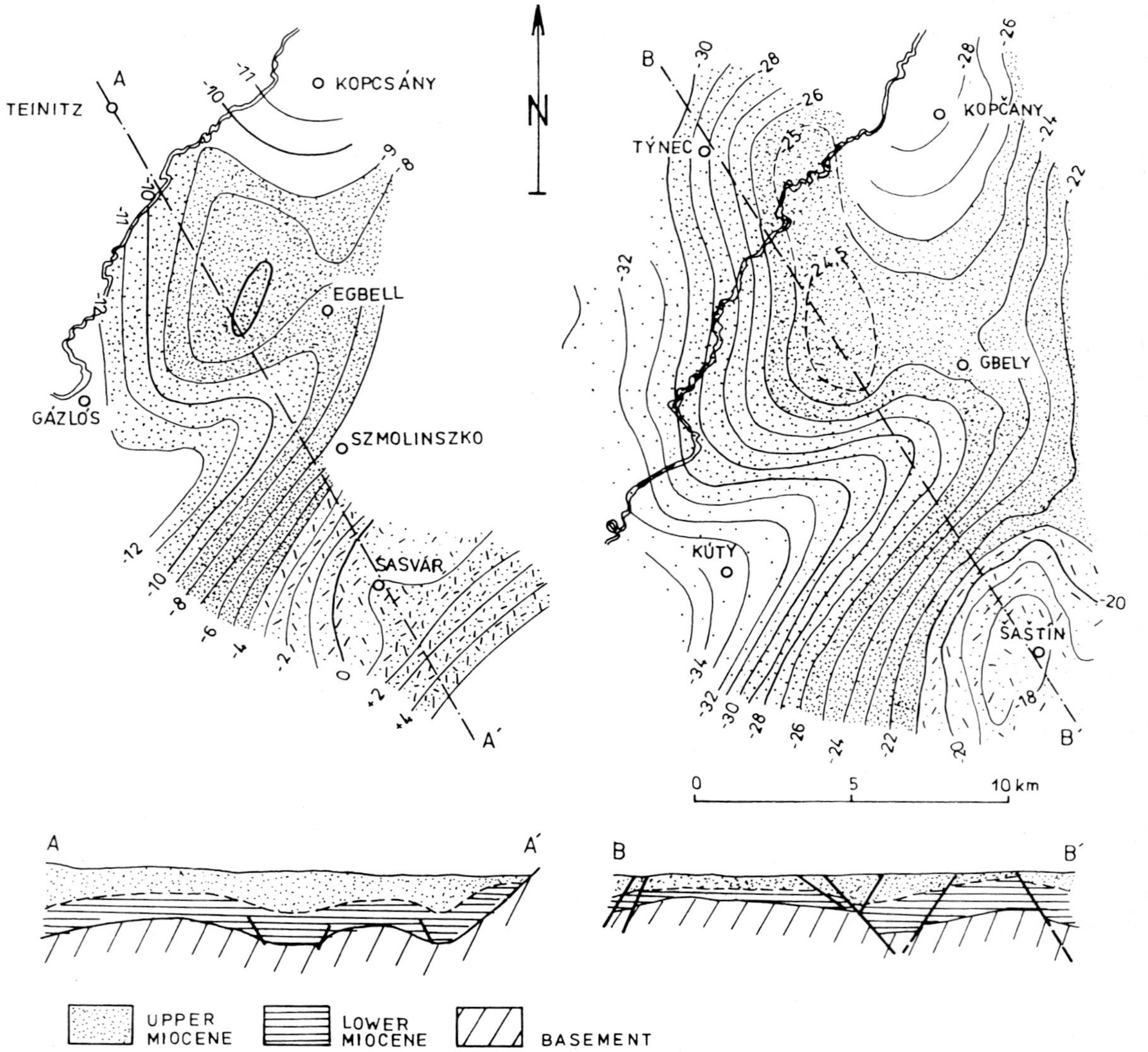

Fig. 3. Comparison of gravity data from the area of Gbely : 1916 (*left*), 1960 (*right*). Interpretation of A-A′ profile, H. Böckh; B-B′ profile, Institute of Applied Geophysics, Brano

accuracy of 3–6 nT was used. The measurement density (Fig. 5) was 1 point per km^2 in the greater part of the Carpathian Foredeep, the western part of the Flysch Belt, the Vienna Basin (A), the Danube Basin (B) and eastern Slovakia (C).

Air-borne geophysical measurements have been conducted since 1957, when aeromagnetic mapping of the Czech Republic and Slovakia at a scale of 1 : 200,000 was initiated. When geophysical methods were updated in the 1970s, mapping at a scale of 1 : 25,000 started in selected areas and is still going on. The distance between profiles is 250 m, clearance is 80 m above the ground, point interval on a profile is 35–40 m. The G-801/38 proton magnetometer and the Di-GRS-800D spectrometer from Geometrics were used. Total gamma-ray activity, delta T and K, U, Th values were obtained. The respective maps are presented at a scale of 1 : 50,000.

Geoelectric and Magnetotelluric. VES measurements (Fig. 6) for oil prospection, with current electrodes up to 20 km apart, have been carried out in the marginal parts of the Neogene Carpathian Foredeep (A), in the Vienna Basin (B), in the Danube Basin (C), and in the East Slovakian Basin (D). The IP method has been used experimentally in the Vienna Basin.

Magnetotelluric measurements have been carried out along two NW-SE-trending lines across

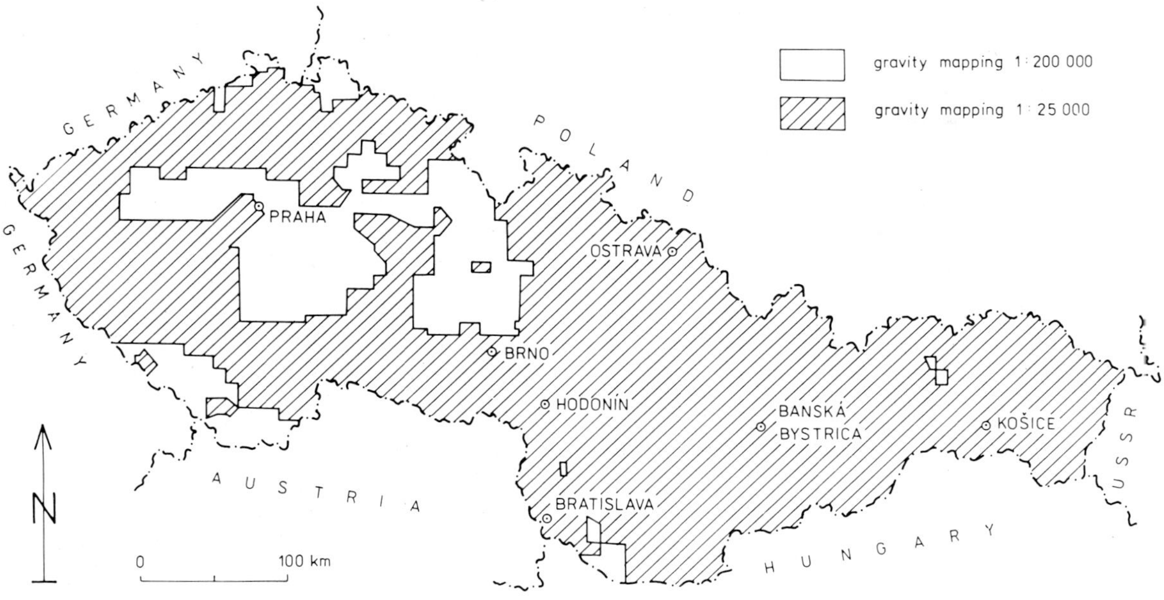

Fig. 4. Gravity mapping of the Czech Republic and Slovakia

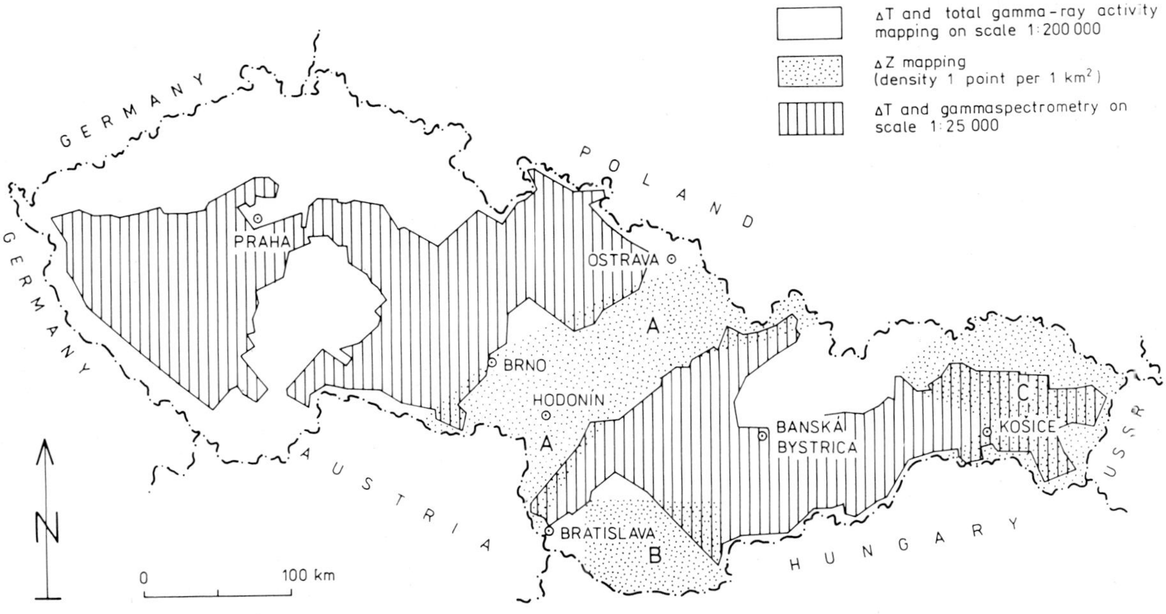

Fig. 5. Airborne magnetic, radiometric and ground magnetometric mapping of the Czech Republic and Slovakia

the Vienna Basin and adjacent areas, and along one line in central Slovakia. Resistivities of rocks in the crystalline basement and thicknesses of sedimentary formations were determined.

Geothermic. Near surface heat-flow values range from 40 to 100 mWm^{-2} (Fig. 7). The highest values have been recorded in the Krusnehory Mts. (A), in the central Slovakian neovolcanites (B), and in eastern Slovakia (C). The lowest values have been observed in the Moldanubicum (I), on the south-eastern slopes of the Bohemian Massif (II), and in the Vienna Basin (III).

Seismic. Reflection and refraction seismic methods have been employed since the early 1950s.

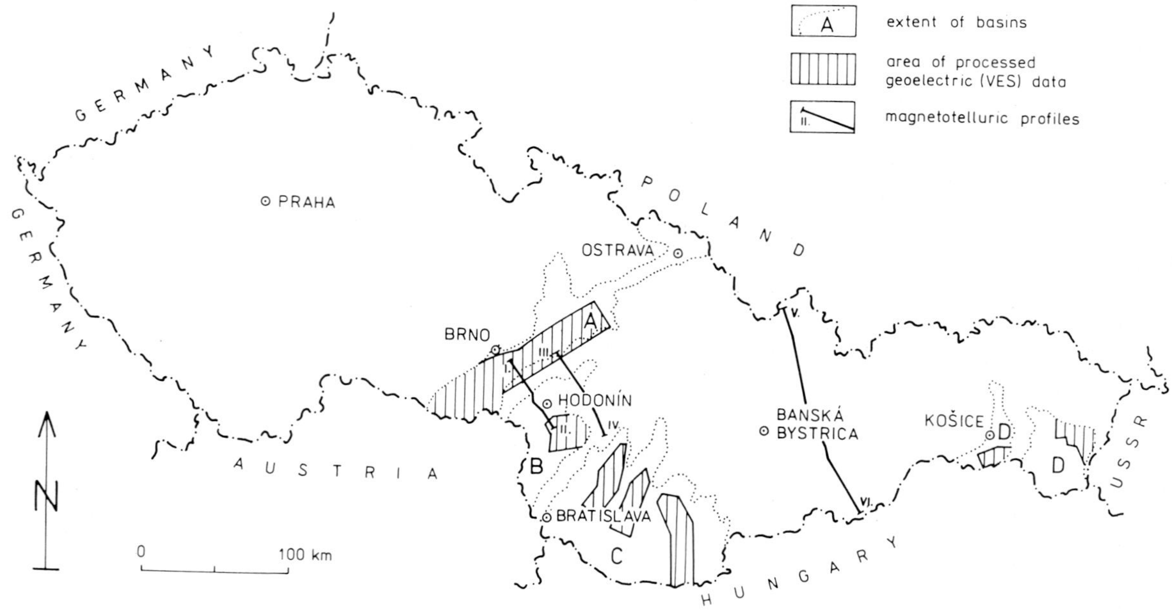

Fig. 6. Geoelectric and magnetotelluric mapping for hydrocarbon exploration in the Czech Republic and Slovakia

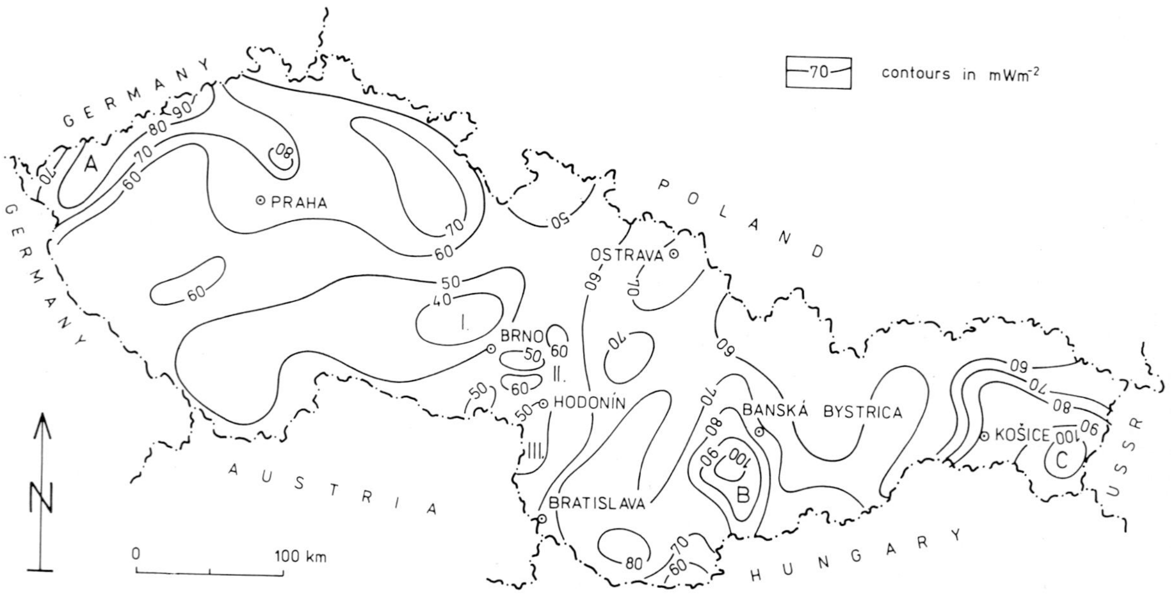

Fig. 7. Heat flow map of the Czech Republic and Slovakia

Refraction measurements have been carried out over 5000 line-km. In subsequent years refraction has focussed on deep seismic sounding.

Reflection seismic has gone from the stage of simple coverage with oscillograph recording to the period of reproducible recording. In the 1970s CDP came into use. Seismic surveys cover 22,000 line-km, with 14,000 line-km in CDP. Geofyzika Brno operated a 96-channel SN 338HR and a 48-channel SN 358HR from Sercel. Dynamite technology and the Vibroseis technology (VVCA vibrators from Prakla Seismos) were used. Measurements have been carried out using 24- and 48-fold coverage. Seismic measurements for 3-D processing have been conducted over 123 km^2 in the Vienna Basin.

In the region of the *Southeastern Slopes of the Bohemian Massif* (Fig. 8) a CDP reflection survey with 12- to 24-fold coverage was carried out along 7500 km of profiles. In addition to basic processing, migration and depth transformation were performed.

In total, 3800 km of profiles and 123 km^2 of 3-D measurements with 12- to 48-fold coverage have been carried out in the *Vienna Basin* (Fig. 8). In addition to basic processing, wave migration and depth transformation, Hilbert transformation has been applied to selected sections. In areas where anomalous high pressure zones were detected, investigation is still in process. Structural maps have been constructed on individual stratigraphic boundaries. Seismic well-logging and vertical seismic profiling have been carried out as well as seismic lithostratigraphy interpretation.

In the *Danube Basin* (Fig. 9), seismic measurements with single coverage have been carried out over the greater part of the region (over 9000 km^2 by 1990). Reliable results were obtained down to the depth of 2500 m. The CDP method has been used on regional profiles with a total length of 500 km.

Owing to the large thickness of the Quaternary cover, the dynamite technique was more successful than Vibroseis. The data was processed with 24-fold coverage. Seismic methods were employed for mapping morphological structures extending from the margins into the deeper parts of the basin. The general structural style and prominent fault lines have been defined.

In *East Slovakia* (Fig. 10), in addition to refraction and reflection seismic methods with single coverage, the CDP method with 12–24-fold coverage has been used for 1646 line-km. The dynamite technique with split-spread arrangement and 25 m distance between geophone groups was used. Besides the basic processing of line sections – Claerbout migration – (depth conversion) and compilation of structural maps, Hilbert transformation and the analysis of dynamic properties along seismic horizons have been carried out. Seismic stratigraphy was also used to evaluate the hydrocarbon potential of the sedimentary fill of the basin.

During the initial stages of exploration, seismic methods helped to explain the basic structural features of the basin, to reveal gas-bearing structures and buried extrusives. Attention is currently centred on zones of facies change and the identification of minor normal faults, volcano-sedimentary layers or anomalous high pressure zones. Seismic measurements have recently revealed the most significant gas accumulation discovered to date, at Senne. Three-dimensional seismic measurements have not yet been carried out in the East Slovakian Basin. They are planned for 1993.

Under the rough conditions of the *Central Carpathian Paleogene* area, seismic measurements have mapped the Mesozoic relief and flat overthrusts. Tectonic zones have been found in intensively folded flysch. The CDP method has been used on profiles over a total length of 500 km. Figure 11 shows the evolution of CDP seismic activity in the Czech Republic and Slovakia.

2.1.2 Drilling

The first reports on oil exploitation in Slovakia and the Czech Republic from pits near Mikova and Turzovka, in the East Slovakian part of the Flysch Belt, date back to 1869. Exploratory drilling was first conducted in the Flysch Belt near Bohuslavice (450 m) and near Vysny Komarnik (820 m) in 1899. Shallow oil accumulations were discovered near Gbely in the Vienna Basin in 1914, and near Hodonin in 1920. Drilling for hydrocarbon, including exploratory, development and service wells, has been continued since and attained all-time records in the 1950s (Fig. 12).

After World War II, exploratory drilling for oil was conducted by the following oil companies: Moravian Oil Company (MND) Hodonin, Nafta Gbely and Geological Survey Ostrava. The companies operate approximately 30 Romanian, Austrian and US-manufactured rigs, of which only five can reach a depth of 6500 m.

Anomalous temperatures (geothermal gradient of 18–20 m/1 °C), and anomalous pressure layers (up to 90% higher than the hydrostatic pressure) have been observed, mainly in the East Slovakian Basin, in the autochthonous Paleogene of the Bohemian Massif (60%), and in the central part of the Vienna Basin (up to 100%). The presence of H_2S has been proven in the Mesozoic formations of the Vienna Basin basement. Salt horizons and volcanites have been revealed in the East Slovakian Neogene, and volcanites in the Danube Basin.

Only five wells deeper than 6000 m have been drilled in the Czech and Slovak republics. The total drilling volume reached 3.7 million meters at the end of 1990.

On the *Southeastern Slopes of the Bohemian Massif* (Fig. 13), during the last few years, some 477,000 m were drilled in 540 wells. Eight drilling rigs have been active with an annual drilled depth

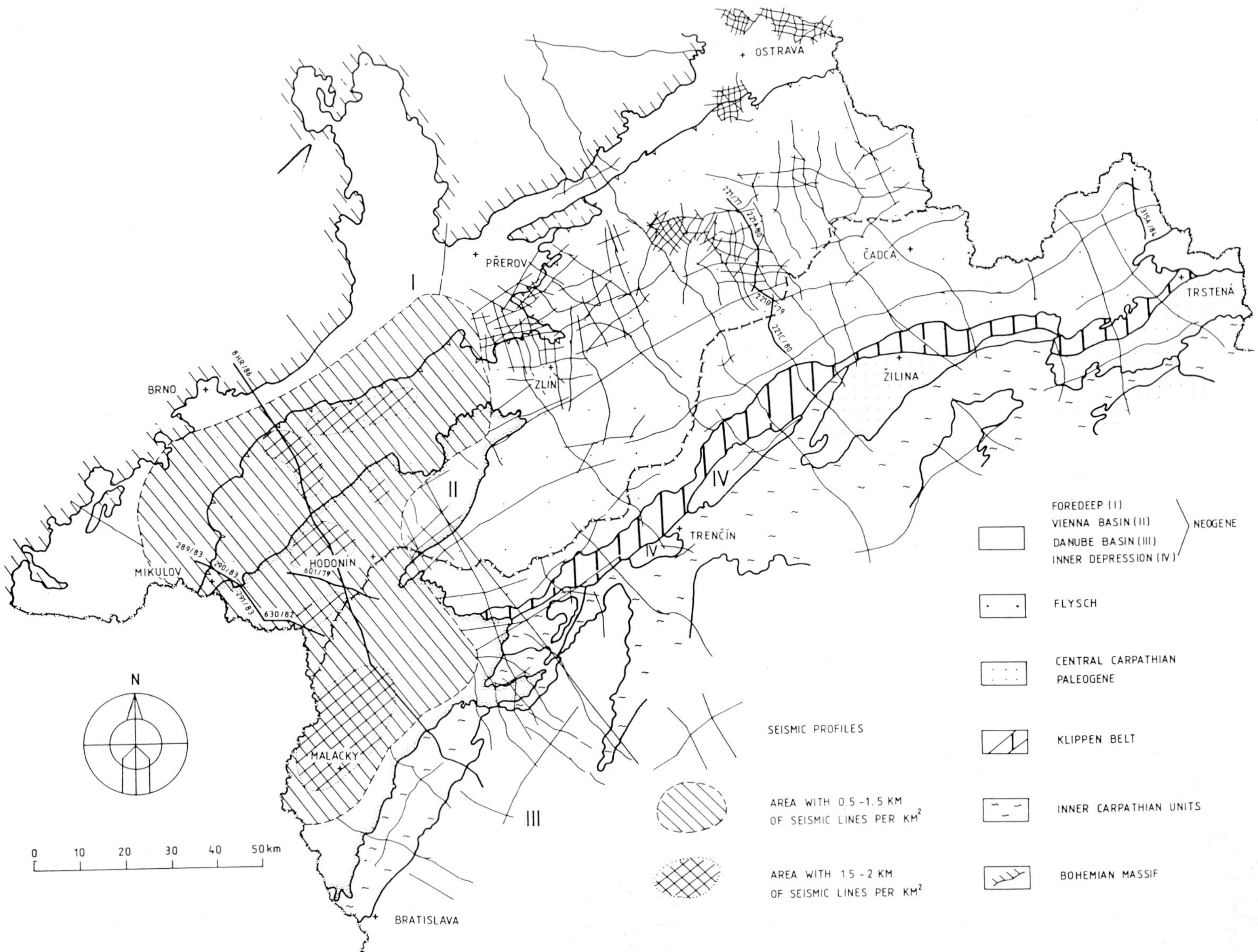

Fig. 8. Seismic profiles on the Southeastern Slopes of the Bohemian Massif and in the Vienna Basin

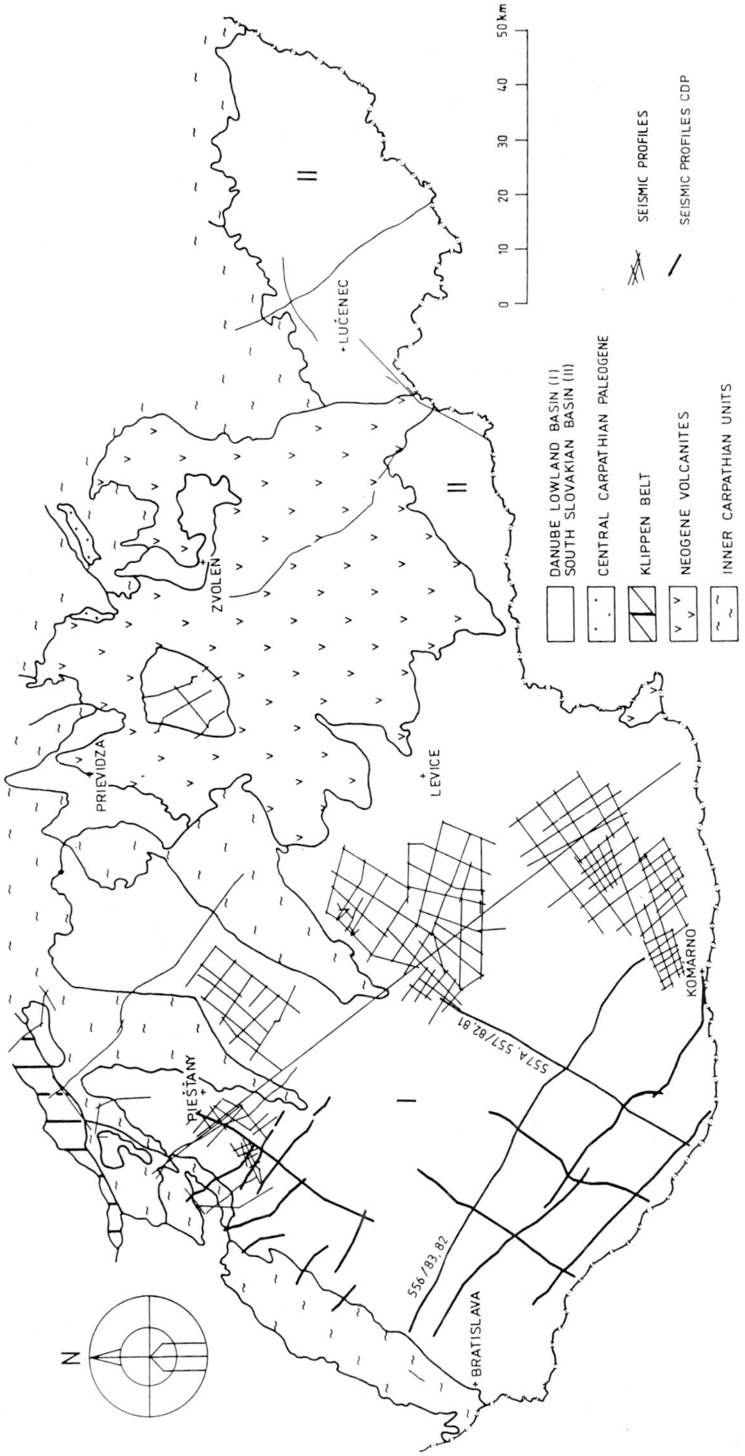

Fig. 9. Seismic profiles in the Danube Basin and in the South Slovakian Basin

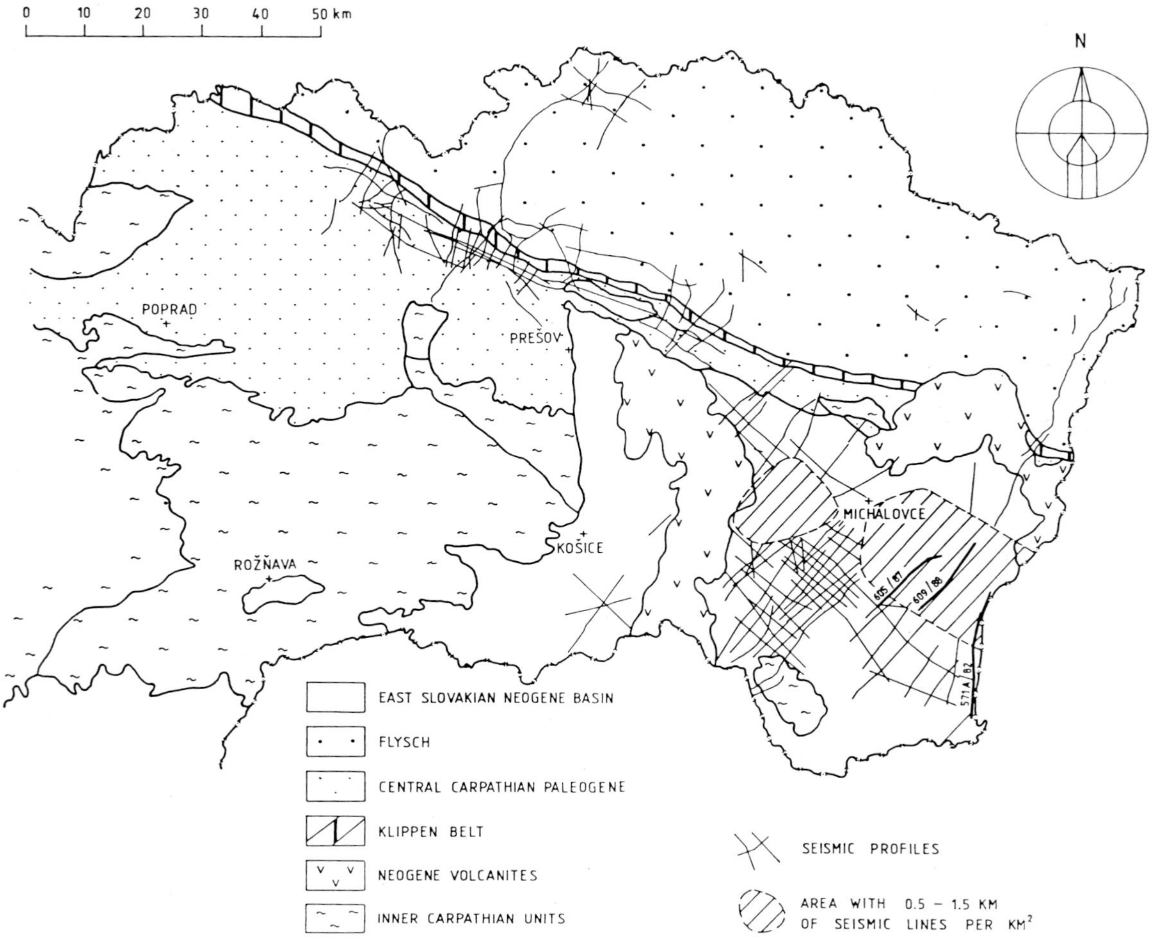

Fig. 10. Seismic profiles in East Slovakia

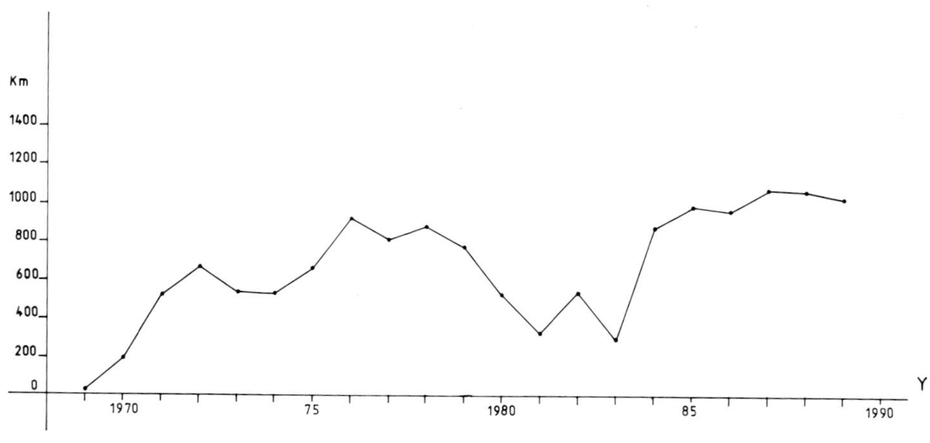

Fig. 11. Evolution of CDP seismic measurements in the Czech Republic and Slovakia

of 23,000. The deepest well, Jablunka-1, was drilled in the Flysch Belt. It penetrated 2900 m of the nappe, then the Paleozoic down to 6315 m and terminated in the crystalline complex at a depth of 6506 m.

In the *Vienna Basin* (Fig. 13), up to 1960, shallow drilling down to 300 m was the main method applied in the search for structural highs. Contrary to other oil prospetive areas, geophysical measurements followed previous intensive drilling period.

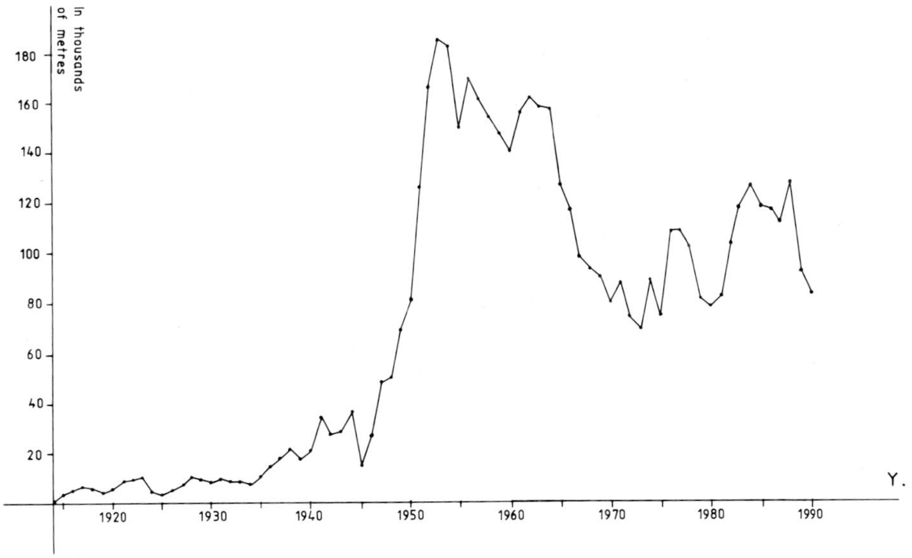

Fig. 12. Evolution of drilling for hydrocarbons in the Czech Republic and Slovakia

Subsequently, complex geophysical measurements together with shallow and deep drilling delimited the main fault systems and identified structures for drilling.

Up to the present, 1724 deep wells totalling 2,343,000 m have been drilled (Fig. 12). The deepest well, Sastin 12, reached the depth of 6505 m in the Frankenfels-Lunz Nappe of the Vienna basin floor. It penetrated 2200 m through the Neogene fill and the remainder in Alpine-Carpathian nappes. Four wells more than 6000 m deep have been drilled so far.

In the *Danube Basin* (Fig. 14) 13 wells totalling 220,000 m have been drilled. The deepest of which is well Sucha-2 (3500 m), in the northwestern part of the basin. Hydrocarbon (gas) traps are sealed tectonically, and less often lithologically. No oil accumulations have been discovered in the basin.

In *East Slovakia* (Fig. 15), drilling is currently under way in the Neogene basin, in the central Carpathian Paleogene, and in the East Slovakian flysch. Five drilling rigs have been operating in the basin, and annual drilled depth has been 20,000–25,000 m. Since 1954, 285 wells with a total drilled depth of 620,000 m have been drilled. The deepest well reached 4207 m (Trhoviste 26). Drilling in the Neogene basin is hampered by thick complexes of volcanic rocks, salt layers, overpressure zones and locally high temperatures.

In the *Central Carpathian Paleogene* and in the *Flysch Belt* of the Inner Carpathians, 13 wells have a total drilled depth of 52,000 m. The deepest of these wells reached 6003 m (Hanusovce-1).

2.2 Petroleum Geology

2.2.1 The Southeastern Slopes of the Bohemian Massif

The Southeastern Slopes of the Bohemian Massif petroleum province cover a region bounded by the eastern outcrops of the Bohemian Massif in the west, and by the Klippen Belt in the east. This area is made up of three distinct sedimentary basins: the crystalline basement and its autochthonous Paleozoic to Early Tertiary sedimentary cover, the Eggenburgian to Badenian Carpathian Foredeep, and the Jurassic-Lower Miocene Carpathian Flysch Nappes thrust over the foredeep and Bohemian Massif sedimentary cover (Figs. 2, 16).

The basin's slope is affected in a NNW/SSE direction by several striking tectonic features, among the most important of which is the platform flexure dividing the area into a shallow part and a deep part (Fig. 17). In a southwestern direction, the slope is divided by the Upper Moravian Basin and the Nesvacilka Graben into three sectors: north, centre and south.

The crystalline basement of the Bohemian Massif slope consists of igneous and metamorphic rocks in the north and of granites and ultrabasics in the

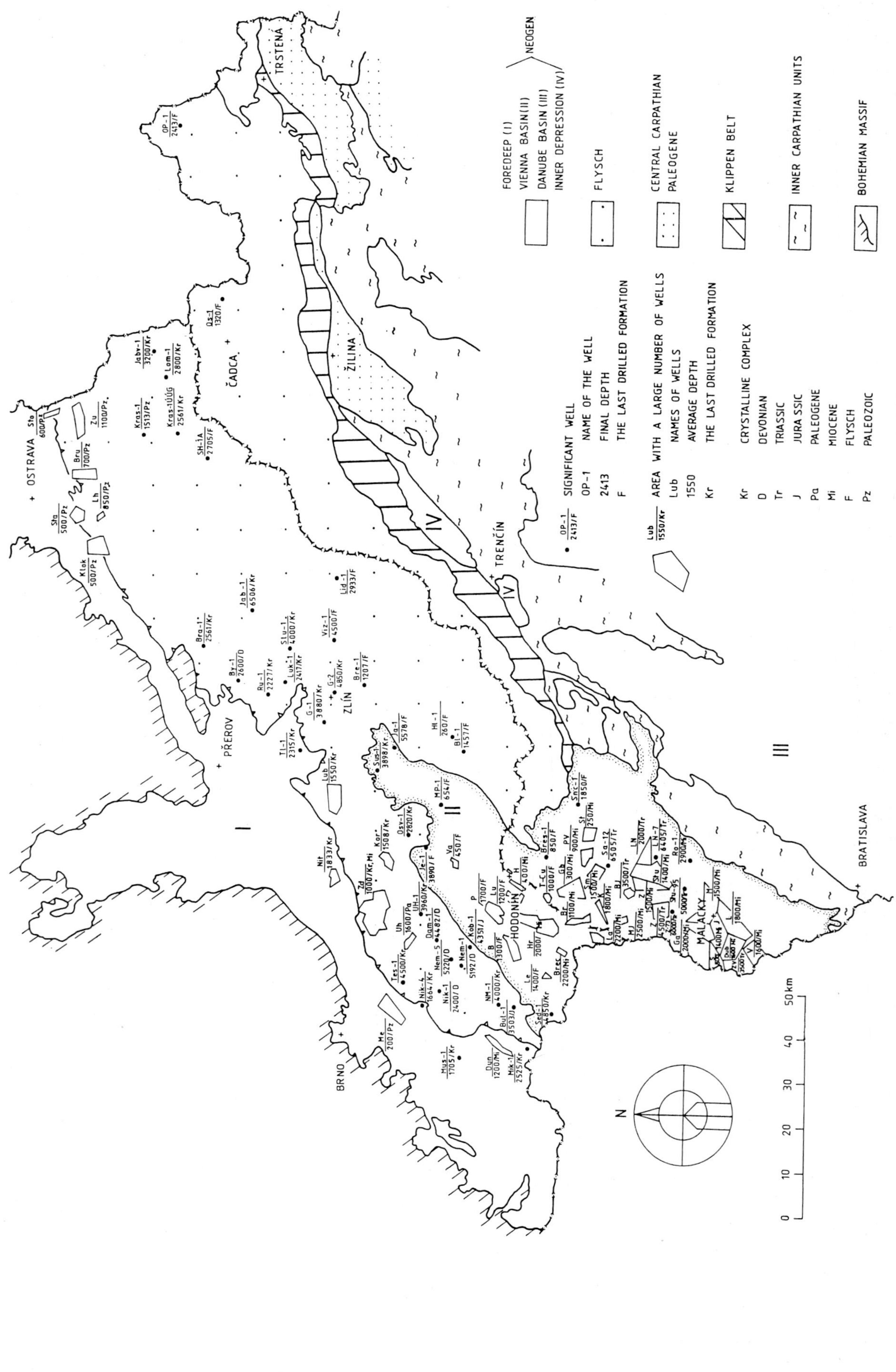

Fig. 13. Location of significant wells drilled on the slopes of the Bohemian Massif and in the Vienna Basin

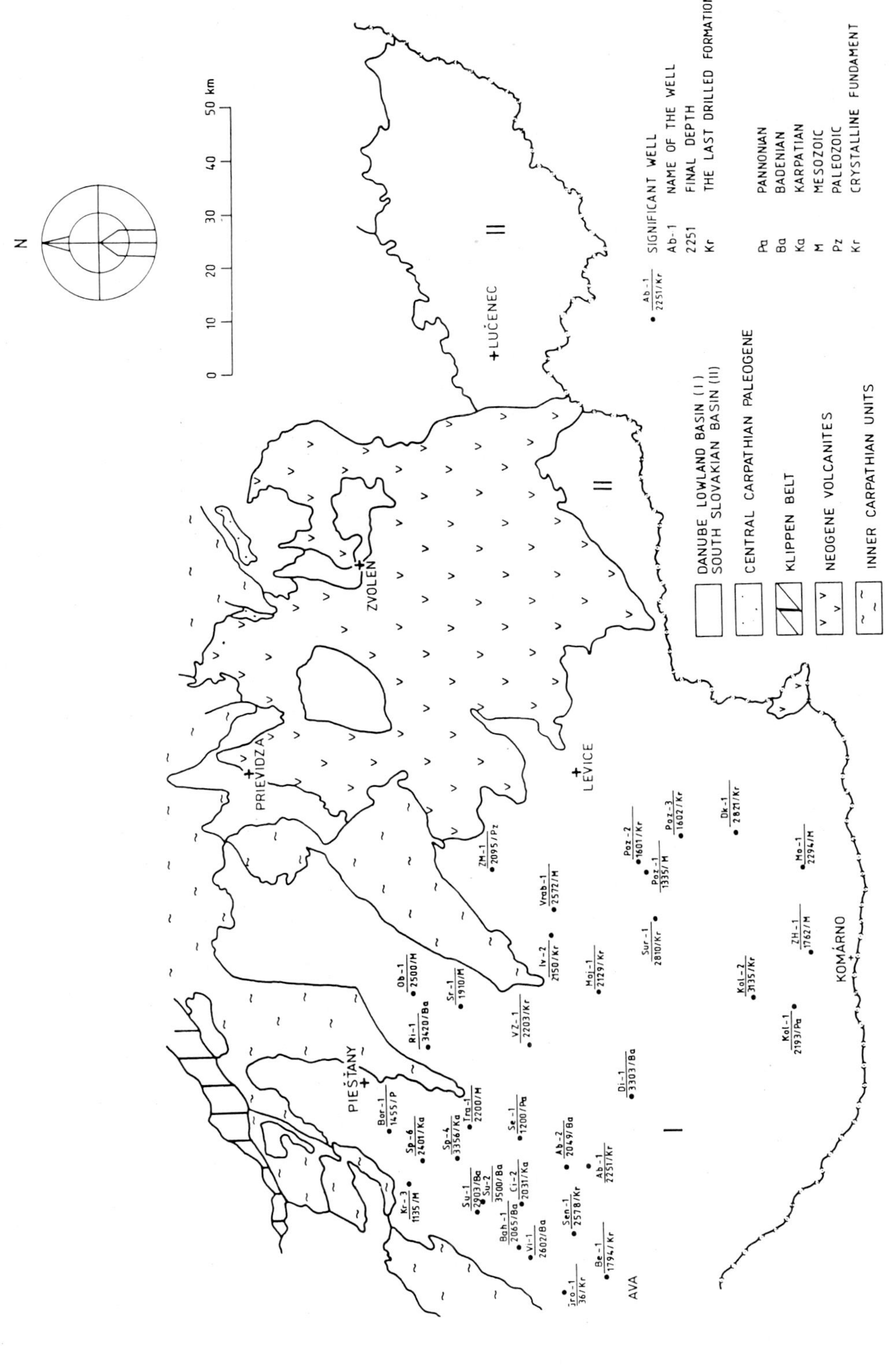

Fig. 14. Location of significant wells drilled in the Danube Basin and in the South Slovakian Basin

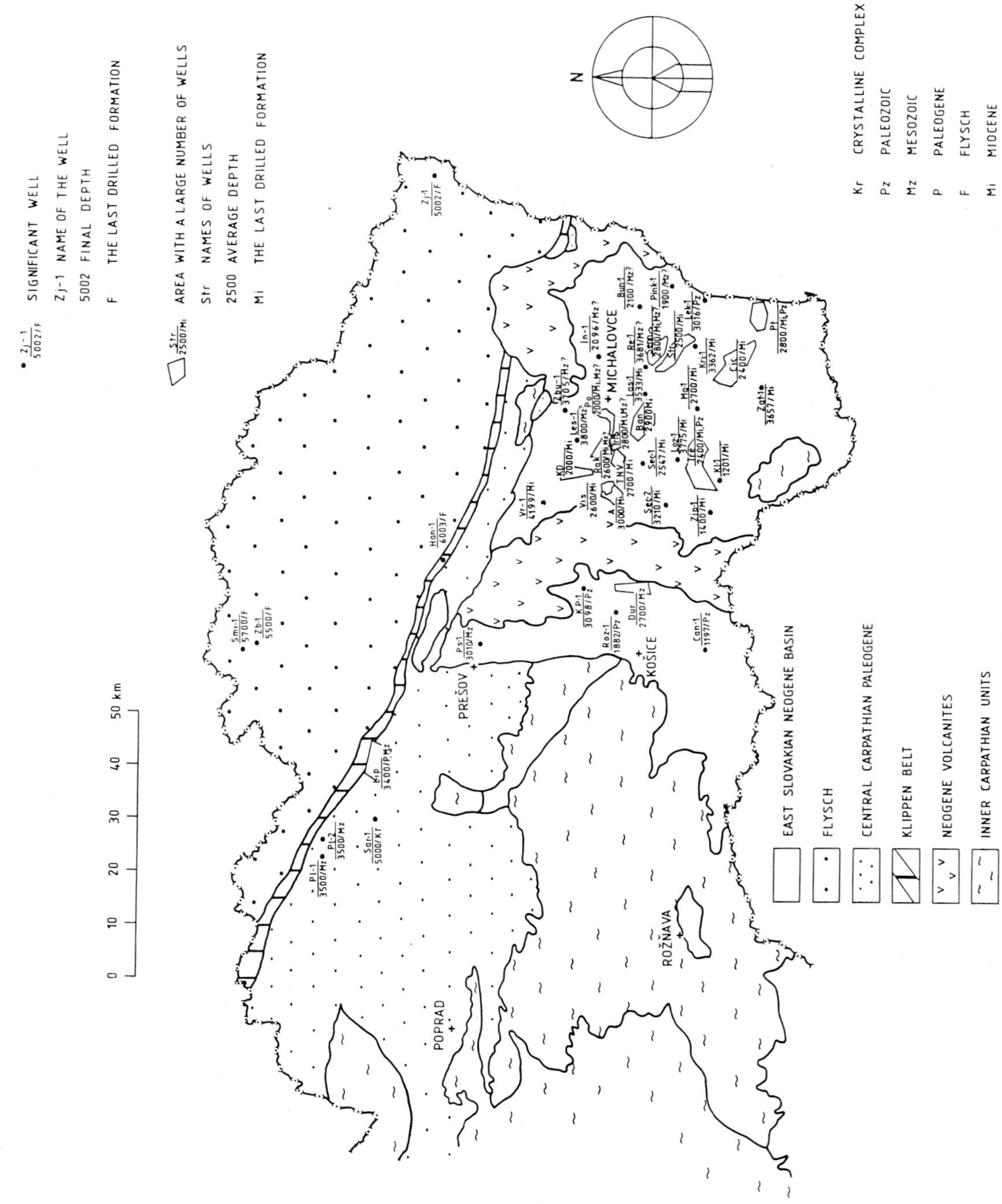

Fig. 15. Location of significant wells drilled in East Slovakia

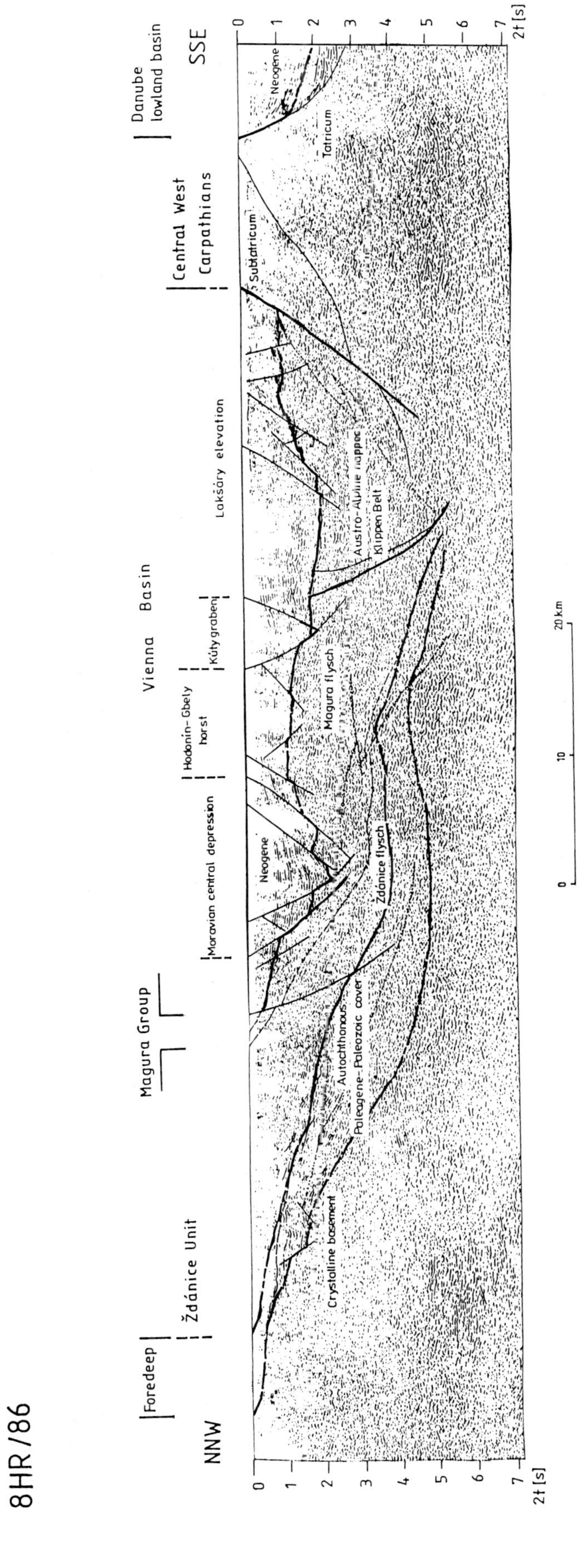

Fig. 16. Seismogeological cross-section 8HR/86. (For line location see Fig. 8)

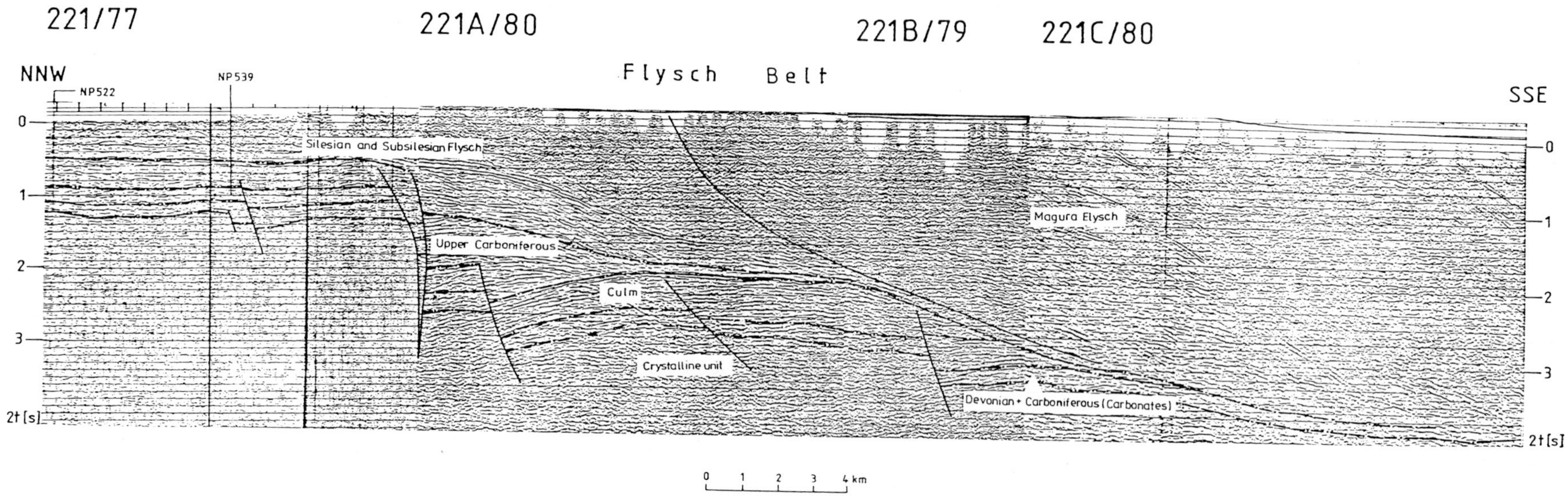

Fig. 17. Seismogeological cross-section 221/77–221C/80. (For line location see Fig. 8)

centre and the south. It dips monoclinally from a depth of 0.5–2 km in the NW, to 12–15 km at the Klippen Belt.

The Autochthonous Cover (Fig. 18) consists of the Lower Devonian Old Red clastics in continental and lagoonal facies, Devonian to Lower Carboniferous platform carbonates, Carboniferous in Culm flysch facies or coal bearing in a paralic and continental molasse facies. The Paleozoic is mainly found in the northern part, in the shallow (NW) and deep (SE) part of the central sector and in the deep part of the southern sector.

Numerous flexures and grabens, e.g. the Jablunkov Graben, the Ostrava-Choryne "morphological" area, and the Jablunkov-Turzovka Fold, disrupt the Paleozoic sequence. In the south, the Nesvacilka-Bilovice Flexure is of the greatest importance. The thickness of the Paleozoic ranges from 2500 to 3000 m. The Mesozoic is represented by Jurassic sediments in a pelitic-carbonate facies and by Cretaceous clastic sediments. The greatest thickness attained is approximately 2000 m.

Autochthonous Paleogene fills the Vranovice and Nesvacilka grabens but also occurs on the related high blocks in the south. It is composed of sandstones and claystones. The assumed thickness is about 1400 m.

The Carpathian Foredeep extends to the northwestern margin of the platform monocline of the southeastern Bohemian Massif slope. From the southeast, the Flysch Belt is thrust over the Foredeep by as much as 30 km in places. The Foredeep is filled with marine Miocene (Eggenburgian to Badenian) sediments (Fig. 18). In the front of the flysch nappes Lower Miocene sequences are incorporated in the nappes. The Eggenburgian-Ottnangian complex developed in the southernmost part and is composed of conglomerates, sandstones

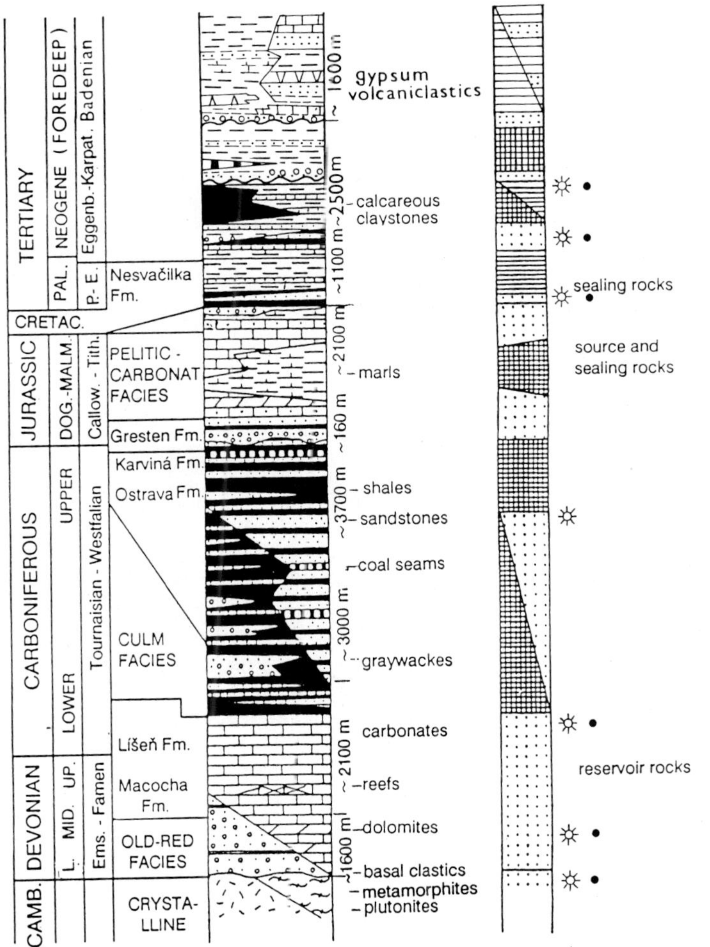

Fig. 18. Generalized lithostratigraphic columns of the Autochthonous Platform cover

and shales. The Karpatian fades away northeastwards and is made up of pelites and sandstones. The Badenian in psephitic and pelitic facies forms a narrow strip along the southeastern margin of the Bohemian Massif. The Neogene, with a total thickness of 1000–1200 m, dips to the southeast beneath the Carpathian nappes.

Hydrocarbon accumulations have been discovered, at shallow depths, in the autochthonous cover of the crystalline complex. Oil and gas accumulations have been found in: the fractured surface of crystalline complexes (Zdanice, East and West Kostelany, Lomna); the Devonian-Lower Carboniferous carbonates (Nikolcice, Uhrice, Nemcicky, Jablunkov); the Jurassic carbonates and clastics (Nikolcice); the Paleogene of the Nesvacilka Graben and associated horsts (Nikolcice, Nemcicky) (Fig. 22, Table 1).

Traps are weathered buried hills, stratigraphic and fault traps sealed by the Neogene rocks of the foredeep and by flysch nappes. Productive horizons occur between depths of 800 to 4000 m. Reservoir rocks are either fractured crystalline rocks, or sandstones with primary porosity (Fig. 23).

The Carpathian Foredeep is a gas-prone area. In the southwestern part, gas reservoirs are in the Eggenburgian to Karpatian clastics, which may be up to 60 m thick at Dolni Dunajovice or at Wildendurnbach in Austria. The Klobucky and Korycany fields contain oil and gas in the 800–1500 m depth

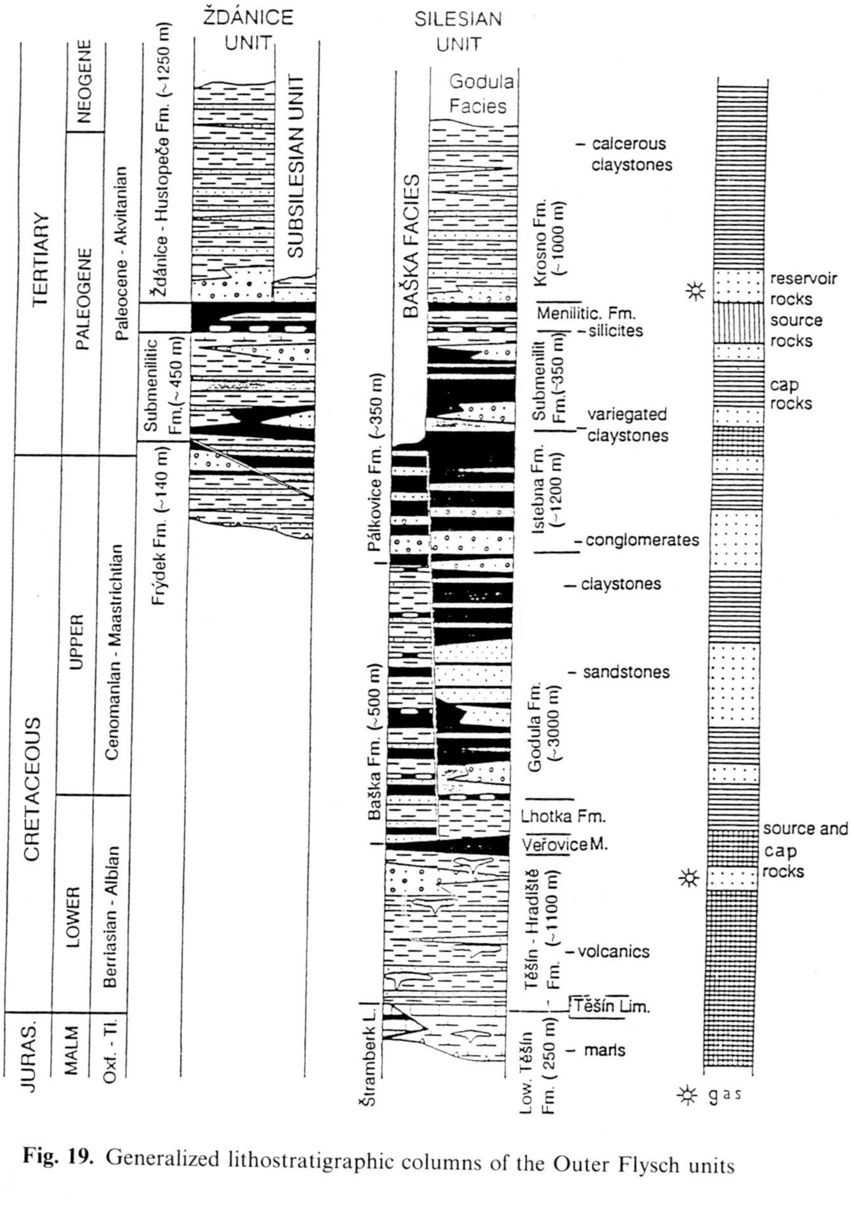

Fig. 19. Generalized lithostratigraphic columns of the Outer Flysch units

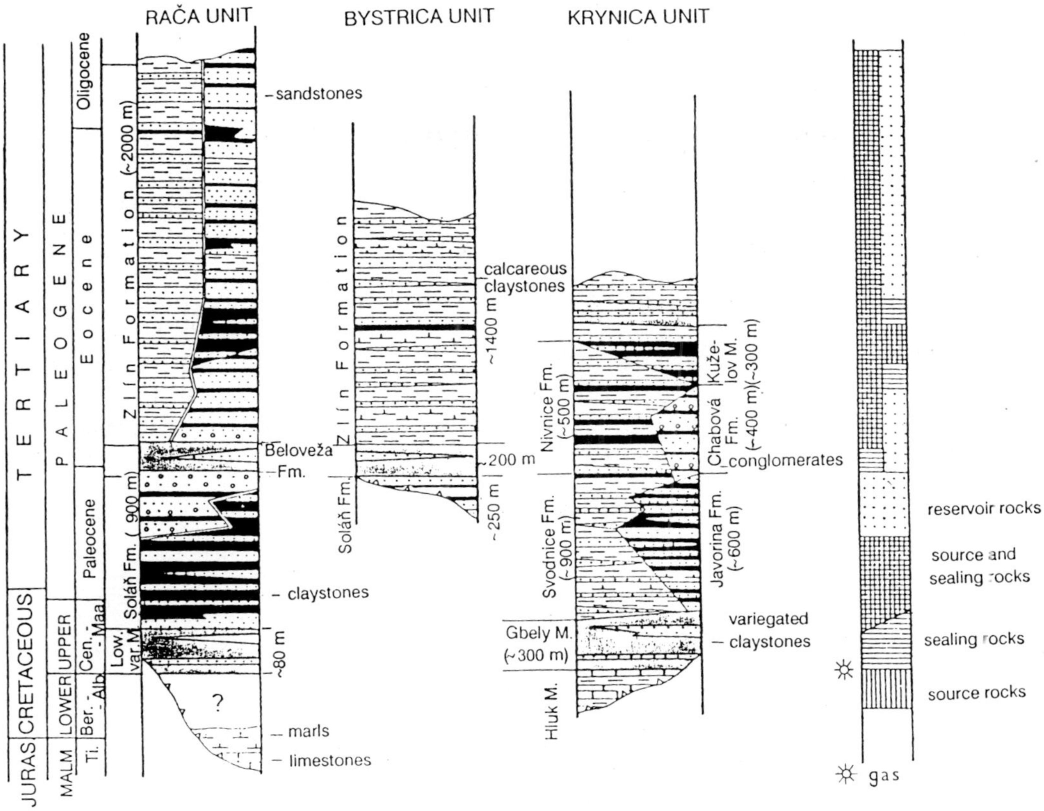

Fig. 20. Generalized lithostratigraphic columns of the Magura Flysch units

interval. In the southeastern part of the foredeep, gas reservoirs have been found mainly in the Karpatian and Badenian sandstones and in the weathered surface of the coal bearing Upper Carboniferous (Fig. 23). Fields are located at depths between 350 and 900 m. The largest of these are Pribor, Klokočov and Horni Zukov. The largest fields are trapped in buried hills while stratigraphic traps occur in drag and half-dome fault structures. Reservoir rocks are clastics with average porosity of 13–15%, and permeability of 100–200 md.

The Flysch Belt is bounded by the Carpathian Foredeep in the northwest and Klippen Belt in the southeast. It is made up of sediments ranging from Upper Jurassic to Lower Miocene (Figs. 19, 20). Their thickness increases towards the Klippen Belt where it may reach over 10,000 m (Fig. 21). The Klippen Belt displays a complicated nappe structure. Movements took place from the Lower Eocene to the Badenian. The Mesozoic is composed of flysch sediments, i.e. sandstones and claystones. The Paleogene formations predominate in the majority of nappe units and are composed of alternating conglomerates, sandstones, marlstones and bituminous shales. The Lower Miocene is in a pelitic-psammitic facies.

In the Flysch Belt, hydrocarbon shows have been encountered in nappe units at Hluk (gas) and at Turzovka and Mikova (oil). Reservoir rocks are mainly Paleogene sandstones sealed by claystones. The average porosity of the flysch sequences is 0.2–5% with permeability ranging from 0.1 to 1.0 md. The average geothermal gradient is 2.85 °C per 100 m. The Flysch Belt is therefore a relatively cool sedimentary province. Pressures are close to hydrostatic (but locally may be up to 60% higher). Crudes are both heavy and light, and contain almost no paraffin (6% at most). As a rule they contain no sulphur. The gas contains some CO_2 (2–5%) and N_2.

Source Rocks. Paleozoic carbonates enriched in organic matter (kerogen type II and III) are considered as oil and gas source rocks. The clastic and Lower Carboniferous (Culm) sedimentary complex, and the coal-bearing Upper Carboniferous contain gas prone kerogen type III.

High source potential has been found in the Upper Jurassic in a pelitic-carbonate facies

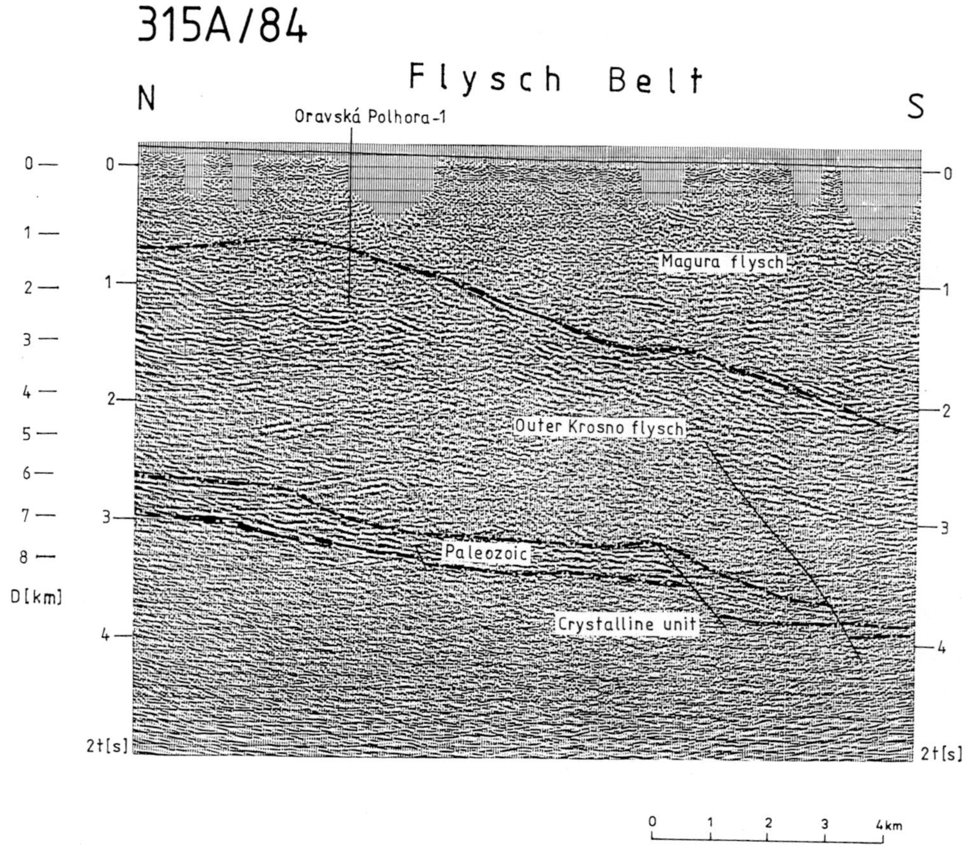

Fig. 21. Seismological cross-section 315A/84. (For line location see Fig. 8)

(kerogen type II–III) which occurs in the southern part of the region and the autochthonous Paleogene (kerogen type III enriched in liptinite) which hypothetically occurs under the Carpathian nappes in the north. Gas generation may also be expected in organic-rich formations and members of the Subsilesian, Silesian and Magura nappes.

Paleozoic source rocks have matured mainly during: the Late Carboniferous in northern Moravia, Jurassic in southern Moravia, Miocene in the area where the total overlying Carpathian nappe was over 2–3 km thick. Present catagenetic temperatures are not sufficient to revive hydrocarbon generation in the Paleozoic source rocks (carbonate and clastics of the Culm facies and coal-bearing Upper Carboniferous) northwest of the Roznow-Valasske Mezirici line. A different situation exists in the northeast, close to Jablunkov, where hydrocarbon generation could have been initiated only after burial by the Carpathian nappes (Fig. 24).

Hydrocarbon generation in the autochthonous Mesozoic and Paleogene rocks in the southern part of the region was initiated in the Savian and Styrian phases of the Alpine orogenesis when the flysch nappes were thrust over the platform cover. Contours of present hydrocarbon generation zones of the Mesozoic and Paleogene are shown in Figs. 25 and 26.

Active liquid hydrocarbon generation from the source rocks of the Bohemian Massif's autochthonous cover, is supposed to occur at depths ranging from 3 to 5 km (oil window), intensive gas generation from 5 to 6–7 km and depleted overmature source rock at 9 km depth and deeper. This vertical catagenetic zonation is valid approximately for the tectonic units of he Carpathian Flysch Belt. In Magura, the Silesian and partially also Subsilesian unit, the observed kerogen maturation level at shallow and medium depth is higher than would correspond to present depth, temperature and burial history during and after the thrusting. It is assumed that the flysch units underwent a pre-thrusting deeper burial with kerogen thermal maturation attaining the threshold of the oil window.

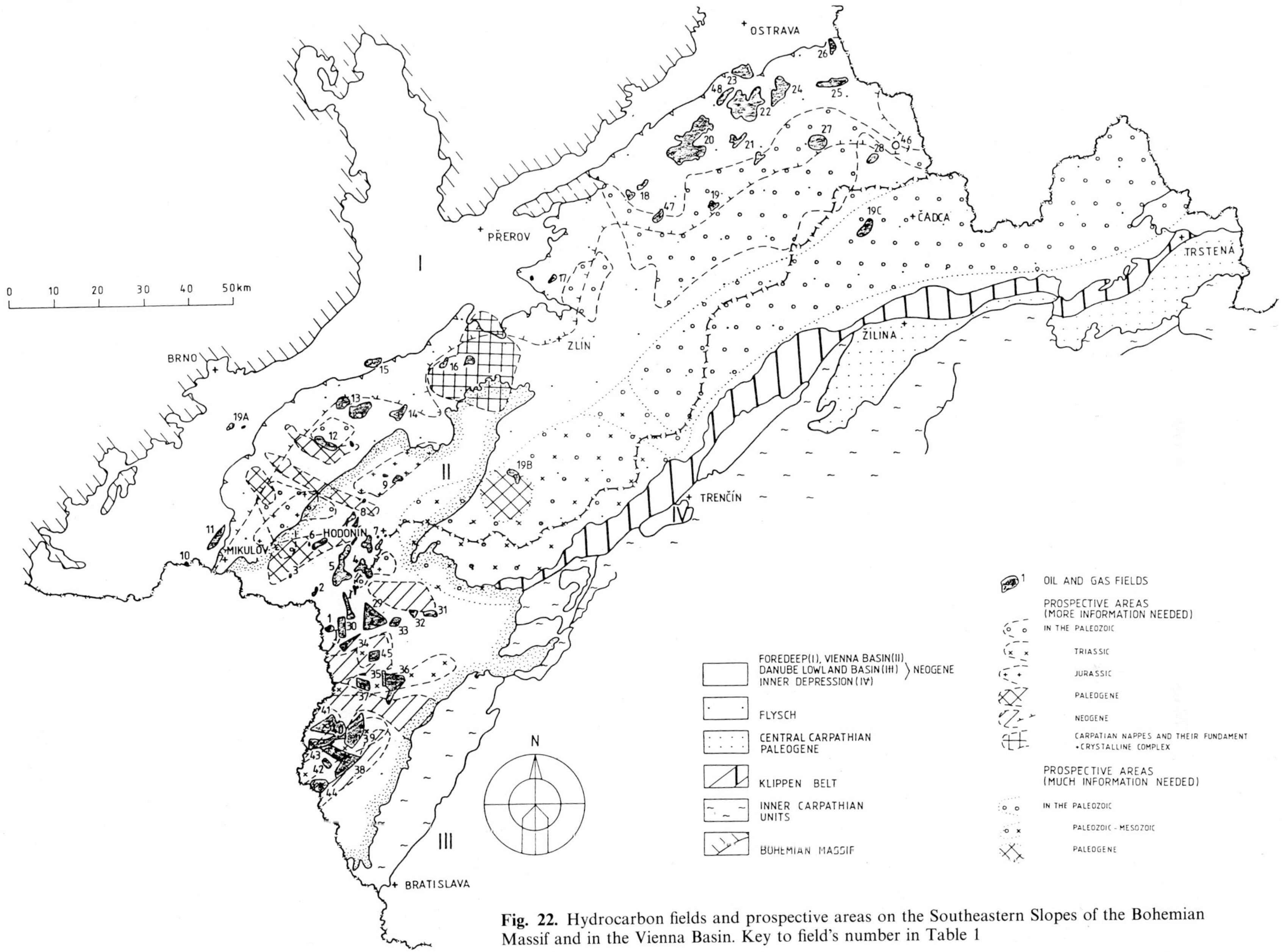

Fig. 22. Hydrocarbon fields and prospective areas on the Southeastern Slopes of the Bohemian Massif and in the Vienna Basin. Key to field's number in Table 1

Table 1. Hydrocarbon fields on the Southeastern Slopes of the Bohemian Massif and in the Vienna Basin

No.	Field name	Year	Depth (m)	Stratigraphy	Reservoir	Porosity (%)	Permeability (md)	Field type	Production status	Reserves[a]
1	Lanžhot	1957	1500–2150	Sarmatian		20	300		Preservation	1
				Badenian	Sandstone	15	100	Oil + gas		
2	Břeclav	1946	1400–1900	Badenian	Sandstone	20	250	Oil + gas	Producing	1
3	Lednice	1979	1150–1350	Badenian	Sandstone	20	300	Gas	Producing	1
4	Týnec-Cunín	1945	300–1300	L. Miocene	Sandstone	15–18	20–100	Oil + gas	Producing	1
		1959		Flysch						
5	Hrušky-Josefov	1959	700–2600	Karpatian		15	150			
		1962		Badenian	Sandstone	20–25	300	Oil + gas	Producing	2
				Sarmatian		25	400			
6	Bílovice-Žižkov	1944	700–2000	Sarmatian	Sandstone	25	400	Oil + gas	Producing	1
				Badenian		20	300			
7	Lužice-Hodonín	1921	300–1700	Sarmatian		20	300			
		1944		Badenian	Sandstone	20	300	Oil + gas	Producing	2
				L. Miocene		15	200			
8	PoddvorOv-	1951	700–2100	Sarmatian	Sandstone	20–25	250–300	Oil + gas	Producing	2
	-Mutěnice			Badenuan		25	400			
9	Vacenovice	1930	150–500	Sarmatian		20	250			
				Flysch	Sandstone	15	100	Oil + gas	Abandoned	1
10	Nový Přerov	1986	800–850	Karpatian	Sandstone	18–25	600–700	Gas	Producing	1
11	Dunajovice	1973	1000–1100	L. Miocene	Sandstone	24	700–900	Gas	Producing	2
12	Uhřice	1982	1500–1800	Oligocene	Sandstone	15–23	1500–2000	Oil + gas	Producing	2
		1978	2400–3100	Devonian	Carbonate	3–4	0–80			
13	Ždánice-	1984	750–1000	L.Miocene	Sandstone	25	200–300	Oil-gas	Producing	2
	-Kloboučky	1983		Old red	Conglomer.	3–5	20–80			
		1973		Cryst. C.	Granite					
14	Koryčany	1978	1400–1500	Karpatian	Sandstone	15–18	200–250	Oil + gas	Development	1
				Cryst. C.	Granite	3–4	0–50			
15	Nítkovice	1971	750–850	Devonian	Carbonate	3	50–100	Gas	Producing	1
16	Kostelany	1968	1300–1500	L. Miocene	Sandstone	15	50–100	Oil + gas	Producing	1
				Crystalline	Granite	3–4	0–50			
17	Rusava	1976	300–500	L. Miocene	Sandstone	12	50–100	Gas	Preservation	1
18	Choryně	1908	400	Carbonifer.	Sandstone	6–20	30–100	Gas	Producing	1
				Karpatian						
19	Rožnov	1983	1650	Paleogene	Sandstone	10	50	Gas	Development	1
19a	Žatčany-Měín	1930	50–80	Karpatian	Sandstone	23	300–400	Oil	Producing	1
		1944		Badenian		30	2500			
19b	Hluk	1943	200–600	Paleogene	Sandstone	2–5	10–20	Gas	Abandoned	1
19c	Turzovka	1902	100–300	Paleogene	Sandstone	2–5	10–20	Oil + gas	Abandoned	1
20	Příbor-Klokočov	1908	260–650	Karpatian	Sandstone	7–20	30–400	Gas	Abandoned	1
		1912		Carbonifer.						
21	Lhotka-Pstruží	1975	600	Paleogene	Sandstone	20	300–400	Gas	Development	1
					Conglom.					

22	Stařič-Lískovec	1913	400–600	Carbonifer.	Sandstone	17	50–400	Gas	Producing	1
				Karpatian	Conglom.					
23	Mitrovice-Paskov	1909	400–900	Badenian	Sandstone	22	50–100	Gas	Abandoned	1
24	Bruzovice-Frýdek	1952	430–630	Carbonifer.	Sandstone	8–20	10–50			
				Karpatian	Conglom.			Gas	Producing	1
25	Horní Žukov	1915	350–450	Badenian	Basal	18–19	100–400	Gas	Producing	1
					Conglom.					
26	Stonava	1952	500	Badenian	Sandstone	20	200–400	Gas	Abandoned	1
27	Krásná Morávka	1980	1500–1600	Devonian	Carbonate	4	0–60	Oil + gas	Development	1
28	Lomná	1984	1950–2050	Crystalline	Crystalline	3–4	0–50	Oil	Development	1
29	Gbely	1914	200–300	Sarmatian	Sandstone	25	500–700	Oil + gas	Producing	2
				Badenian		19	80			
30	Brodské	1951	1000–1100	Badenian	Sandstone	20	60–100	Oil	Preservation	1
31	Štefanov	1948	250–260	Sarmatian	Sandstone	25	500–700	Oil	Producing	1
32	Petrova Ves	1954	850–900	Karpatian	Sandstone	18–20	10–80	Oil	Producing	1
33	Smolinské	1957	1400–1500	Karpatian	Sandstone	18–20	10–80	Gas + oil	Abandoned	1
		(1979)								
34	Kúty	1974	1700–1800	Badenian	Sandstone	22	150–200	Gas	Producing	1
35	Závod	1960	1450–1500	Badenian	Sandstone	23	500–700	Oil + gas	Producing	1
36	Studienka	1959	600–700	Sarmatian	Sandstone	25	500–700	Oil + gas	Producing	1
			1350–1450	Badenian		23	580–700			
37	Závod Fundament	1977	4000–4500	Triassic	Dolomite	3	10–100	Gas + H_2S	Producing	2
38	Láb	1951	1750–1800	Badenian	Sandstone	22	200–400	Oil + gas	Producing	2
39	Malacky	1955	600–700	Sarmatian	Sandstone	25	500–700	Gas	Producing	1
				Badenian		22	200–400			
40	Suchohrad-Gajary	1956	750–800	Panonian	Sandstone	33	980	Gas	Producing	2
41	Gajary-Badenian	1951	1800–2000	Badenian	Sandstone	23	350–500	Oil + gas	Producing	2
		(1980)								
42	Dúbrava	1979	1600	Badenian	Sandstone	15	20–50	Oil	Producing	1
43	Jakubov	1955	1400	Sarmatian	Sandstone	25	500–700	Gas	Producing	1
				Badenian		22	200–400			
44	Vysoká	1955	1600	Badenian	Sandstone	25	200–400	Gas	Producing	3
45	Borský Jur	1979	3300–3500	Triassic	Dolomite	4	20–120	Gas + H_2S	Development	2
46	Jablunkov	1984	2900	Devonian	Dolomite	3–4	20–50	Gas	Development	1
47	Valaš. Meziříčí	1986	2800–2900	Devonian	Dolomite	3–4	20–50	Gas	Development	1
48	Janovice	1967	850–900	Karpatian	Sandstone	20	300	Gas	Development	1
				Carbonifer.		7				

[a] Reserves up to 0.5×10^6 t oil or 0.5×10^9 m^3 gas = 1; reserves up to 5×10^6 t oil or 5×10^9 m^3 gas = 2; reserves over 5×10^6 t oil or 5×10^9 m^3 gas = 3.

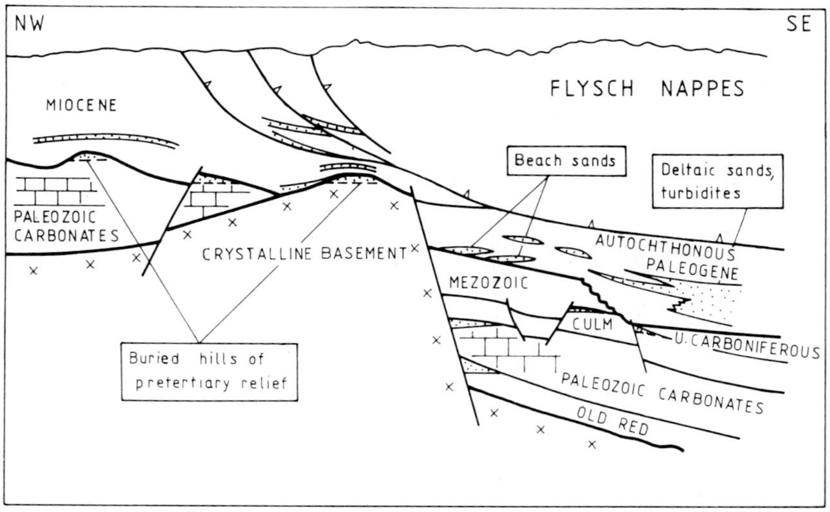

Fig. 23. The Southeastern Slopes of the Bohemian Massif – play types

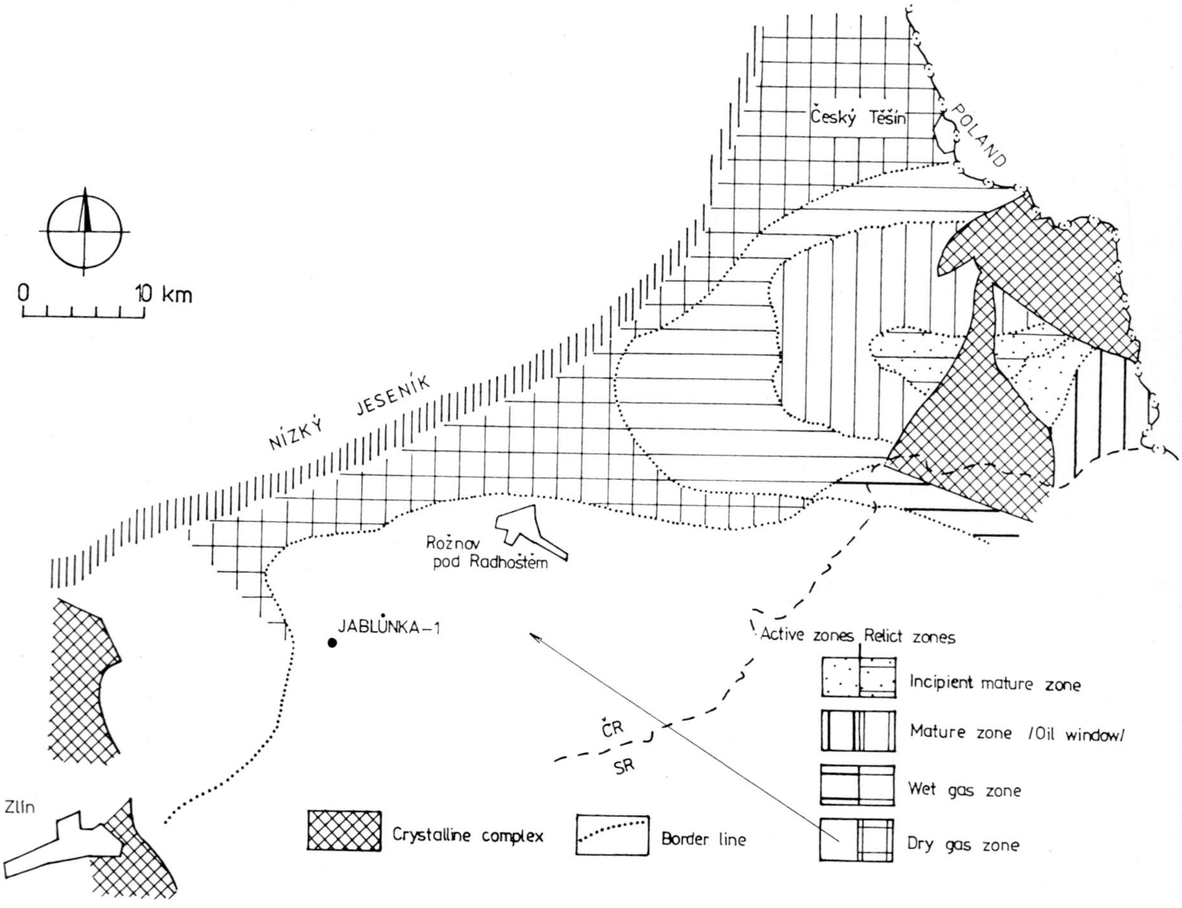

Fig. 24. Maturation map of the carbonate Paleozoic formations, N. Moravia

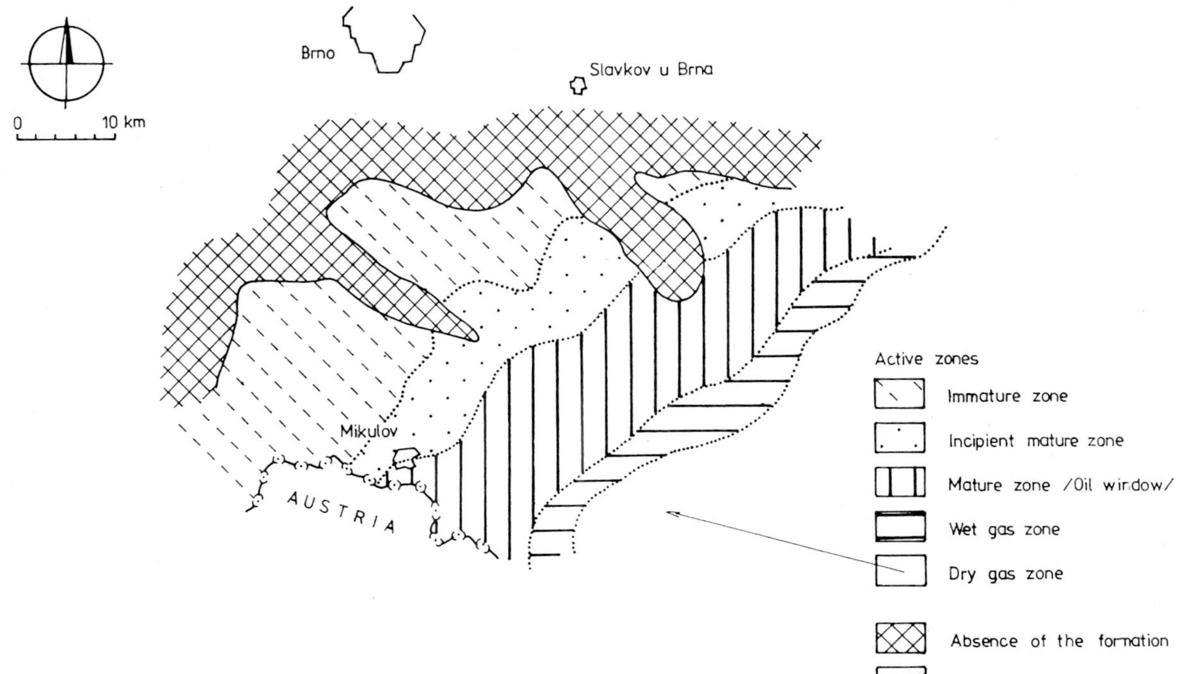

Fig. 25. Maturation map of the Autochthonous Mesozoic

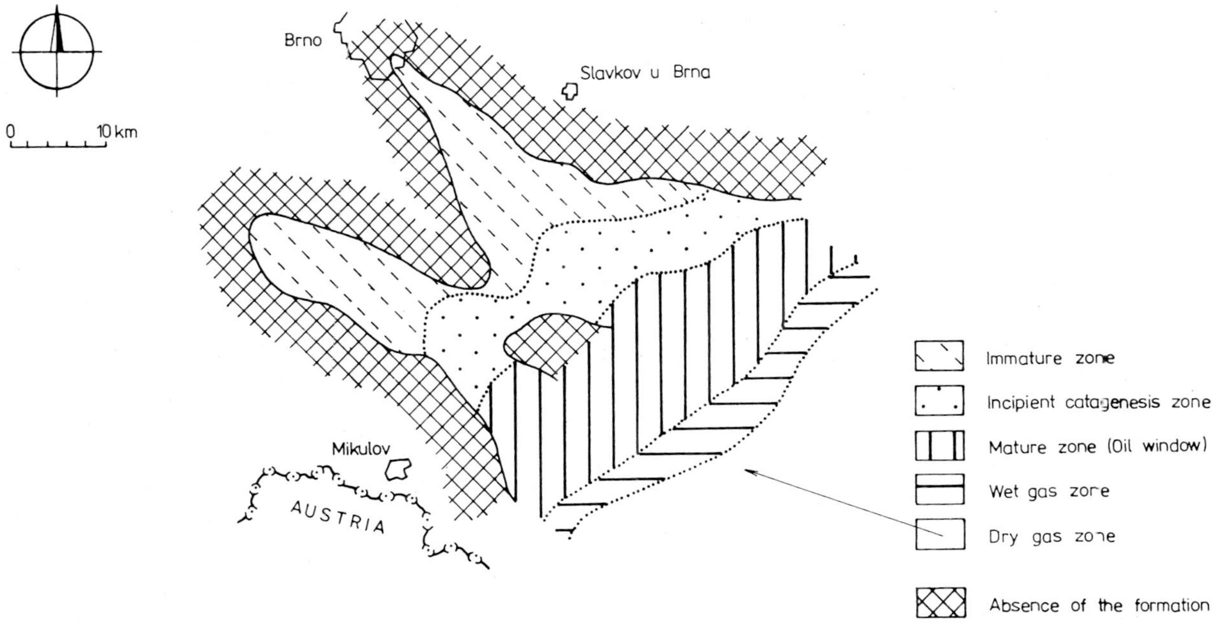

Fig. 26. Maturation map of the Autochthonous Paleogene

However, post-thrusting burial and thermal exposure has been sufficient to revive the hydrocarbon generation in the deeper flysch source rocks.

Paleothermal history and kerogen-to-hydrocarbon conversion is distinct for each lithostratigraphic unit. Fluid mobilization, migration, hydrocarbon accumulation is linked in the whole region with the geological evolution, from the nappe thrusting to present.

The overthrust plane of the Carpathian nappes is considered to be the main migration path of liquid and gaseous hydrocarbons. The fracture zones along the faults and rocks with higher pore and fracture permeability, have acted up to now, as the probable feeding canals for hydrocarbons joining the main migration front from the source rocks to reservoir and traps.

2.2.2 The Vienna Basin

In the Czech Republic and Slovakia, the Vienna Basin (Fig. 2) originated by extension along the boundary of the Alps and Carpathians during Late Alpine orogenic stages. It is a graben depression elongated in a NE-SW direction, and bounded by large faults with amplitude of several hundred to over 1000 (Fig. 27).

The Neogene fill is composed of an Eggenburgian to Pontian sequence which transgresses onto the Alpine-Carpathian units in the southeast, and flysch nappes in the northwest. The depositional axis and extent of the Vienna Basin has changed. In the Lower Miocene, the basin extended to the west. It developed its present shape in the Badenian. During the development of the basin, marine, brackish, freshwater, lagoonal, limnic and deltaic sedimentation stages succeeded one another. The depocentre shifted from north to south. The maximum thickness of the fill is approximately 5500 m. Eustatic and tectonic movements created large lithological facies change. The Lower Miocene comprises conglomerates, sandstones, marine shales, ("Schlier" facies) and freshwater sandy pelites.

The Middle Miocene is made up of conglomerates, shales, sandstones and occasional limestones. In its southern portion abundant deltaic sediments are present. The Upper Miocene is made up of brackish and freshwater pelites with sandy layers, paleodeltaic sediments and lignite. The Pliocene is represented by a fresh water pelitic-psammitic series.

Three sedimentary complexes have been identified beneath the Neogene basin. The lowermost complex is composed of autochthonous Paleozoic and Mesozoic-Paleogene of the slopes of the Bohemian Massif. The intermediate complex is composed of the nappe folded Flysch and Klippen Belt rocks (Jurassic-Egerian). The uppermost complex is made up of the Triassic-Paleogene Central Carpathian and East Alpine nappe units.

The surface of the crystalline complex is assumed to be within a depth range of 7–15 km. In the northwest it is composed of granitoids and meta-

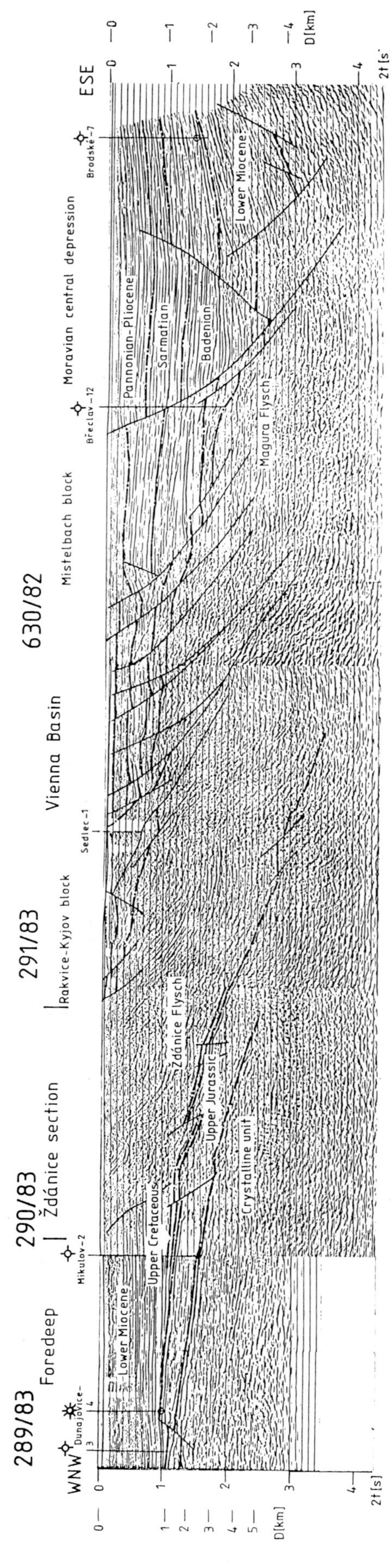

Fig. 27. Seismogeological cross-section 289/83–630/82. (For location see Fig. 8)

morphites of the Bohemian Massif, and in the southeast of crystalline rocks of the Central Carpathians. The basement is unconformably covered by the Paleozoic, Mesozoic and Paleogene sedimentary sequences. The Flysch and Klippen Belt complex is composed of nappes, with rhythmically alternating Paleogene sandstones and claystones. The East Alpine and Central Carpathians Nappe units are mainly composed of Triassic dolomites, shales, limestones, sandstones, with evaporites, Jurassic, Cretaceous, and Paleogene pelites and clastics (Fig. 28).

Commercial oil and gas accumulations occur throughout the Miocene in the depth interval of 150 to 2000 m, the most favourable are however the Badenian, Sarmatian and Pannonian sediments (Fig. 28). Reservoir rocks are sandstone and conglomerates 2 to 30 m, and exceptionally, up to 60 m thick.

In the south, the limestones may reach a thickness of 110 m. Porosity of sandy layers ranges from 10 to 29%, permeability from 50 to 250 md. Clay is the seal rock. The most significant reservoir rocks are the Lab sands, 20 to 25 m thick and widespread throughout the basin. Gas accumulations have been proven in deltaic sediments of the Badenian and Pannonian (Fig. 22, Table 1).

Hydrocarbon *traps* in the Neogene fill have developed mostly in drag folds, half-dome structures and in faulted brachyanticlines. Non-structural traps such as Lithothamniumn reefs (Leitha facies) and traps related to unconformities have also been encountered. Hydrocarbon accumulations may be trapped in blocks created by the main fault systems, horsts, grabens and regional arches (Fig. 29).

Oil and gas occurrences in the flysch floor of the Neogene Vienna Basin are known along the northwest margin of the basin and in the Tynec-Cunin structural element. Traps here are complex. Reservoir rocks are usually fractured sandstones with variable permeability. In the portion of the basin where the basement is composed of Mesozoic nappe units, gas accumulations relate to the erosional tectonic relief and are sealed by overlying Neogene. These nappes are mostly composed of Triassic dolomites, shales, limestones, sandstones with evaporites, Jurassic, Cretaceous and Paleogene pelites and clastics. At the Zavod field, gas was discovered at 4200 m, and at Borsky Jur, a producing horizon occurs at a depth of 3040 m. The reservoir rock is a fractured Upper Triassic dolomite with a porosity of up to 3%, and permeability of 3–5 md. In the Austrian portion of the basin, hydrocarbon fields also occur within the folded nappe structures.

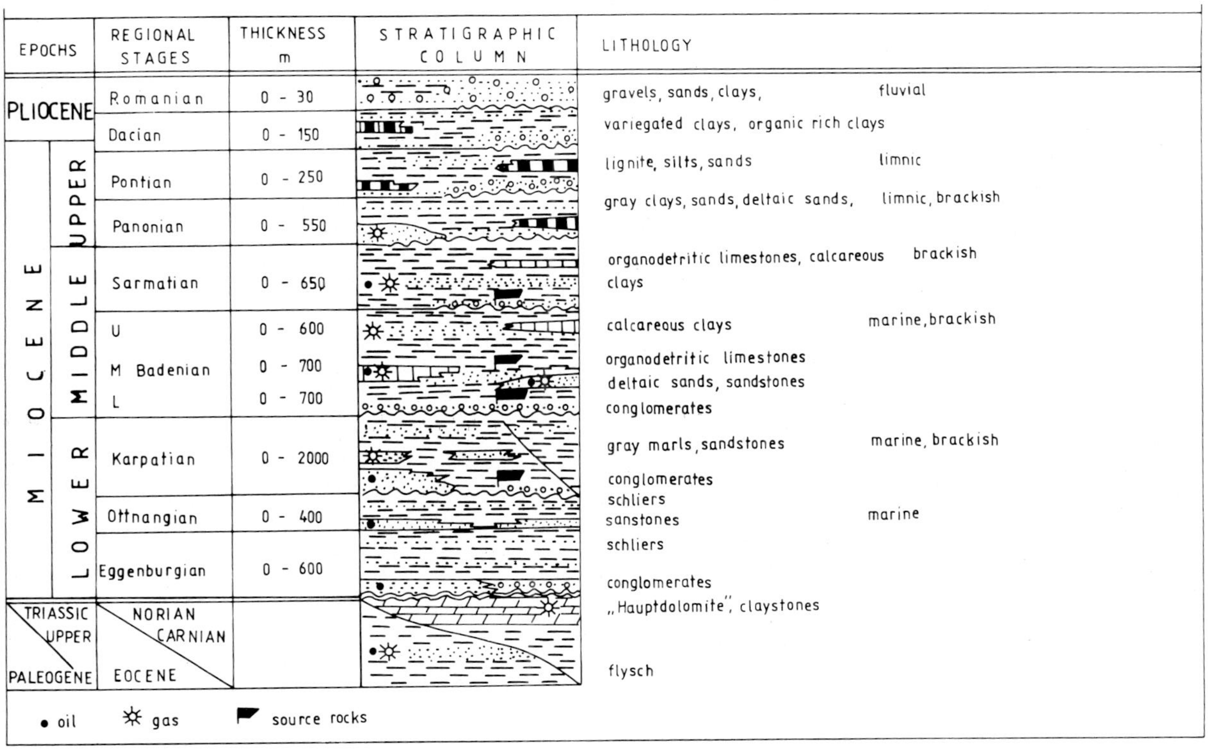

Fig. 28. Generalized stratigraphy of the Vienna Basin

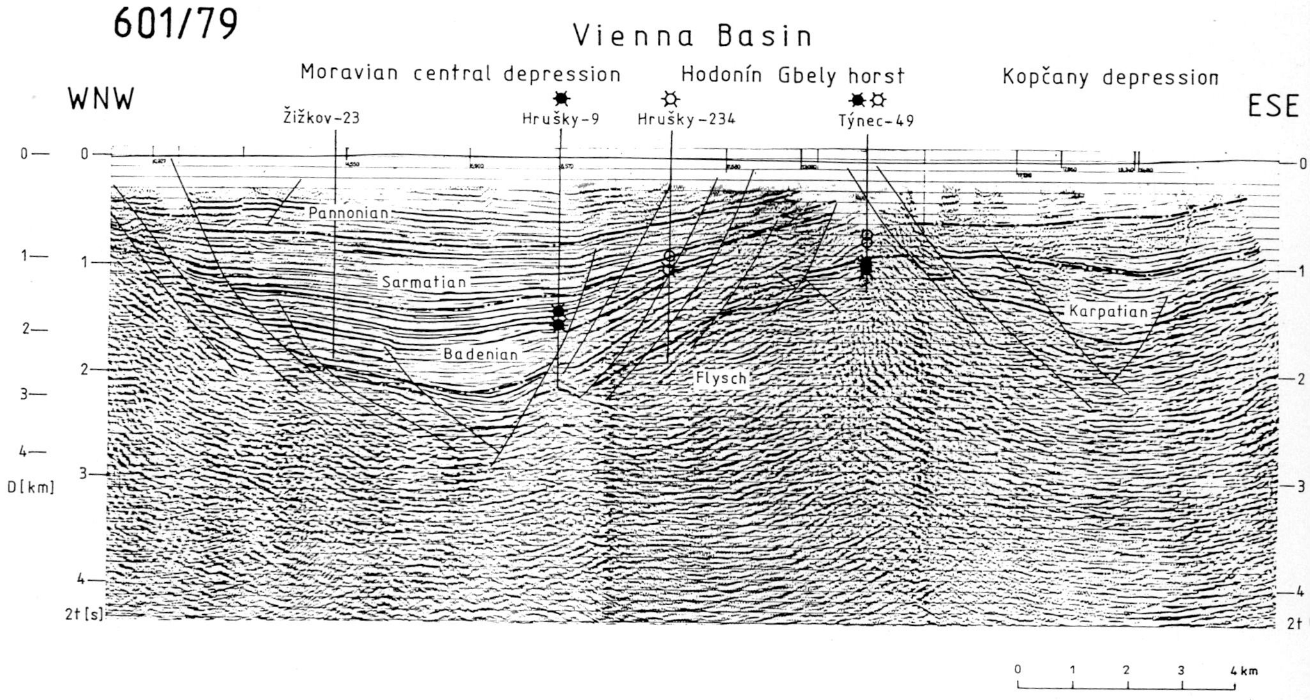

Fig. 29. Seismogeological cross-section 601/79. (For location see Fig. 8)

Source rocks occur in the Middle and Lower Miocene. Due to low organic carbon content (average TOC is less than 1%), and humic type of kerogen their source potential is rather low (less than 2 kg HC/mt of rock).

The Vienna Basin is an area of rather low heat flow and the average geothermal gradient in the depth interval 0–3 km is 2.8–3.0 °C/100 m. Based on vitrinite reflectance, Rock-Eval pyrolysis and modelling the zone of early generation of oil and gas occurs at depths of 3 to 4 km, and the zone of main oil generation (oil window) at 4 to 6 km depth, which overwhelmingly includes the pre-Neogene units (Fig. 30).

A relatively low initial source potential, and only incipient conversion of kerogen to hydrocarbons in the Neogene basin fill, suggest that most of the oils originated from the underlying pre-Neogene formations. These include the Flysch, Carpathian-Alpine nappes and Paleozoic to Paleogene autochthonous sediments of the Bohemian Massif. The latter are considered as the most important source rocks.

Oils in the Neogene reservoirs show decreasing density with depth. They are paraffinic-naphthenic in the Lower Miocene, naphthenic in the Sarmatian and paraffinic in the Badenian and Paleogene pools. Oils at depths of less than 1000 m are depleted of light paraffins due to biodegradation which disappears at about 2000 m. Based on biological markers, the oils are derived from kerogen of mixed humic and marine type, and their maturity is typically higher than that of the adjacent Neogene source rocks.

Gases contain up to 91–97% methane, and often a few percent of gasoline. Mixing with biogenic, isotopically light methane increases with decreasing depth. Formation pressure is mostly hydrostatic. In the deepest parts of the basin only, the pressures can be 30–80% higher than the hydrostatic pressure.

2.2.3 The Danube Basin and the South Slovakian Basin

The Danube Basin originated as a subbasin of the Paratethys (Figs. 2, 31). The basement of the basin is made up of the tectonic units of the Inner Carpathians, with prevailing crystalline and Paleozoic rocks. The Mesozoic in the northern and northeastern parts of the basin belongs to the cover unit of mountain ranges and to remnants of the Veporic nappes. In the southeastern part of the basin, drilling has encountered the Mesozoic of the Hungarian Central Range and the Paleogene, which is also known in the northwestern part of the basin.

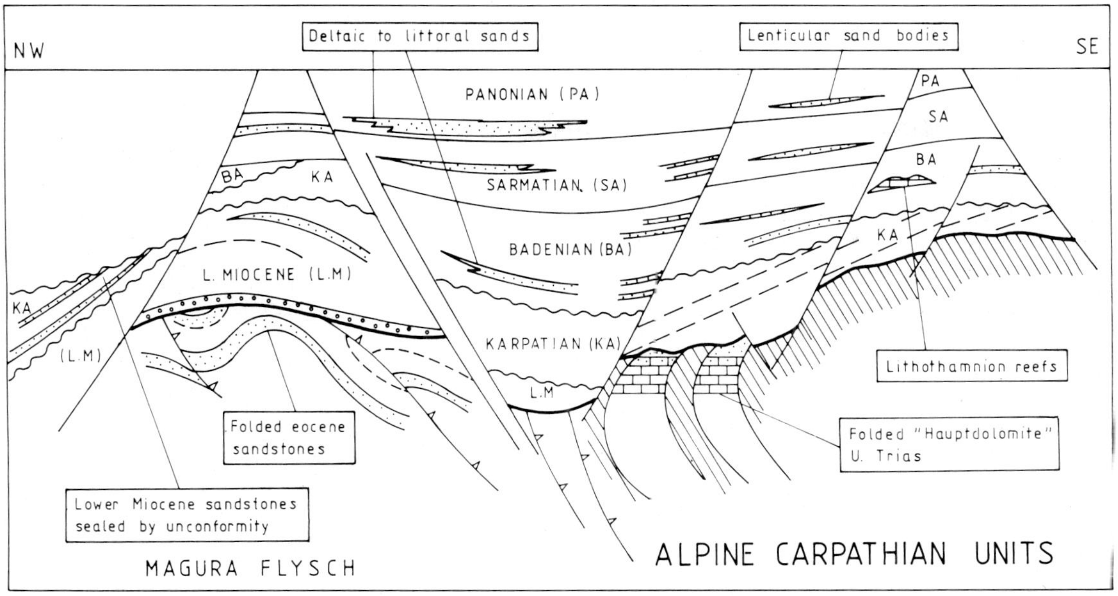

Fig. 30. Vienna Basin—play types

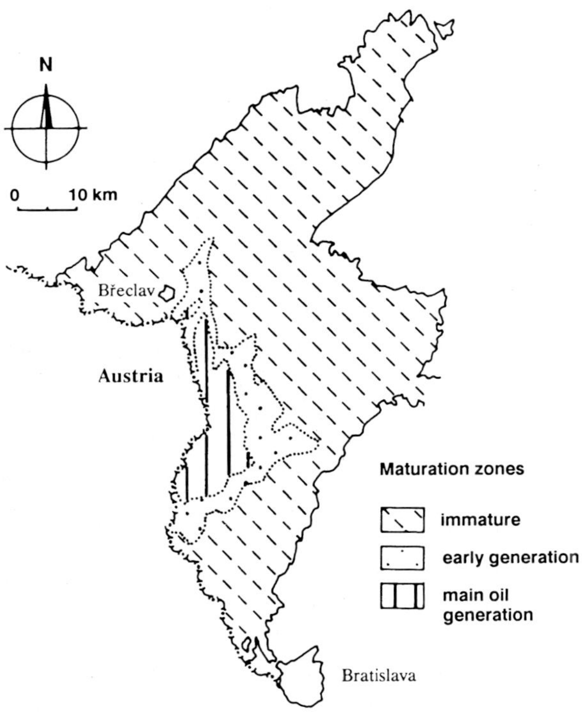

Fig. 31. Organic maturation zones in the Vienna Basin

In its central part, the basin has a depth in excess of 6 km (Fig. 32). In the Middle Badenian, the centre of the basin was to the north, later it shifted south. Sedimentation took place in intramontane seas and lakes, which resulted in diversified facies development, a variable thickness of sediments and numerous hiatuses.

The basin achieved its present-day structure only in the Sarmatian-Pliocene. Upward fining clastics with interbeds of lignite in younger series of strata form the main sedimentary fill. The presence of volcanites and volcanic material is typical in the Lower Badenian and Sarmatian of the centre and east of the basin (Figs. 32, 33).

The principal feature of Neogene tectonics is the normal faulting. Highs confined by faults are the dominant structures in the basin (Fig. 34).

Hydrocarbon *traps* are sealed tectonically and less often lithologically (Figs. 35, 36, Table 2). Commercial accumulations have not yet been found in the Danube Basin. Sporadic oil shows have been noticed in the Sarmatian of the Kolarovo High, and in shallow boreholes in the Komarno area. In the Trnava Bay, at 3000 m depth, methane gas has been encountered. In shallow reservoirs and in the horst areas, biogenic nitrogen and volcanogenic CO_2 are the dominant gas components. Gases are dry with no condensate fraction.

The Neogene *source rocks* contain mainly humic type kerogen in average concentration lower than 1% TOC. The initial source potential of immature shales and marls does not exceed 2 kg HC/mt of rock and is relatively higher in the Badenian and Karpatian rocks.

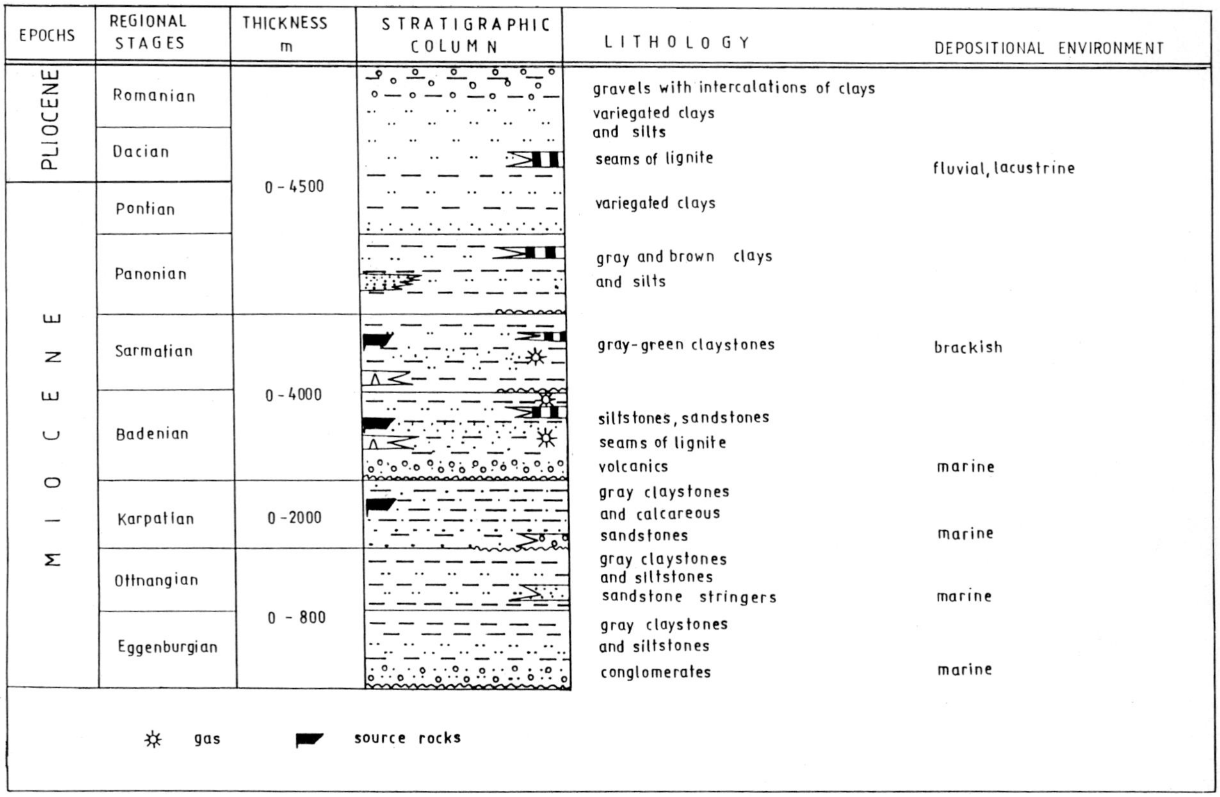

Fig. 32. Generalized stratigraphy of the Danube Basin

Geothermal gradients (in depth range from 0 to 3 km) vary from 3.3 °C/100 m in the colder northern and western marginal areas, to 4.3 °C/100 m in the eastern and central parts of the basin, where the diagenetic zones are shifted to shallower depths. Down to 2.7 km, the source rocks are at an immature stage. The zone of early hydrocarbon generation occurs in the depth range from 2.7 to 3.2 km, and the zone of main oil generation from 3.2 to 4 km in the central part (Fig. 37). Modelling predicts the zone of dry gas below 4.5 km.

There are no data on the underlying pre-Neogene formation source rock potential in the deeper parts of the basin, as they have not been sufficiently cored. It is worth mentioning that the total calculated amount of hydrocarbons formed in the Neogene of the Danube Basin is comparable in order of magnitude to that of the East Slovakian Basin or even higher. However, the absence of commercial hydrocarbon accumulations in the Danube Basin contradicts this conclusion. The reason may be unfavourable trapping conditions in the shallower parts and the little-known geological knowledge of the deeper parts of the basin.

The South Slovakian Basin is a projection of the Pannonian Basin (the Buda basin and the Novohradska Paleogene basin in Hungary). The pre-Tertiary basement is composed of rocks of the Central Carpathians, predominantly Early Paleozoic crystalline schists which in places is overlain by Permian and Triassic, and also sporadically by Jurassic sediments. In the eastern part of the basin, Triassic rocks, mainly carbonates and Paleozoic and Mesozoic rocks of a low metamorphic grade rest on the weakly metamorphosed Upper Paleozoic units of the Inner Carpathians (Fig. 38).

The South Slovakian Basin is filled with Kiscellian and Egerian marine calcareous siltstones and claystones. Thin layers of coarse clastics, organogenic limestones, and carbonaceous clays with coal seams are to be found in the lower part of the section.

In the central and western parts of the basin, Eggenburgian-Karpatian sediments predominate with sandstones, siltstones and claystones with coal seams, layers of rhyodacite tuffs and tuffites which originated both in the sea and on land. The basin margins (particularly the southern one) are covered

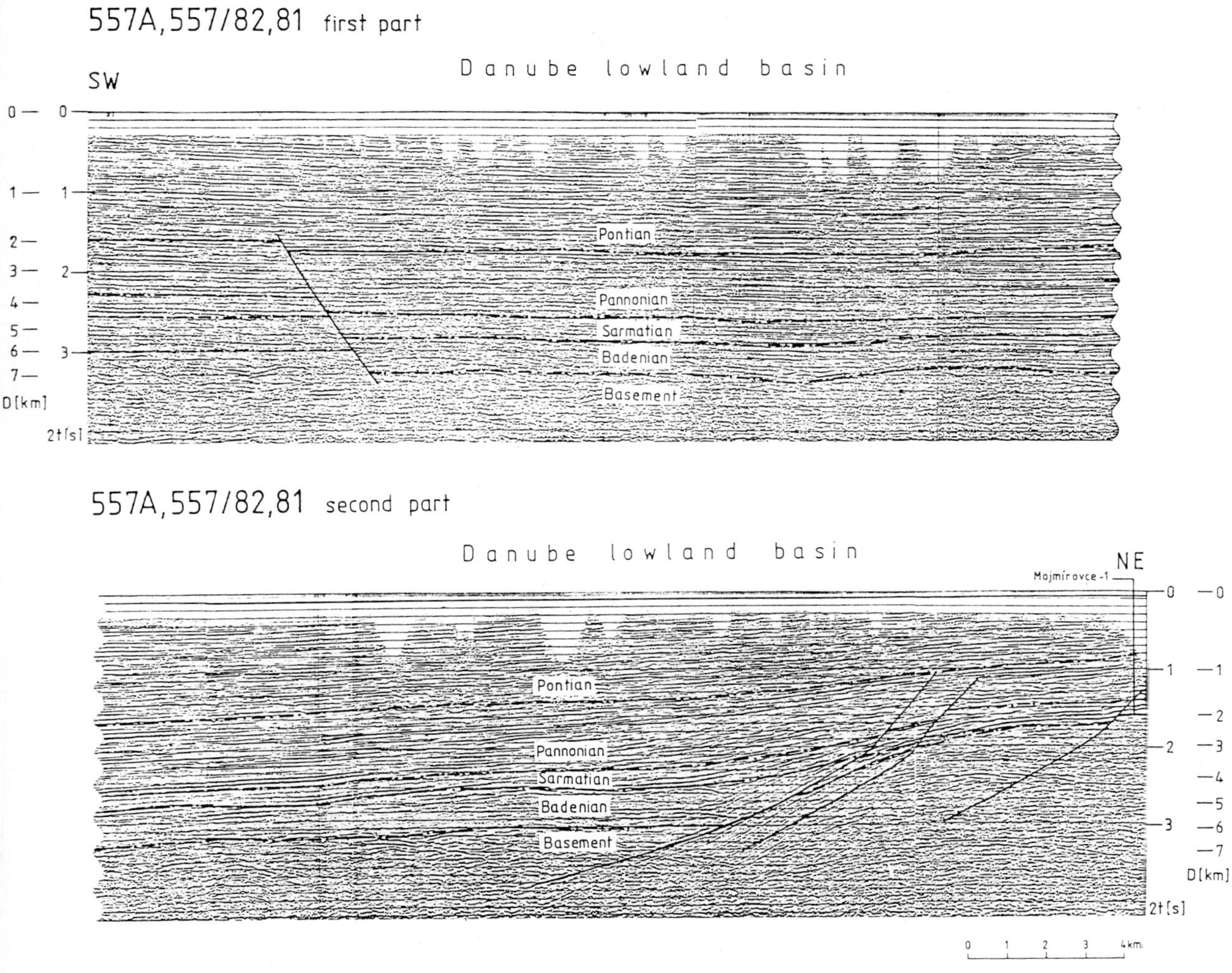

Fig. 33. Seismogeological cross-section 557A, 557/82, 81. (For location see Fig. 8)

with Middle Miocene andesite volcanoclastics. In the northeast, fluvial to lagoonal Pontian sediments occur. The centre is composed of basalt flows of Pliocene to Pleistocene age.

The basement and basin fill are affected by two fault systems, one northeastern and another northwestern.

Since the South Slovakian Basin sediments in Slovakian territory are not thick enough, they did not create favourable conditions for hydrocarbon generation.

2.2.4 The East Slovakian Basin, the Central Carpathian Paleogene and the East Slovakian Flysch Belt

The East Slovakian Basin is a sub-unit of the greater Pannonian basin. It is a pull-apart basin which started its evolution in the northern part of Paratethys, on a basement composed of Mesozoic and Paleozoic nappes of the Inner Carpathians. The marginal normal faulting took place along the older thrust. The depositional environment of the lowermost Miocene sequences was controlled by the mobile zone of the Klippen Belt in the north. In the Badenian and Upper Miocene depocentre gradually shifted to the south, on a NW-SE trend.

The thickness of the Neogene fill in the deepest parts of the basin reaches 7000 m (Fig. 39) The principal lithological types are clays and sands of shallow marine origin with evaporite intercallations which document episodes of arid environment during the Middle Karpatian and during uppermost Middle Badenian. The thickness of the Middle Badenian salt-bearing sequences reaches 300 m. During the Upper Badenian clays of a deeper sea environment were deposited in the south

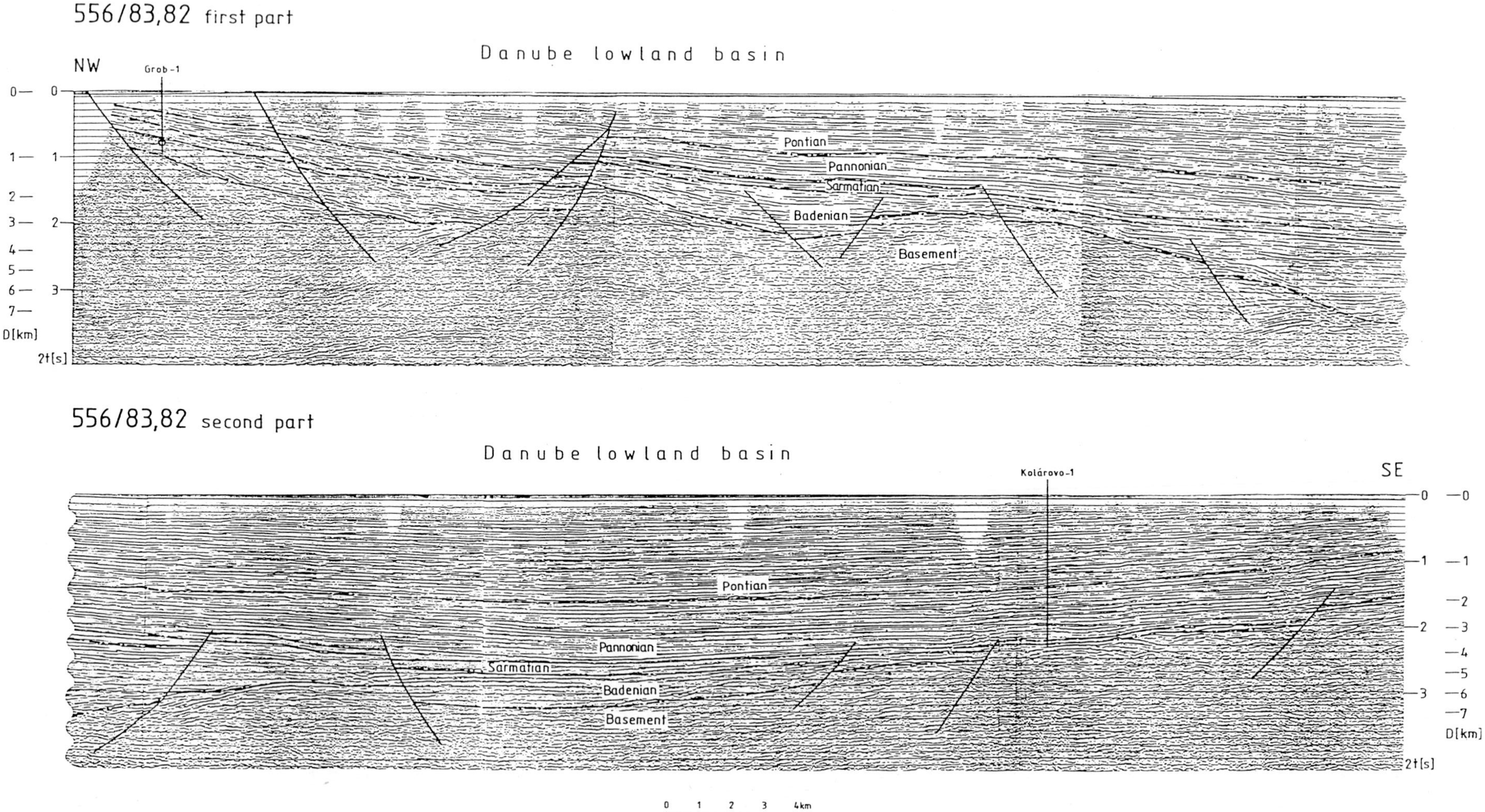

Fig. 34. Seismological cross-section 556/83, 82. (For location see Fig. 9)

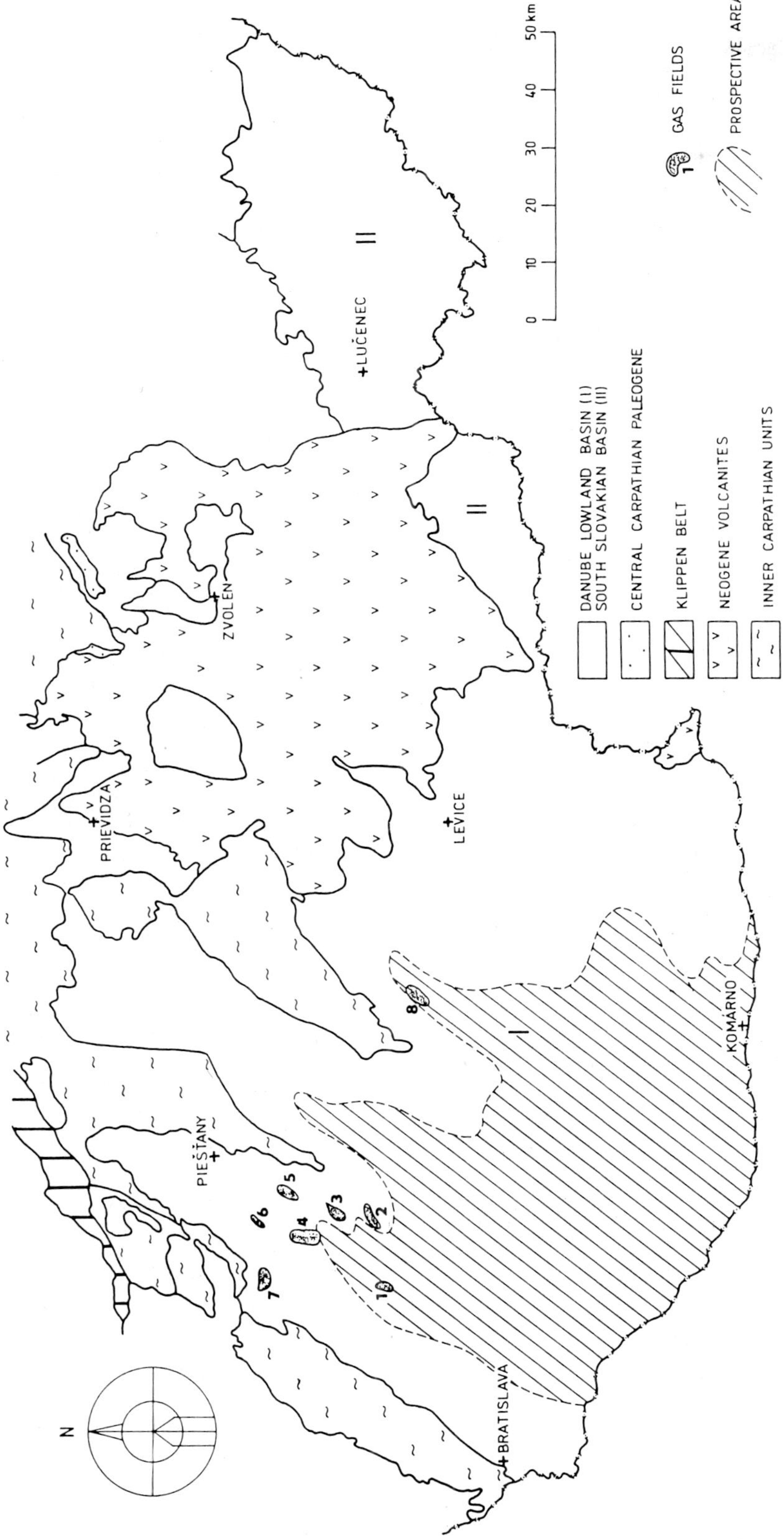

Fig. 35. Hydrocarbon fields and prospective areas in the Danube Basin and in the South Slovakian Basin

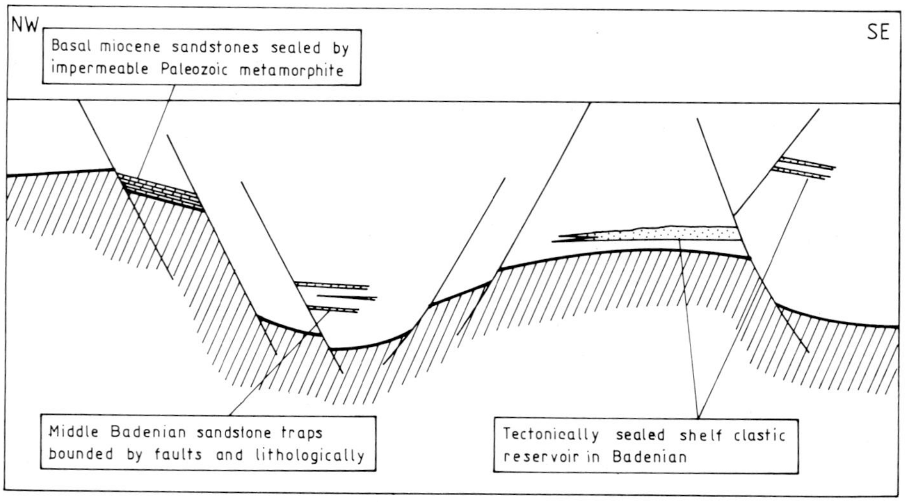

Fig. 36. Danube Basin – play types

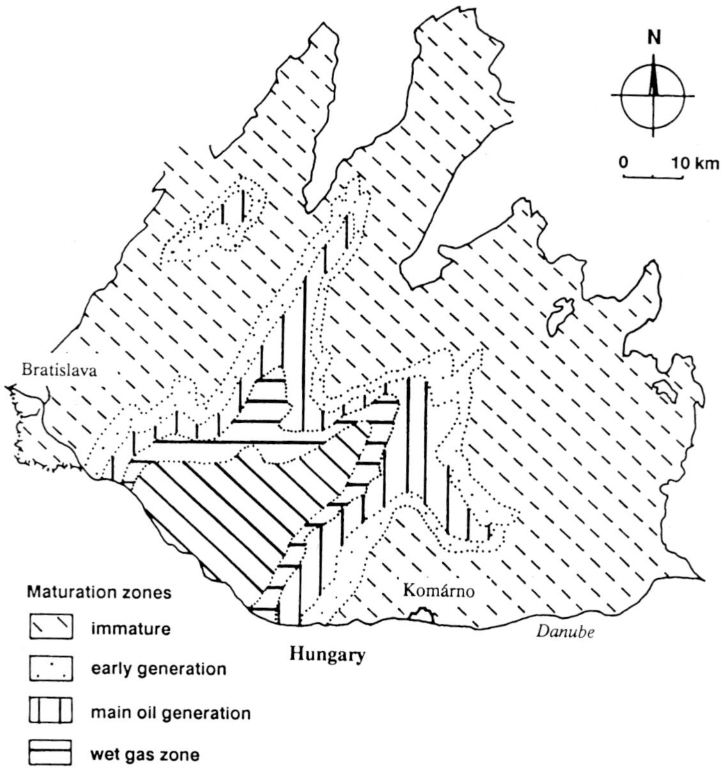

Fig. 37. Organic maturation zones in the Danube Basin

and centre of the basin, while on the northern margin some 1500 m of river and deltaic sediments were deposited. The porosity of deltaic sands is high, reaching 25%.

The most important gas-bearing rocks are found in the Lower Sarmatian series, i.e. 1500-m-thick clayey-sandy sediments of shallow sea origin, including sands with porosity of up to 20%.

In the upper Sarmatian, the subsidence is considerably lower. The brackish to freshwater clays and sands reach only 500 m at the depocentre. Gradual uplift of Carpathian Flysch resulted in

Table 2. Hydrocarbon fields in the Danube Basin

No.	Field name	Year	Depth (m)	Stratigraphy	Reservoir	Porosity (%)	Permeability (md)	Field type	Production status	Reserves[a]
1	Cifer	1956	620,1430	Sarmatian Badenian	Sandstone	18–20	Up to 100	N_2 + gas	Preservation	1
2	Sereď	1964	1000	Badenian	Sandstone	25	Up to 100	CO_2 + gas	Preservation	2
3	Trakovice	1967	950	Badenian	Sandstone	17–19	Up to 100	Gas	Producing	1
4	Špačince	1958	1800–2850	Badenian	Sandstone	15–18	Up to 100	Gas	Preservation	1
5	Madunice	1968	550, 1180	Badenian	Sandstone	18	Up to 100	Gas	Preservation	1
6	Nižná	1967	600–800	Badenian	Sandstone	22	Up to 100	Gas	Preservation	1
7	Krupá	1969	200–300	Badenian	Sandstone	20	Up to 100	Gas	Preservation	1
8	Ivánka	1971	1800	Sarmatian	Sandstone	22	Up to 100	$N_2 + CO_2$ + gas	Preservation	1

[a] Reserves up to 0.5 ×10^9 m^3 gas = 1.
Reserves up to 5 ×10^9 m^3 gas = 2.

Table 3. Hydrocarbon fields in the East Slovakia

No.	Field name	Year	Depth (m)	Stratigraphy	Reservoir	Porosity (%)	Permeability (md)	Field type	Production status	Reserves[a]
1	Ptrukša	1964	1500–1700	L Sarmatian U Badenian	Sandstone	13–24	10–100	Gas + cond	Producing	2
2	Stretava	1963	1150–1500	L Sarmatian U Badenian	Sandstone	17	61	Gas + cond	Producing	2
3	Trhoviště -Podišovce	1958	500–1500	U Badenian	Sandstone	14–29	9–300	Gas + cond	Producing	2
4	Bánovce	1965	1100–2000	U Badenian	Sandstone	8–21	18–120	Gas + cond	Producing	2
5	Rakovec	1986	1000–1150	U Badenian	Sandstone	15	200	Gas	Development	1
6	Senné	1984	1040–1940	L Sarmatian U Badenian	Sandstone	13–22	10–300	Gas + cond	Producing	2
7	Višňov	1984	1450	U Badenian	Sandstone	15	up to 200	Gas	Preservation	1
8	Kolčovo Dlhé	1962	1925	L Badenian	Sandstone	12	up to 100	CO_2	Preservation	1
9	Kecerovské Pekl'any	1973	2160–2705	Mesozoic	Dolomite	5		CO_2	Preservation	2
10	Lipany	1978	1900–2450	Paleogene	Sandstone	11		Gas + oil	Preservation	1
11	Miková	1850	100–800	Paleogene	Sandstone	5–10		Oil	Abandoned	1

[a] Reserves up to 0.5 ×10^6 t oil or 0,5 ×10^9 m^3 gas = 1.
Reserves up to 5 ×10^6t oil or 5 ×10^9 m^3 gas = 2.
L = Lower; U = Upper; Cond = Condensate

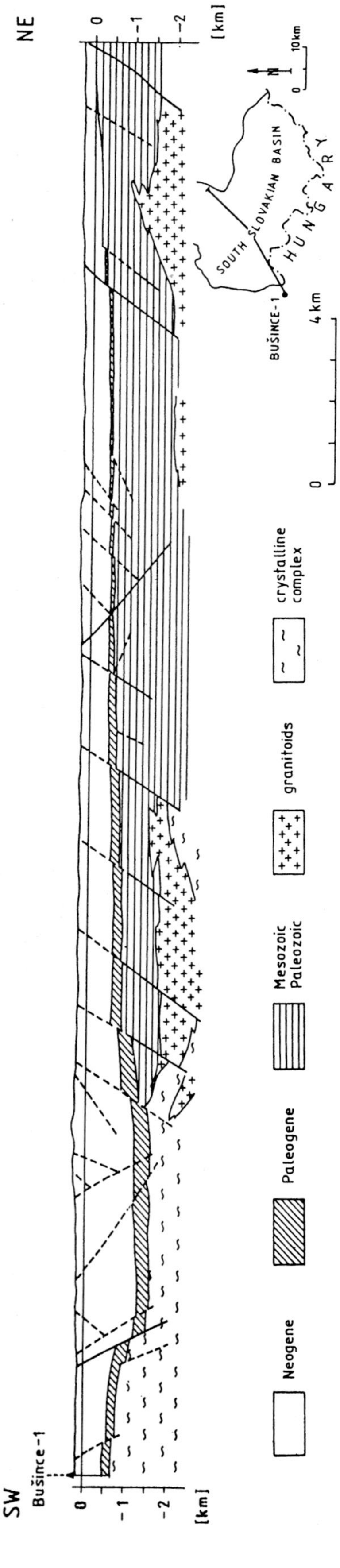

Fig. 38. Geological cross-section of the South Slovakian Basin

shallowing and subaerial exposure of the basin during the Pannonian to Romanian. The maximum thickness of these continental sediments is around 1000 m.

Subsequent volcanism is significant for the development of the East Slovakian Basin. Rhyolites have been encountered from Eggenburgian to Middle Badenian. Intermediate volcanites-andesites prevail in the Upper Badenian-Pannonian period (Fig. 39).

The general development of the basin has been affected mainly by synsedimentary faults and by volcanic activity. The compressive movements in the Pannonian were of importance and probably caused the origin of a striking feature, the Stretava-Banovce Ridge (Figs. 40, 41) which contains gas accumulations. The structure of the basin was also largely affected by the presence of salt layers and by differential compaction (Table 3).

Hydrocarbons have been discovered in various play types (Fig. 42), in all Neogene intervals except the Pannonian. Lithological traps prevail at e.g. Senne, Stretava, Ptruksa (Fig. 43), and Visnov. Reservoirs could have been sealed by normal faults: e.g. at Banovce, Trhoviste, and Pozdisovce (Fig. 44). The usual reservoir rocks are sandstones but in the centre of the basin, weathered extrusive reservoirs sealed by younger Neogene sediments may also be found.

Reservoirs are covered by thick complexes of clays or salty layers. In the structural traps at the localities Kecerovske Peklany and Kolcovo Dlhe hydrocarbons are accompanied by CO_2.

Source rocks are of Early Karpatian, Badenian and Early Sarmatian ages. Average concentration of organic matter is lower than 1 to 1.3%, and kerogen is mainly humic gas prone similar to occurrences in the Vienna and Danube basins.

Wet gas with condensate was found in commercial accumulations in the eastern East Slovakian basin. The distribution of light hydrocarbons in the condensates suggests that the generation depth is at about 2.3 to 3 km. Probably all the hydrocarbons are derived from the Tertiary source rocks as the Pre-Tertiary rocks were overmature prior to the Paleogene. Gases contain higher hydrocarbons and are isotopically heavier, hence suggesting their catagenic origin in the oil and gas generation zones. Biogenic methane occurs only in shallow reservoirs with a biochemical cut off point at depths of about 1.2 km.

The high diagenetic rate with depth is due to high geothermal gradient, which ranges, in the

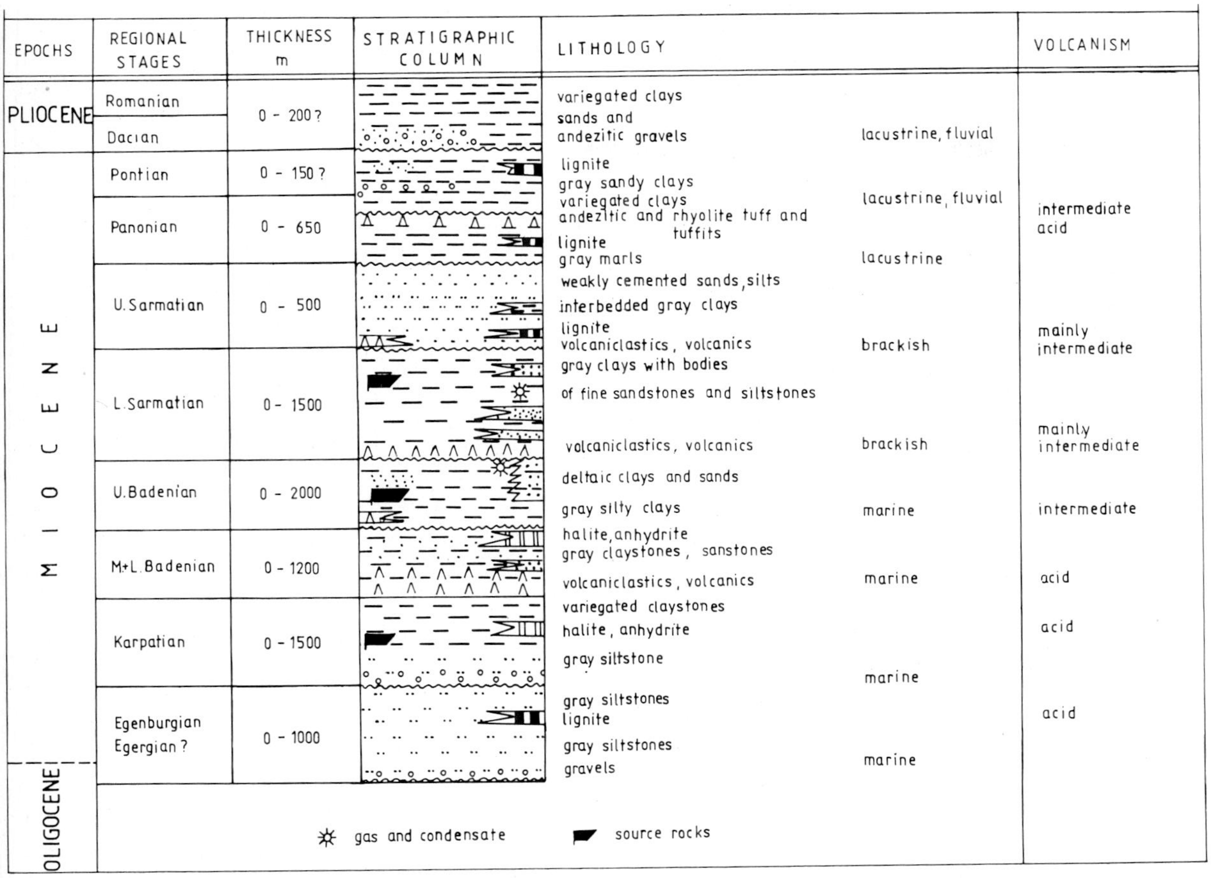

Fig. 39. Generalized stratigraphy of the East Slovakian Basin

depth interval of 0 to 3 km, from 3.7 °C/100 m in the northwestern part to 5.0 °C/100 m in the central, southern and eastern parts of the basin. The immature zone occurs down to 1.7 km: from 1.7 to 2.0 km early generation occurs followed by the main oil generation from 2 to 2.8 km, wet gas from 2.8 to 3.4 km, and dry gas over 3.4 km (Fig. 45). Most of the kerogen to hydrocarbon conversion takes place between 2 and 3 km. In the central part of the basin it was accomplished prior to the Pliocene sedimentation.

The geochemical characteristics of the hydrocarbons is in good agreement with the concept of a Neogene source rocks origin. Studies of pre-Neogene source rocks suggest that hydrocarbon generation has occurred only during the pre-Neogene evolution.

The Central Carpathian Paleogene and the *East Slovakian Flysch Belt* constitute another potential hydrocarbon prospective area. The Klippen Belt separates the more deformed southwestern flysch part from the less tectonized Central Carpathian Paleogene. The thickness of Paleogene resting on the Mesozoic-Paleozoic units of the Central Carpathians reaches 3000 m towards the Klippen Belt. Northwards along the Slovak-Polish frontier, the assumed thickness of the flysch may reach 12,000 m.

At Lipany, in the Central Carpathian Paleogene zone, fractured claystones and marlstones yield a small amount of oil and gas. Potential reservoirs are fractured sandstones with a low primary porosity. The structure is located on a Mesozoic basement high. Oil in the Lipany oil and gas discovery is light paraffinic with high amounts of gasoline (density from 0.81 to 0.85 g cm^{-3}). Gas dissolved in the oil is composed by 96–99% methane, in other horizons by 50% methane and 50% nitrogen. In the uppermost horizon gas with 92% CO_2 and 8% nitrogen occurs. Drilling in the Carpathian Flysch Belt has also detected minor gas shows.

Source rocks in the inner Carpathian Paleogene have up to 1.5% TOC, and their source potential is mostly depleted due to high diagenetic/catagenic

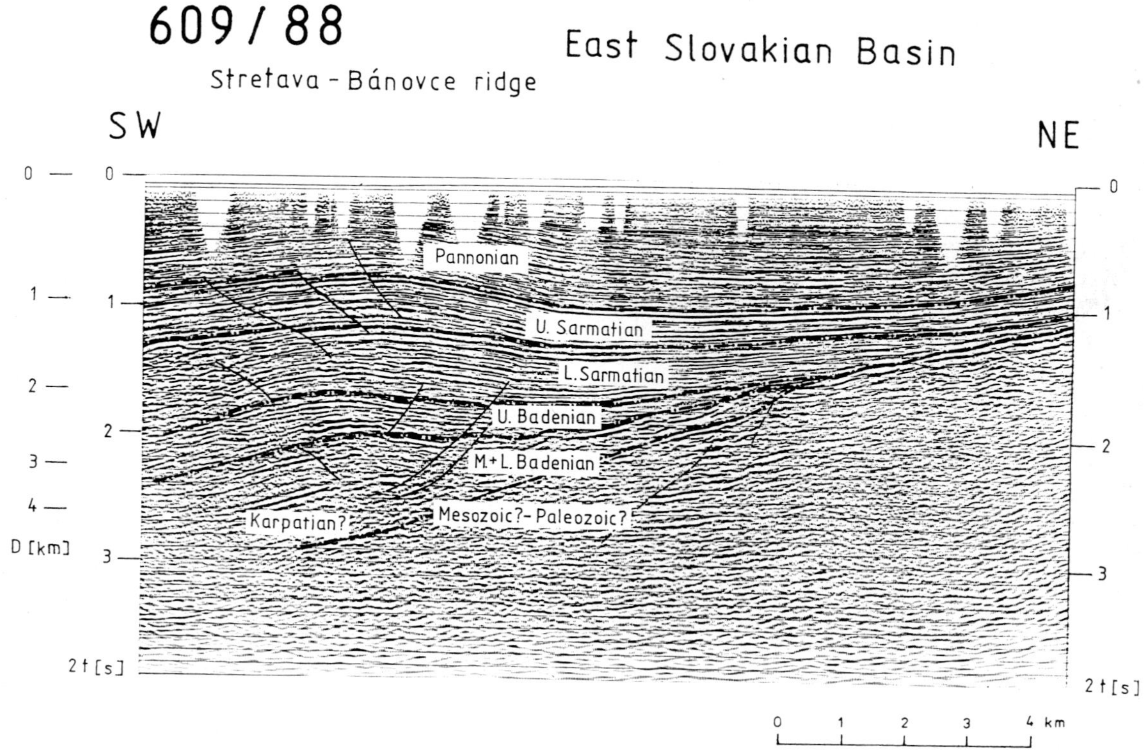

Fig. 40. Seismogeological cross-section 609/88. (For location see Fig. 10)

stages. In the peri-Klippen area the Paleogene kerogen maturity corresponds to the end of oil generation and wet gas stages (vitrinite reflectance R_o is over 1.2% at the surface and reaches 1.8 at 1 km depth).

The underlying Mesozoic rocks of the Krizna Nappe are in both the wet and dry gas window. The present catagenic degree of alteration is a relict one and is a result of Oligocene/early Miocene maximum burial when the oil was generated.

2.3 Production and Future Potential

2.3.1 The Southeastern Slopes of the Bohemian Massif

Hydrocarbon accumulations occur mainly in the: Middle Devonian-Lower Carboniferous carbonates, Paleogene terrigenous sediments of Bohemian Massif slope, Neogene clastic series of the Carpathian Foredeep, Upper Carboniferous Molasse terrigenous and carbonate Mesozoic, sandstones series of the Carpathian Flysch, as well as in the weathered surface of crystalline basement.

Forty structures have been selected for exploration: 30 fields (gas fields prevail) have been discovered, some with very small reserves of up to 0.5 billion m^3 (17.5 BCF) of gas. Only one field has larger reserves. Two fields yield over 500,000 tons (3 MMbbl) of oil. Altogether, 40% of the wells drilled are producing.

Most hydrocarbon accumulations have been found at a depth of 3000 m in the northwest part of the province where the platform is at shallow depth. Future potential could be in the deeper part of the platform, in the depth range from 5 to 7 km, where the depositional environment is favourable to hydrocarbon accumulation and sourcing. The largest reserves are expected in the Paleozoic of the northern sector, in the Jablunkov-Turzovka depression and in the Choryne-Ostrava area and in the southern sector, in the Nesvacilka-Bilovice area. In the Flysch Belt, hydrocarbons may be found within the Paleogene series of the Inner Flysch Nappes and in front of the Klippen Belt (Fig. 22). A good indication might be the high content of dissolved methane in natrium carbonate fluids tested in wells which penetrated Flysch Nappes and the Paleozoic-Mesozoic cover of the Bohemian Massif, at the platform/crystalline contact.

The fields so far discovered occur in traps of structural type as well as in weathered basement. Exploration should concentrate on traps of

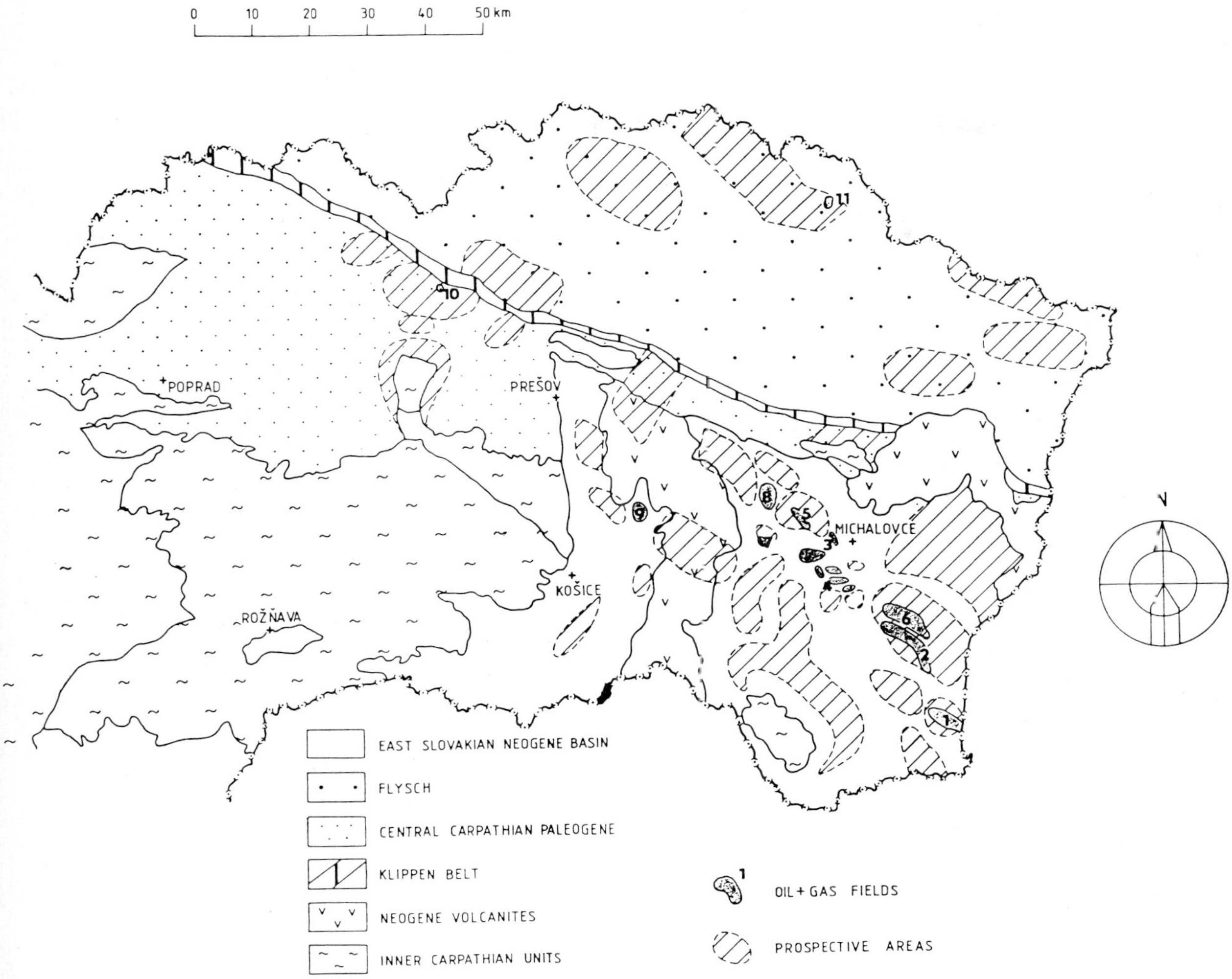

Fig. 41. Hydrocarbon fields and prospective areas in East Slovakia

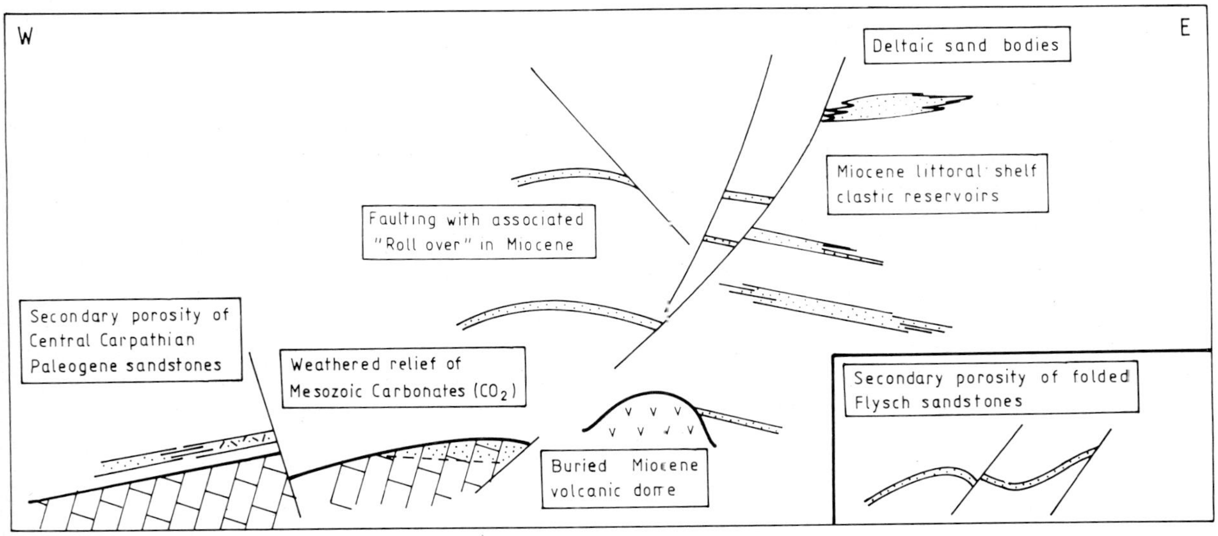

Fig. 42. East Slovakia – play types

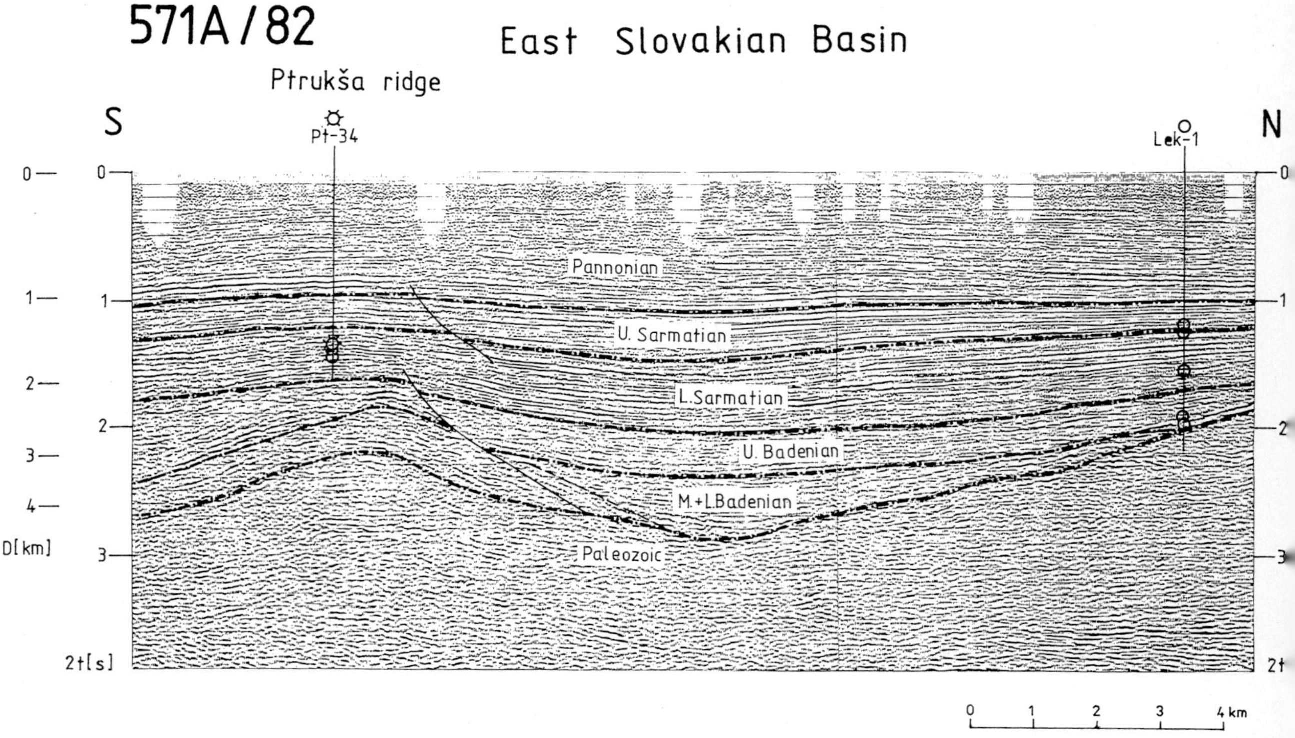

Fig. 43. Seismogeological cross-section 571A/82. (For location see Fig. 10)

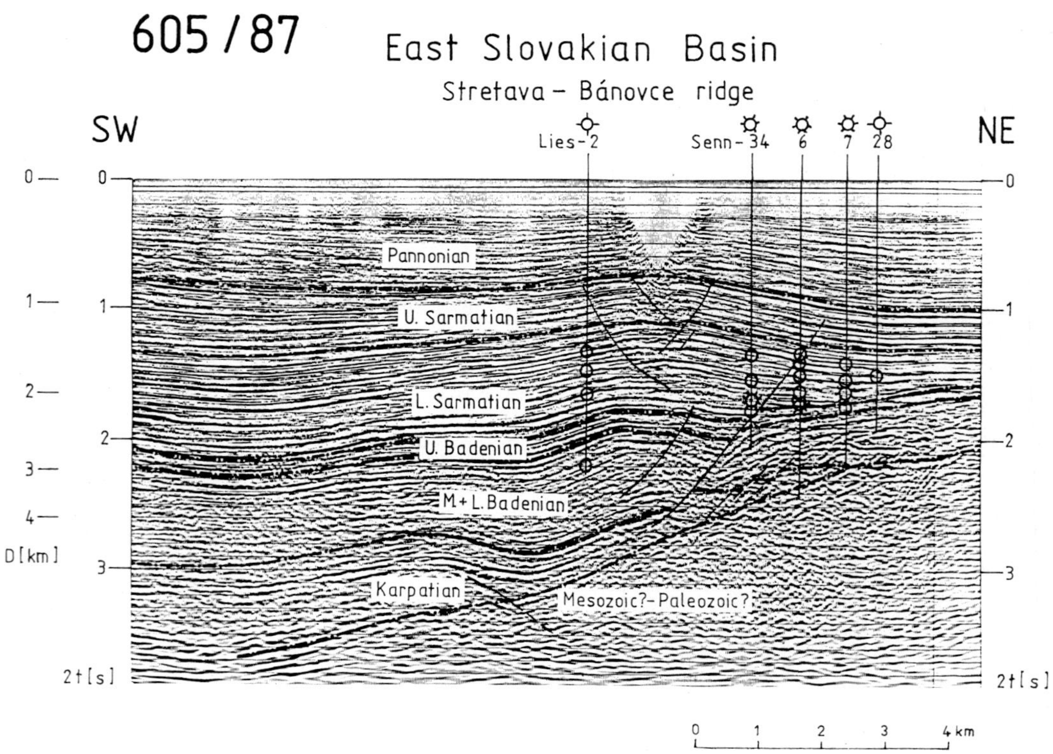

Fig. 44. Seismogeological cross-section 605/87. (For location see Fig. 10)

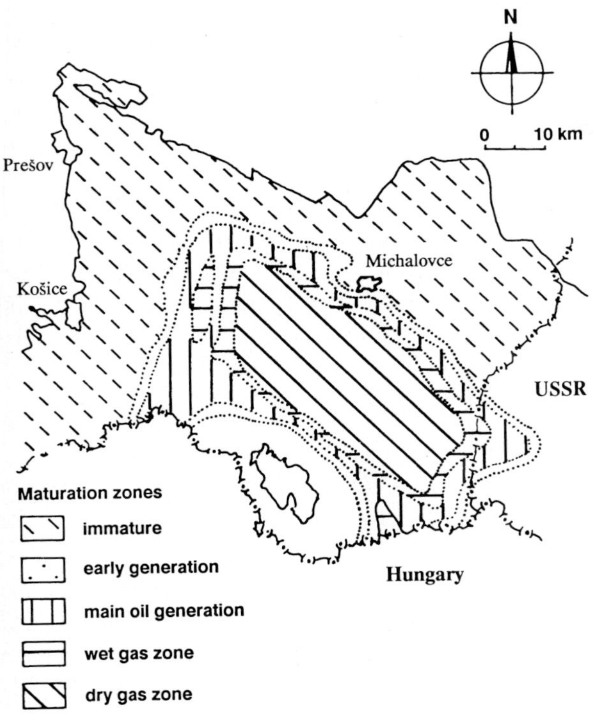

Fig. 45. Organic maturation zones in the East Slovakian Basin

lithological type and on fractured zones related to major fault systems.

2.3.2 The Vienna Basin

Commercial reserves have been found throughout the Miocene and in the underlying Flysch or Alpine-Central Carpathian units. Of the 44 explored structures, 34 yielded hydrocarbons. Exploratory drilling has an overall 57% success ratio.

Despite the very good knowledge of the basin, new prospects can still be found there. In the well-explored Neogene, for instance, Lower Miocene sequences of the slopes of interior depressions, Badenian, Sarmatian and Pannonian paleodeltaic facies and sandy reservoirs at all stratigraphic levels must be taken into account.

However, the principal area of interest is the folded Alpine-Carpathian Mesozoic basement of the basin. At Borsky Jur and Zavod, oil and gas have been discovered in the "basement" at depth intervals of 3000 to 5000 m. The objectives are the frontal parts of the individual Alpine-Carpathian nappes in the south and centre of the basin. In the north, the autochthonous Paleozoic-Paleogene cover of the Bohemian Massif, assumed at a depth interval of 7–10 km on the Tynec-Cunin and Skalice-Radosovce highs, has been selected for further exploration (Fig. 22).

2.3.3 The Danube Basin

Up to now, only gas has been found in the basin. The gas tested in the Middle Badenian of the northwest part of the basin contains 85–95% of CH_4. However, in places (Cifer, Sered, Ivanka), abundant admixtures of CO_2 and N_2 even prevail over hydrocarbons. A large accumulation near Sered, with reserves of 4.7 billion m^3 of gas, contains approximately 10% of CH_4 and 90% of CO_2.

Here reservoirs are sandstones with volcanic admixture in places. They have large variations in grain sizes and cementation. Hydrocarbons have not yet been detected in the Neogene basement (Fig. 36). Deep drilling was carried out on 29 structures, eight of which yielded hydrocarbon accumulations but the reserves are small. At the Trakovice field approximately 100 million m^3 of gas has been produced and the field has been abandoned.

Although no direct data, mainly from the deeper parts of the basin, are available, the conditions in the Danube Basin are favourable for generation, migration and accumulation of hydrocarbons. The presence of source rocks, reservoirs and seals has been confirmed by previous exploration. More information is still required to find out whether the source rock thickness in the deeper parts of the Neogene fill is sufficient for generation of commercial hydrocarbon accumulations, as has been suggested by previous data analysis.

The search for non-structural traps in the deep, central part of the basin is another task. Indications of prograding clinoforms in the Badenian and Sarmatian have been observed on seismic lines in the neighbourhood of the Surany faults. The Lower Pannonian could be a good seal. Even the small amount of exploration work so far accomplished in the Danube Basin has shown that commercial hydrocarbon deposits can most likely be found in the deeper parts of the basin, at intervals of 1500 m to 6000 m.

2.3.4 The East Slovakian Basin, the Central Carpathian Paleogene and the East Slovakian Flysch Belt

Hydrocarbons have been encountered in all pre-Pannonian Neogene reservoirs. Commercial deposits have been found in the Lower Sarmatian and in the Upper Badenian. Of the 40 structures originally selected, only 20 have been explored. Commercial accumulations have been found in seven small fields. Two fields contain reserves of over 2 billion m^3 of gas. The majority of reserves occur in the Lower Sarmatian and in the Upper Badenian, where 50% of the wells drilled are producing.

In the East Slovakian Basin, hydrocarbon accumulations may be found on the flanks of highs (Stretava, Ptruksa) where exploration has concentrated on their uppermost parts. The traps could be facial closures of sandstone layers in the Middle Badenian to upper Sarmatian at depths of 1200 to 1900 m. Another prospective area is the entire northeastern margin of the basin, where the Karpatian, Middle and Upper Badenian, and especially Lower Sarmatian, gradually pinches out from the southwest to the northeast. Hydrocarbon traps may be found in bar sand bodies, at depths between 1200 and 2200 m.

Traps can also be found in the uplifted northern part of the basin filled by huge Upper Badenian deltaic deposits. In this part of the basin, note must also be made of the basement highs which originated along pre-Neogene faults which were also active in the Karpatian and Badenian. Shows of hydrocarbons have also been found on the slopes of buried stratovolcanoes.

The deepest parts of the Meso-Paleozoic "basement" have not yet been explored. CO_2 gas, with small amounts of hydrocarbons has been found on the margins. Gas production from low permeability reservoirs of the southern part of the basin presents an unsolved technical production problem.

At Lipany, in the Central Carpathian Paleogene, fractured claystones and marlstones yield 220 t of oil and 6000 m^3 gas (Fig. 41). This region, as well as the zone of folded flysch, should be included in the potentially prospective regions, where for economic reasons exploration work was interrupted.

3 Sedimentary Provinces with Unexplored Potential (Fig. 46)

Sedimentary sequences in which exploration for oil has not yet been carried out lie on the Bohemian Massif. They include: the Prague Basin, the Lower Silesian Basin, the Moravian Paleozoic Basin and the Central Carpathians.

The Prague Basin (Fig. 47) is filled with Ordovician-Devonian (Barrandian) sedimentary rocks approximately 3000 m thick. The basement is composed of weakly metamorphosed Proterozoic and Cambrian rocks. A gravity survey has been carried out and a reflection seismic survey conducted along

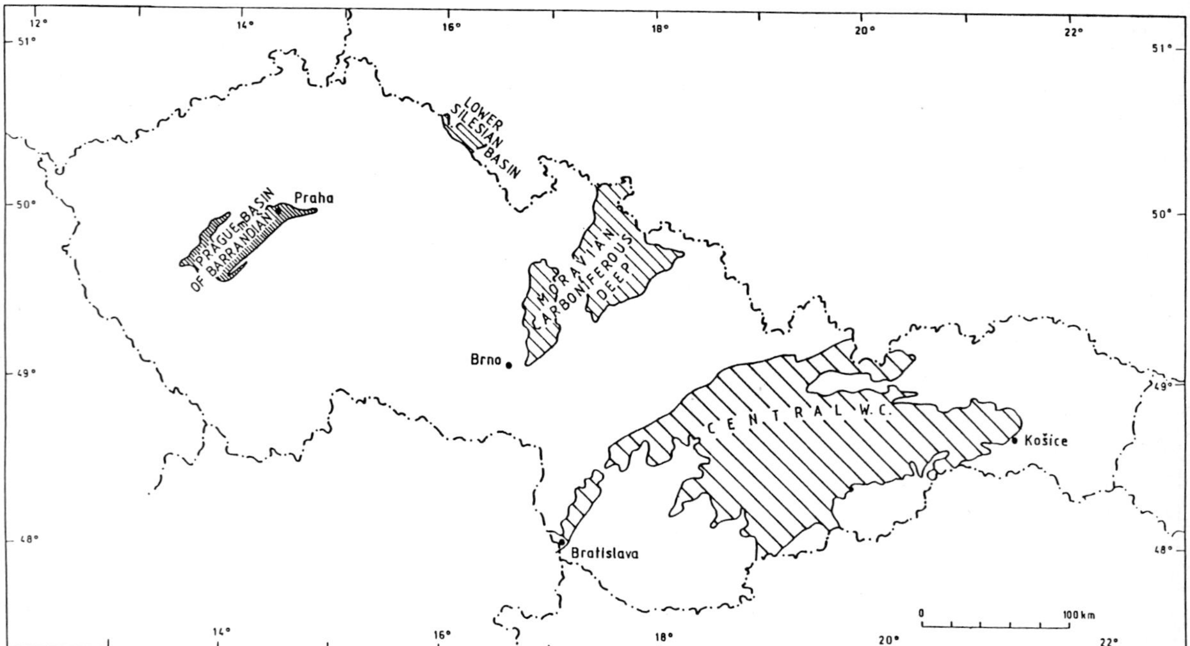

Fig. 46. Sketch of the sedimentary provinces with unexplored potential

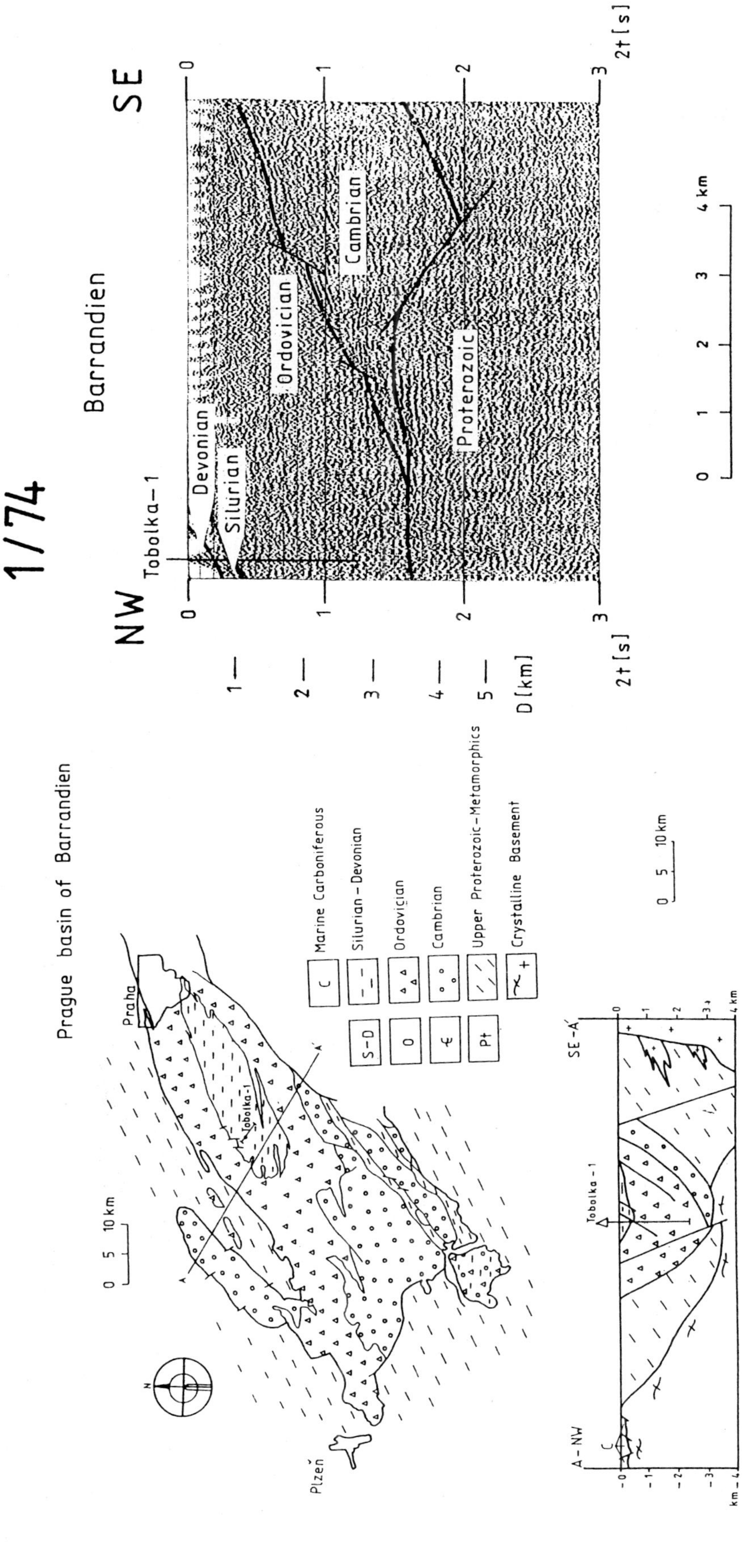

Fig. 47. Prague Basin

one profile. The well Tobolka-1 was drilled down to 2712 m and encountered a flow of strongly mineralized NaCl water saturated with dissolved methane gas and nitrogen. Porosity of reservoirs varies from 9 to 30%. Stratigraphic and lithological traps bounded by faults may be found. The structures are mainly tilted blocks with reservoirs of Ordovician-Silurian layers of volcanogenic origin.

The Lower Silesian Basin has the shape of an elongated ellipse. It is filled with Permo-Carbonife-

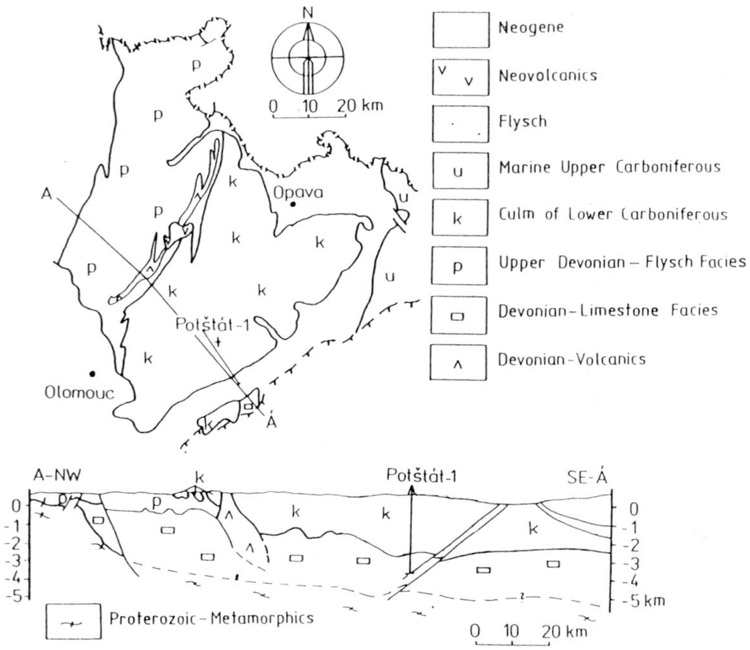

Fig. 48. Northern Moravia Paleozoic Basin

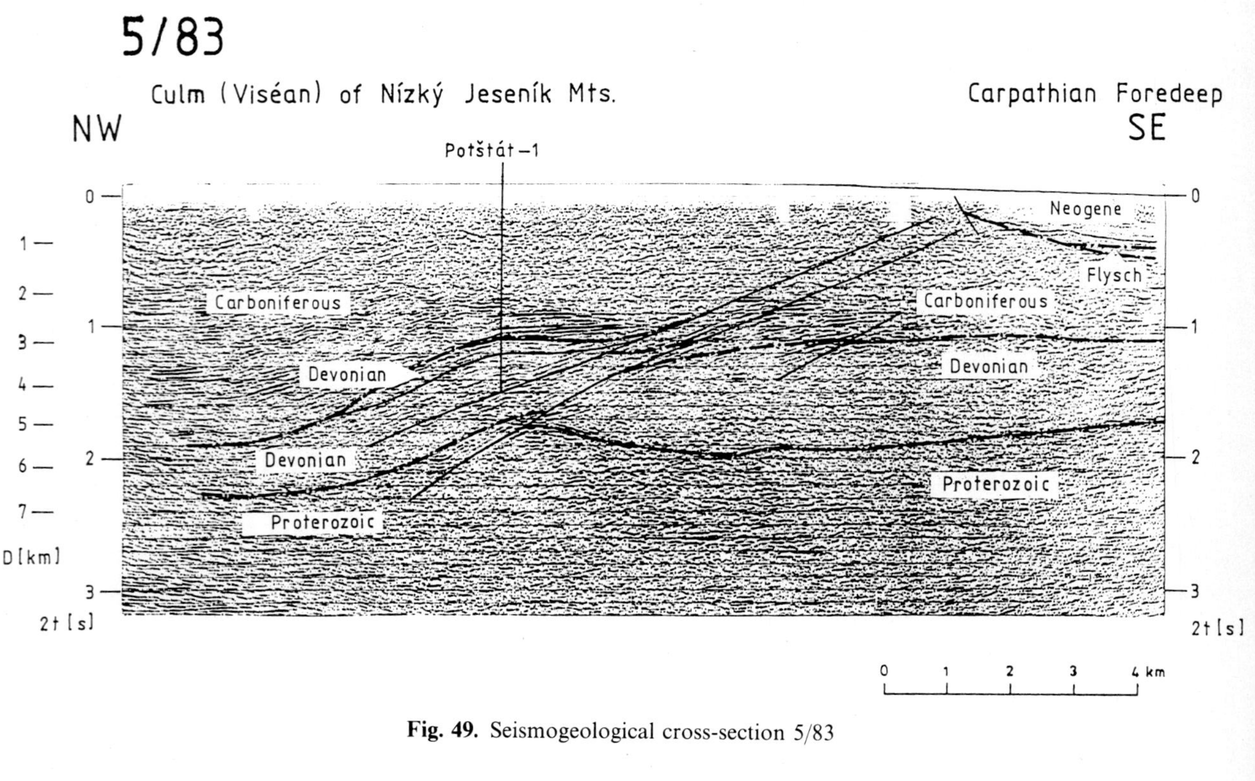

Fig. 49. Seismogeological cross-section 5/83

rous arkoses, shales, sandstones, coal seams and volcanogenic series. The basement is composed of crystalline rocks. The thickness of the fill exceeds 3000 m. No geophysical surveys for oil exploration have yet been carried out in this area. Some indications of methane have been found there: well Broumov-1, 2628 m deep, encountered the Permo-Carboniferous at a depth of 2516 m, where some coal bed methane was tested. Reservoirs are molasse of Upper Carboniferous and Permain age. The traps are of the stratigraphic type, lithological or tectonic.

The Moravian Paleozoic Basin (Figs. 48, 49) lies on the northeastern margin of the Bohemian Massif. Seismic measurements have been only locally conducted. The typical rock sequence includes: conglomerates, greywackes and aleuropelites in Culm facies. The basement is composed of folded, weakly metamorphosed Devonian. Well Potstat-1 was abandoned dry at a depth of 4100 m. Reservoirs in Culm facies were found to have a fair secondary porosity.

The Central Carpathians (Fig. 50) extend over a large area in Central Slovakia. They have a complex nappe structure which includes a crystalline core with its original autochthonous Paleozoic to Early Mesozoic cover. Numerous hydrocarbon seeps have been noticed. A systematic exploration for hydrocarbons has not been conducted in this region. Reservoirs could be sandstones or fractured rocks. Stratigraphic and massive structural traps are likely.

4 Conclusion

Commercial accumulations of hydrocarbons within the territory of Slovakia and the Czech Republic

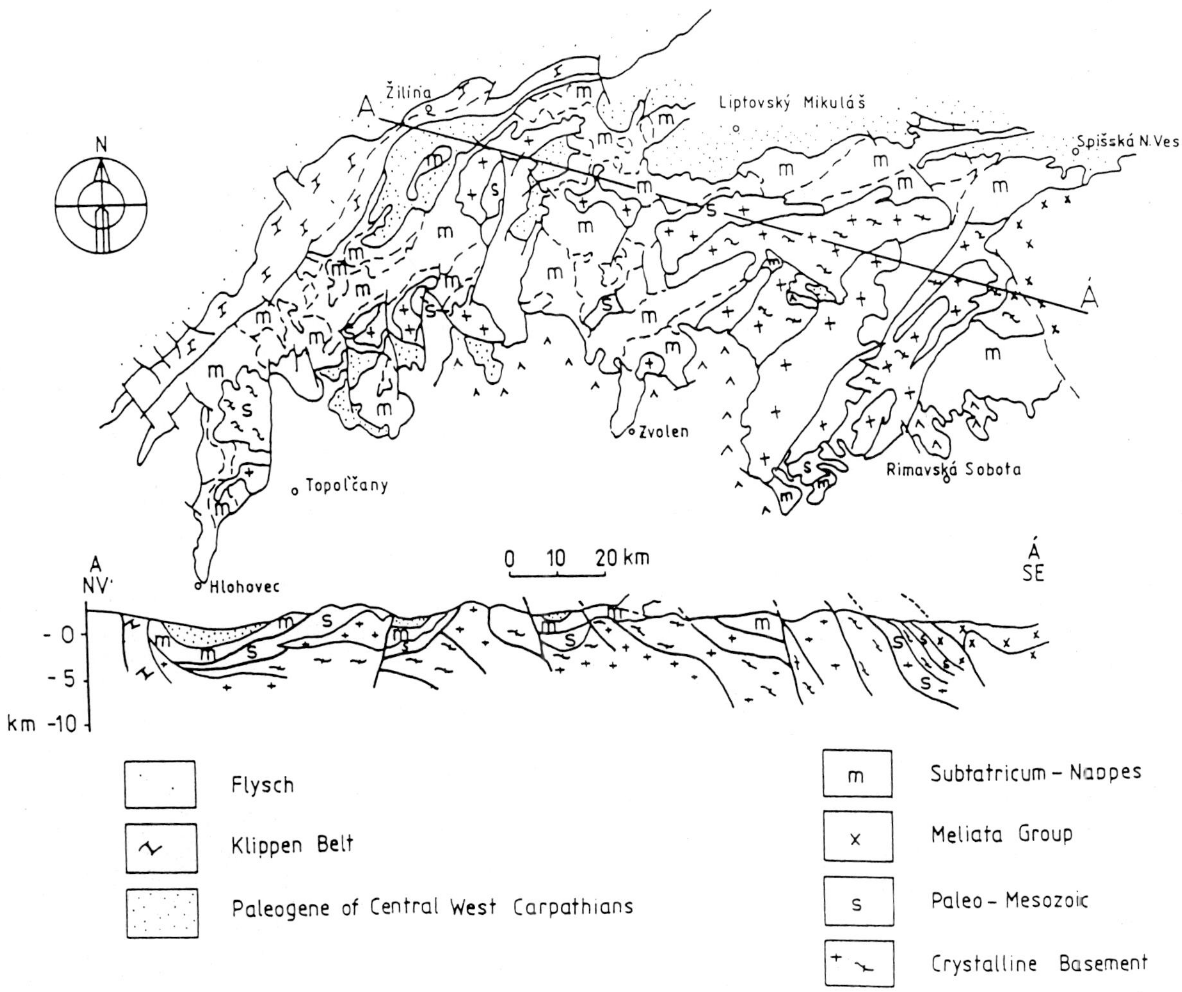

Fig. 50. Schematic geological cross-section through the Central West Carpathians

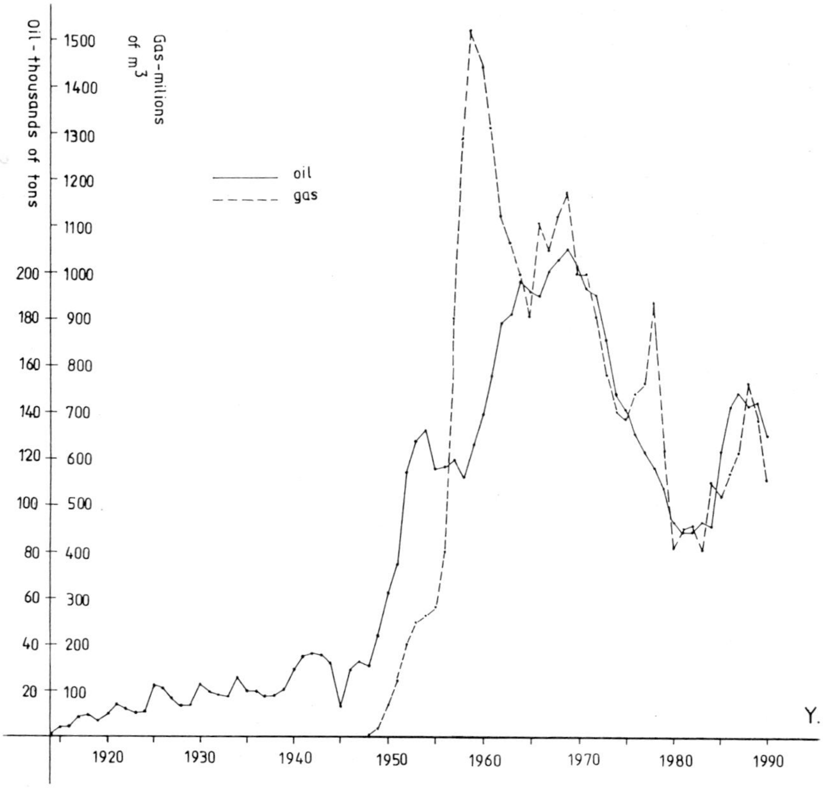

Fig. 51. Oil and gas production in Slovakia and the Czech Republic

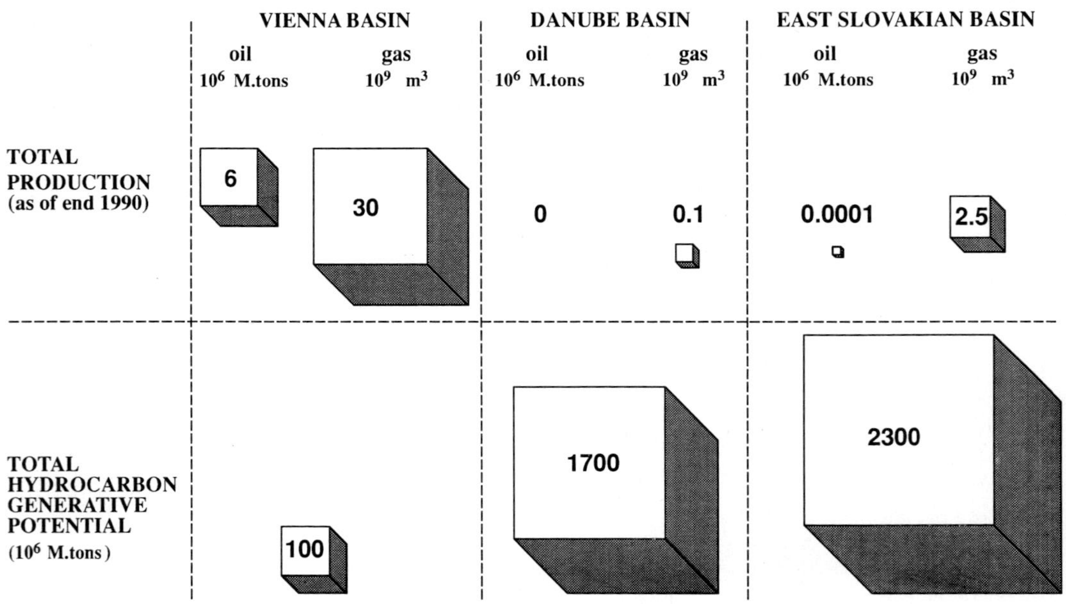

Fig. 52. Cumulative hydrocarbon production in the Neogene basins and the calculated source potential of the Neogene

have been found in the region of the Alpine-Carpathian orogen and in its autochthonous basement. Neogene depressions of the Paratethys with sedimentary fill up to 7000 m together with the Carpathian Foredeep have yielded the most important part of the hydrocarbon production. The evolution of the oil production in the Czech Republic and Slovakia is shown in Fig. 51. Up to now, the main structural features of the Neogene basins have been explored. Studies of kerogen to hydrocarbon conversion show that total estimated potential of the Neogene sediments decreases from the East Slovakian to the Danube and Vienna basins (Fig. 52).

At present, exploration is aimed at deeper structures located both in the basement of the Neogene basins and in the platform cover of the Bohemian Massif, part of the Central European Platform, where significant economic hydrocarbon accumulations have been proved.

Greater production of hydrocarbons may result from more intensive exploration of non-structural traps in the Neogene and earlier formations.

Other prospects are related to the deeper parts of the whole Alpine-Carpathian region, and to the areas where favourable thermobaric conditions may be expected: the Carpathian Foredeep, the sedimentary cover of the Bohemian Massif, and the overthrust flysch nappes, the Vienna Basin, the East Slovakian Flysch and the Central Carpathian Paleogene. Prospects for hydrocarbon generation and preservation in the pre-Neogene rocks increase in a westerly direction and are best in the Vienna Basin.

An exploration programme with deeper targets will require further geophysical investigation and additional drilling.

Acknowledgements. The authors thank many anonymous geoscientists for their outstanding contribution to the understanding of petroleum geology of the Czech and Slovakia republics. Most of their work is in unpublished reports.

The authors also appreciate the contribution of Dr. Bogdan Popescu, Petroconsultants, Geneva, for the professional editing of the text.

4 Development of Oil and Gas Exploration in the Eastern Part of Germany

E.P. MÜLLER[1]

CONTENTS

1 Introduction

Although the presence of gas fields in eastern Germany, formerly GDR, was proved as early as the beginning of the century, an aggressive hydrocarbon exploration policy including geological, geophysical and drilling activities was implemented only in the 1950s. A resulting volume of 150,000 m drilled per year was achieved in the early 1960s. The Mesozoic of the North German-Polish Basin and the Zechstein of the Thuringia Basin were primary target horizons. Zechstein discoveries in the Thuringia Basin and disappointing results of Mesozoic exploration shifted the emphasis to the exploration of pre-Jurassic objectives in the following years. The oil field Reinkenhagen (Zechstein) in Mecklenburg-Vorpommern was discovered in 1961 and the gas field Salzwedel-Peckensen (Rotliegendes) in the Altmark region in 1968. Both are located in the North German-Polish Basin.

About 4.3 million exploratory and development metres were drilled from 1950 to 1991. The successful exploration, including drilling of more than 1500 boreholes, led to the discovery of 297.5 billion m^3 of gas reserves in the Permian (Zechstein and Rotliegendes) and 4.3 million tonnes of oil reserves in the Zechstein reservoirs as of 1.1.1991. Cumulative gas production from the Permian amounted to 180 billion m^3 and the Zechstein oil production to 2.5 million tonnes as of 1.1.1991.

This paper presents the results of geological, geophysical and exploratory drilling of Permian plays: Zechstein carbonates and Rotliegendes clastics. We anticipate that, in addition to a continuously refined Permian exploration, the Devonian and the Lower Carboniferous will be the "frontier" area of the following years.

2 Geologic Framework

The investigation area lies between the Baltic Sea and the Thuringia Forest, and is part of the Middle European Depression, a vast epicontinental basin which had developed southwest of the Eastern European Platform. Main petroliferous provinces are the North German-Polish and Thuringia basins.

The basement of this area consolidated during different phases. While the Variscan (Hercynian) folded basement rock dominates in the south – the Thuringia Forest area, the Harz Mountains and in the block of Calvoerde – a young Caledonian basement develops in the north (Fig. 1). In the central part of the North German-Polish Basin, a Dalslandian-Baikalian basement is supposed to exist. The basement may lie at depths up to 10,000 m, as proved by deep seismic investigations and by deep wells drilled in the northern and southern basin areas.

The subsidence of this epicontinental basin began in the Permian period after the Late Carboniferous folding, uplifting and erosion of the basement. The basin's development was in four phases:

[1]Erdöl-Erdgas Gommern GmbH, Postfach 21, 3304 Gommern, Germany

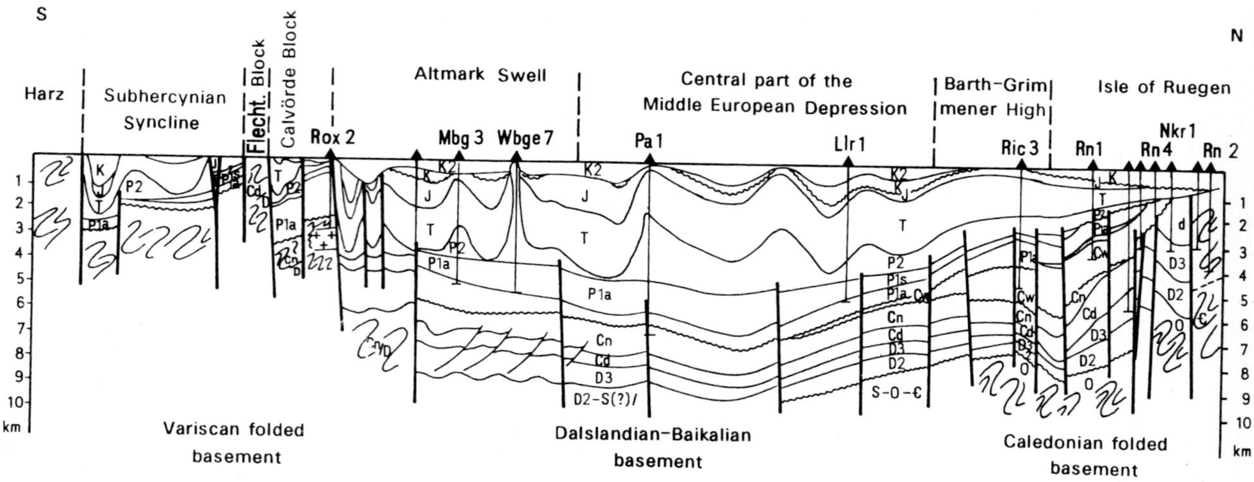

Fig. 1. Schematic cross-section between Harz Mountains and Isle of Ruegen

1. Taphrogenetic stage (Autunian-Saxonian I).
2. Main subsidence stage (Saxonian II-Middle Keuper).
3. Differentiation stage (Upper Keuper-Lower Cretaceous).
4. Stabilization stage (Upper Cretaceous-Cenozoic).

Transgressions occurred in the Zechstein, Muschelkalk, Lower and Middle Jurassic periods and in the Hauterivian stage. The oil and gas deposits are located with Rotliegendes (Saxonian), Zechstein (Stassfurt Carbonate) and partly in Triassic reservoir rocks (Fig. 3).

The *taphrogenetic stage* (Autunian-Saxonian I) is characterized by active fracture tectonics accompanied by intensive volcanism. The Autunian effusives are of basic, intermediate or acid types in caldera and ignimbrite complexes.

The *main subsidence stage* (Saxonian II-Middle Keuper) covers the major sedimentation cycle of the Subhercynian and Thuringia basins in the study area. The main reservoir rocks for natural gas deposits in the area are clastics. The gas traps were sealed by the Zechstein transgression.

Here, "paleoaccumulations" were formed in the Rotliegendes by the migration of CH_4 from Carboniferous source rocks. Nitrogen gas and CH_4 were generated and could migrate into these paleotraps much later. In the Zechstein Stassfurt Carbonate, oil and natural gas were penecontemporaneously expelled from the organic matter.

During the *differentiation stage* (Upper Keuper-Lower Cretaceous) subsidence was much slower over large areas. During the Early Kimmeridgian vertical movements reached a climax with the formation of regional highs (e.g. the Altmark Swell) and ended with local trap development in the Austrian phase. Local subsidence might have initiated secondary migration.

During the *stabilization stage* (Upper Cretaceous-Cenozoic), inversion of troughs is characteristic. The formation of brachyanticlines also took place during this stage. It initiated a secondary migration of natural gas from the Rotliegendes "paleoaccumulations". Secondary migration of oil and natural gas fields was also typical for the Stassfurt Carbonate. These phenomena led to the current distribution of the eastern German hydrocarbon fields (Figs. 2, 3).

3 Exploration History

The hydrocarbon exploration within the study area can be traced back to 1907, when the first gas flow was recorded from the potassium well Graefentonna/Langensalza (Triassic) in the Thuringia Depression.

Subsequently, the following discoveries were made:

- 1930, the Volkenroda oil field in a potassium salt mine.
- 1932, the Muehlhausen natural gas field.
- 1934, the Fallstein oil field.
- 1935, the Langensalza natural gas field in the Stassfurt Carbonate Formation.

However, intensive hydrocarbon exploration integrating geological, geophysical and drilling activities started after 1950. At the beginning of the

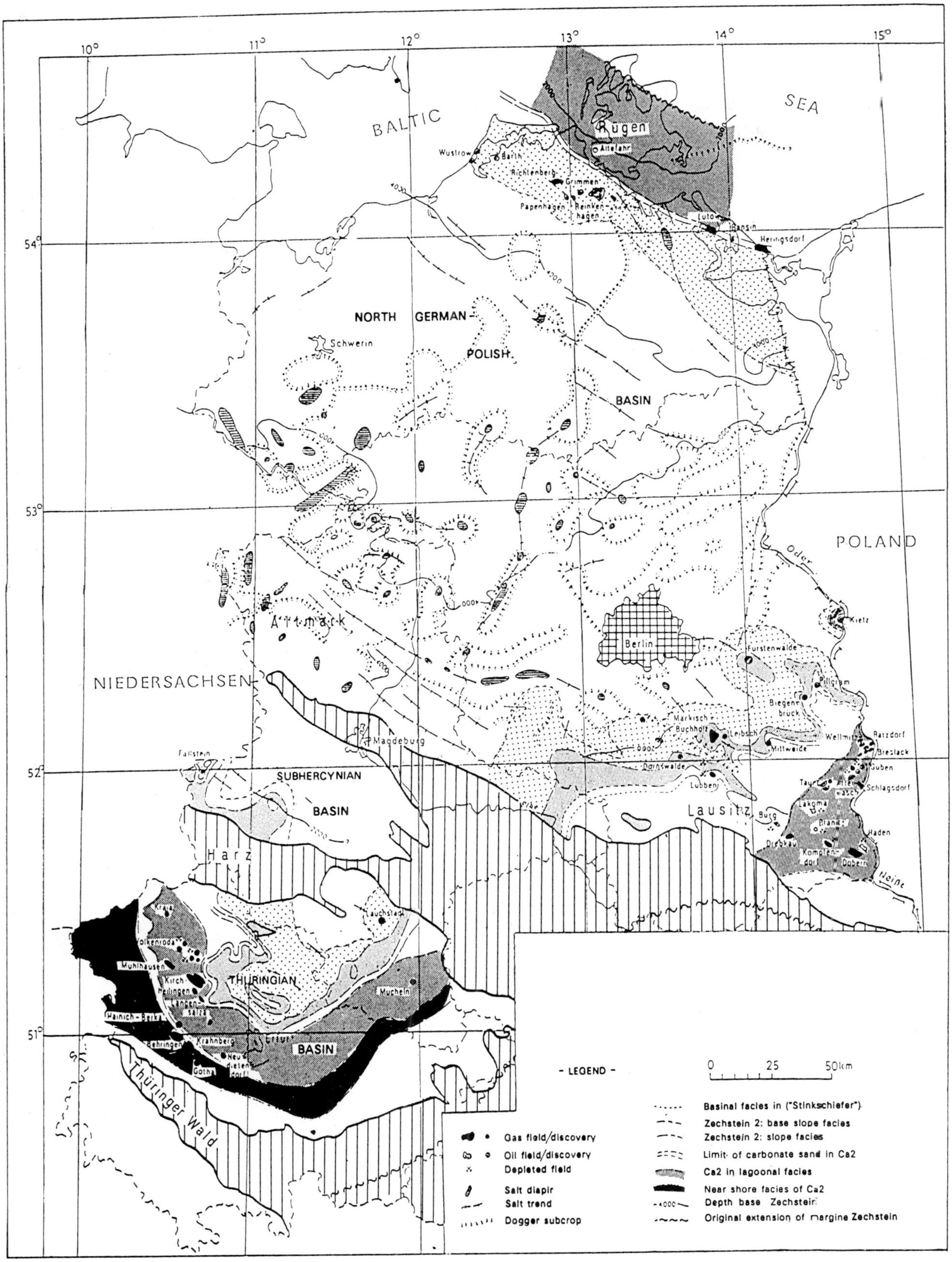

Fig. 2. Map of oil and gas fields (Zechstein) of eastern Germany

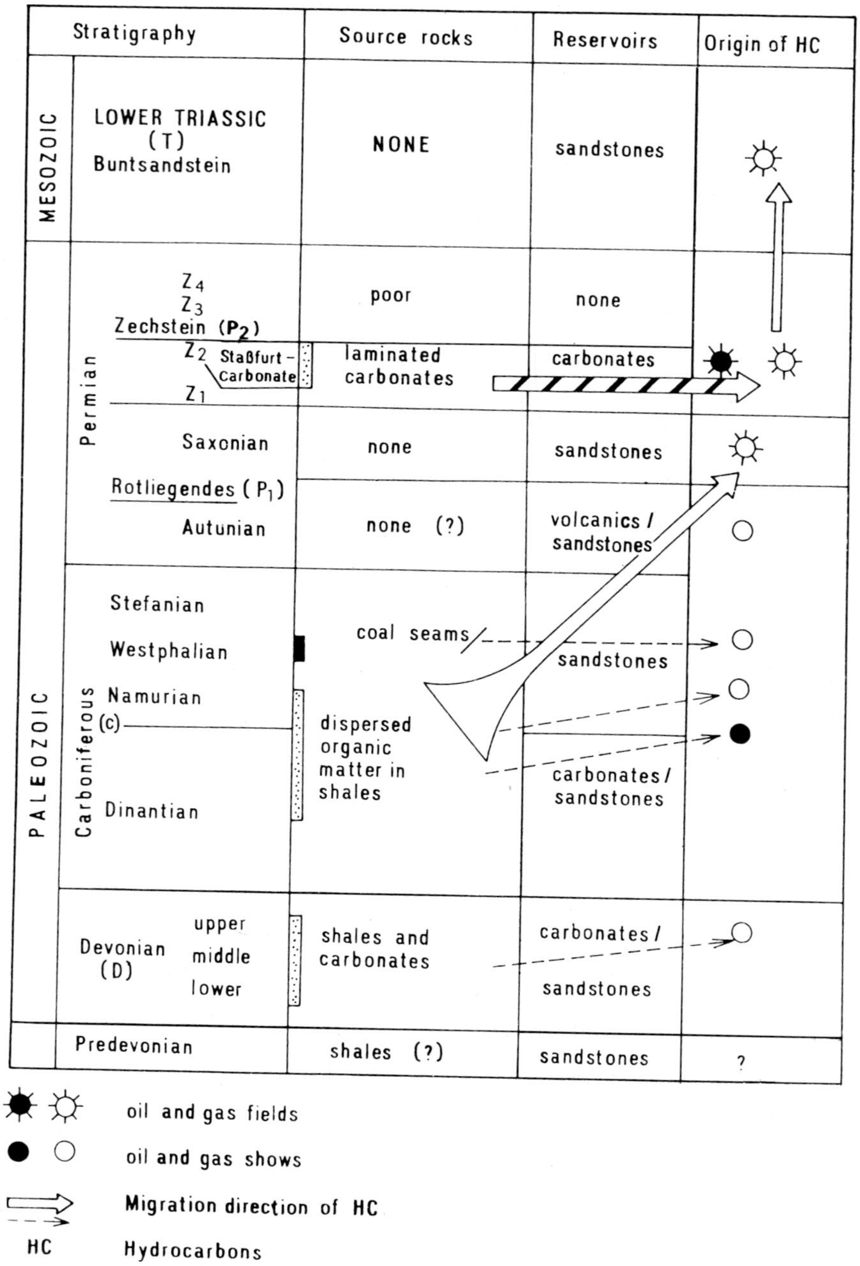

Fig. 3. Stratigraphic distribution and genesis of oil and gas fields in NE Germany and Thuringia Basin

1960s, it had already reached volumes of more than 150,000 m a year (Fig. 4). At that time, targets were the Mesozoic of the North German-Polish Basin and the Zechstein of the Thuringia Basin. The exploration of the Mesozoic was disappointing: oil and natural gas shows were tested in more than 425 wells but no economic flows have been encountered so far. Oil and gas reserves were first proved in the Zechstein of the Thuringia Basin (Kirchenheiligen-Allmenhausen, Krahnberg, Behringen and others). In 1961, the oil prospectivity of the Zechstein in the northern part of the study area was substantiated by the discovery of the Reinkenhagen oil field in the Mecklenburg/Vorpommern area. The exploration of the Rotliegendes play resulted in the discovery of the Salzwedel/Peckensen gas field in the Altmark region in 1968. The discovery of oil and natural gas fields in Zechstein and Rotliegendes are the result

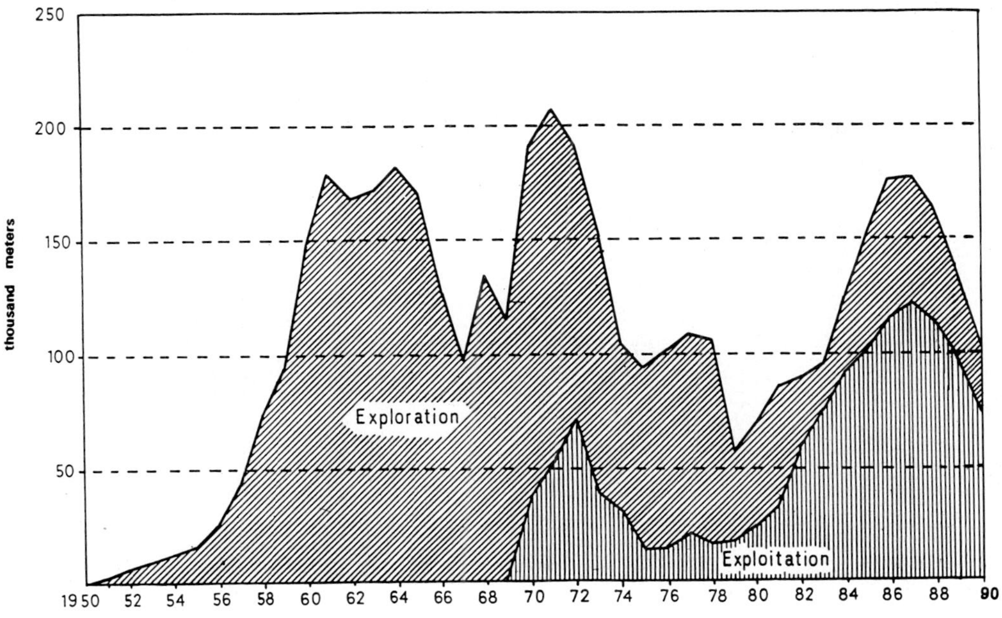

Fig. 4. Drilling metres per year (1950–1990)

of the application of modern geophysical and geochemical methods. The following prospective areas were intensively explored over the past 25 years:

North German-Polish Basin

- Mecklenburg/Vorpommern (Zechstein and Rotliegendes plays)
- Mecklenburg/Brandenburg (deeper basin Rotliegendes plays)
- Altmark (Rotliegendes plays)
- Brandenburg/Lausitz (Zechstein and Rotliegendes plays)
- Subhercynian Basin (Rotliegendes and Zechstein plays)
- Thuringia Basin (Zechstein and Triassic plays)

So far, in eastern Germany, two oil- and gas-producing provinces have been discovered:

- The North German-Polish Basin (North East German Sub-basin) and
- The Thuringia and Subhercynian basins.

The oil and gas fields are related to anticlinal and combined traps of the Saxonian in the Rotliegendes and to the carbonate facies of the Stassfurt Series in the Zechstein. In the Jurassic and other Mesozoic stages, only oil shows have been found so far. Oil shows were also tested from the Carboniferous /Devonian of the Isle of Ruegen.

3.1 Surface Exploration

The hydrocarbon fields in the Zechstein and Rotliegendes are controlled mainly by structural factors. However, combined lithological-structural traps are also common in this stratigraphic interval. These traps were mainly identified using digital-seismic (DS) methods (Fig. 5). Two maxima are apparent: a sharp increase in 1968–1972 when the main target was the Rotliegendes of the Altmark region (5500 line-km/year) and 1982–1988, reflecting intensive exploration of the Rotliegendes, Zechstein and pre-Permian (2500–3000 line-km/year).

Until 1970, analogue recording and refraction were the most used seismic methods, then digital recording and CDP techniques were employed.

The highest line density (Fig. 6) was achieved in the region of the Altmark Swell with 8 line-km/km^2 including 3-D seismics. The central basin areas of the Rotliegendes in West Mecklenburg-Brandenburg show an average density of 0.5–0.6 line-km/km^2, while shallow areas have only 0.1 line-km/km^2. The Zechstein plays in the Mecklenburg-Vorpommern and Brandenburg/Lausitz areas have densities of 1.5 line-km/km^2. Comprehensive refraction-seismic, magnetic, gravimetric (500 m point spacing covering the whole area) and magnetotelluric measurements were additional requirements for the preparation of Rotliegendes and pre-Permian leads.

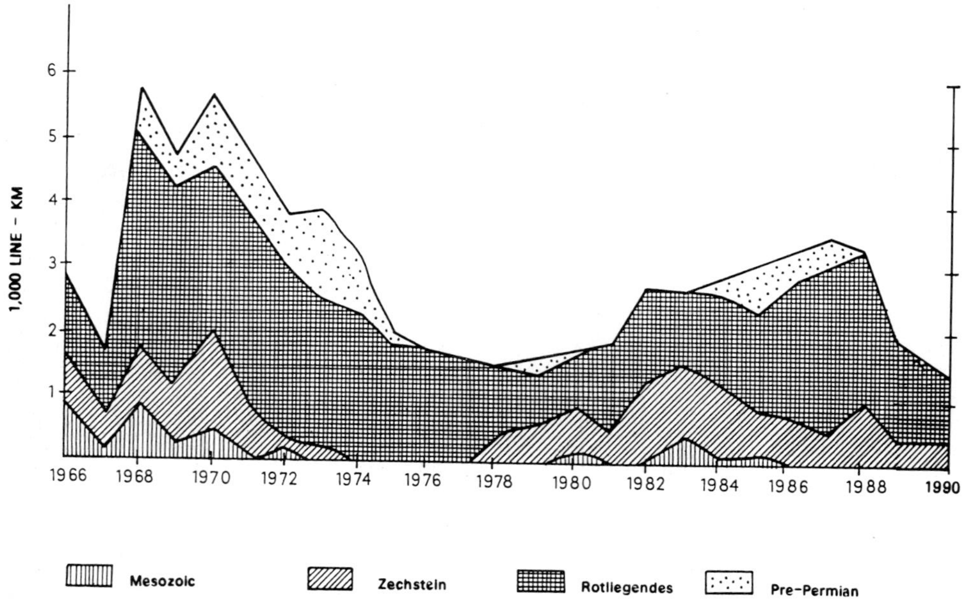

Fig. 5. Annual digital seismic recording per target (1966–1990)

3.2 Drilling and Reserves

A total of 4.3 million m of exploration and development hole was drilled for oil and gas from 1950 to 1991. The drilling of stratigraphic tests, rank wildcats, exploration and development wells have had average depths of about 3500 m for the Rotliegendes objectives, and 2300 m for Zechstein objectives. The exploration meterage reached peak values of more than 150,000 m a year/1960–66 and 1970 (Figs. 4, 7). Drilling for pre-Permain and Zechstein was highest in 1970 with 20,000 and 35,000 m drilled respectively. The development meterage reached maximal values of 75,000 m in 1972 to 120,000 m in 1989.

The highest drilling density was reached in the Altmark Swell area (Fig. 8) with up to 144 wells per 125 km^2. Not surprisingly, high activity was also recorded in the western part of the Thuringia Basin, and in the Cottbus and Stralsund areas of the North German-Polish Basin, which are also main producing areas.

Exploration effort in the post-World War II period, included more than 1500 wells drilled, which proved 381.3 billion m^3 of natural gas in the Permian and 13.831 million tonnes of oil in the Zechstein, both in place reserves, by the end of 1990 (recoverable reserves in Fig. 9). Up to the end of 1990, the cumulative gas production amounted to 184.7 billion m^3 gas and 2.574 million tonnes of crude oil. At the same date, proven reserves were 46.98 billion m^3 of gas and 0.86 million tonnes of oil (Table 1). Gas has been produced from the Zechstein and Rotliegendes and oil from the Zechstein. Exploratory drilling of the pre-Permian (Lower Carboniferous/Devonian) was only a research project that has not proved the existence of commercial hydrocarbon accumulations.

3.3 Deep Well Drilling

Deep wells have been drilled with Soviet rigs of the 3D-67 type from 1950 to 1962 then, with modern Romanian rigs of 4DH-315 and 3DH-400 types. The drilling of 27 wildcats with depths of more than 5000 m, of which seven reached depths of more than 7000 m, was characterized by the following problems, which were successfully controlled:

- Drilling through the Permian plastic salt with high pressures. Measures to avoid casing deformation consisted in the close control of the influence of the Mesozoic faulting and of the pre-Permian compressional tectonics.
- Reduction of mud losses while drilling low-pressured reservoir horizons, by addition of special chemicals.
- Control of the bottom temperatures up to 235 °C with salt-saturated, water-based muds.

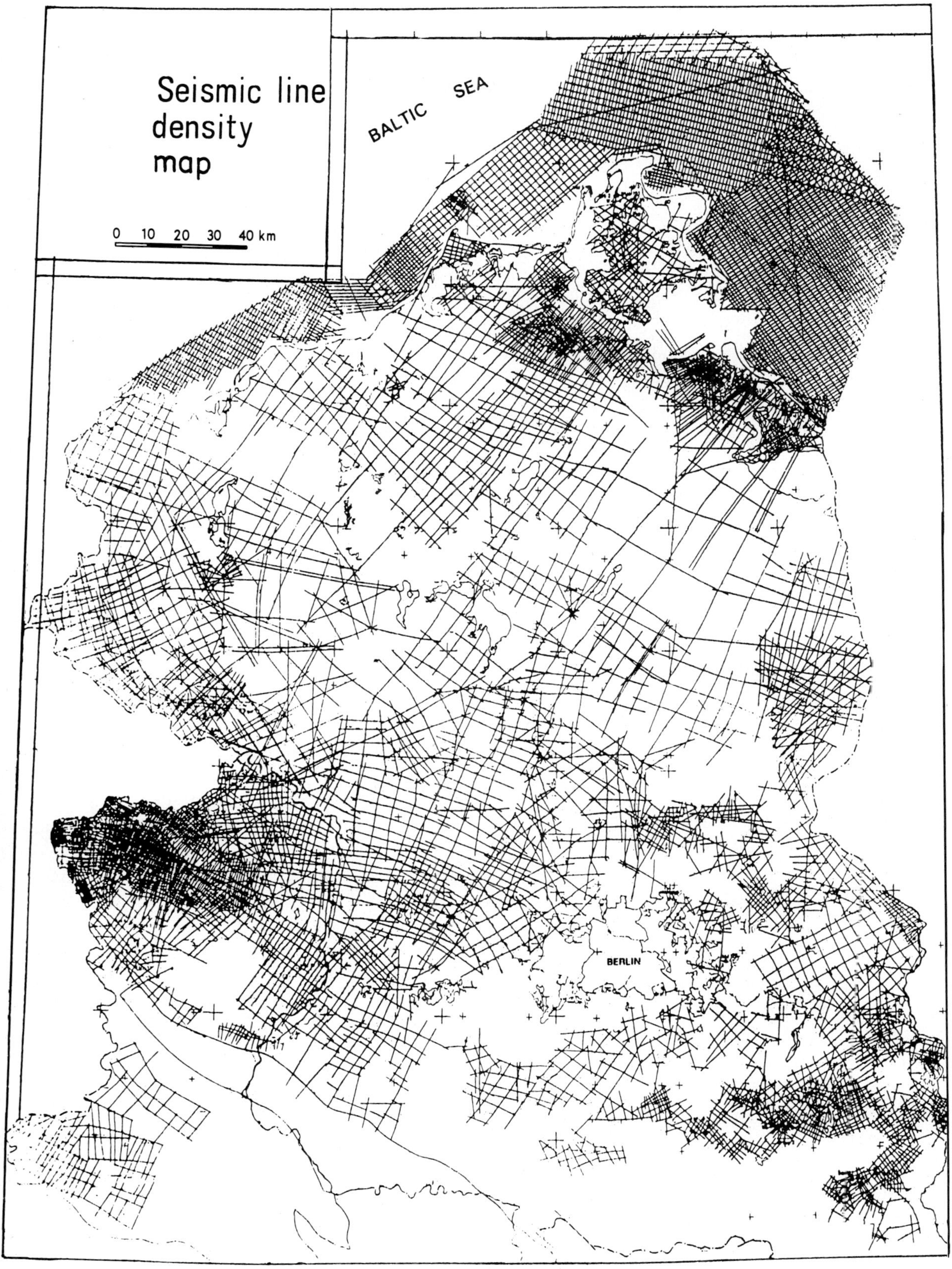

Fig. 6. Seismic line density map

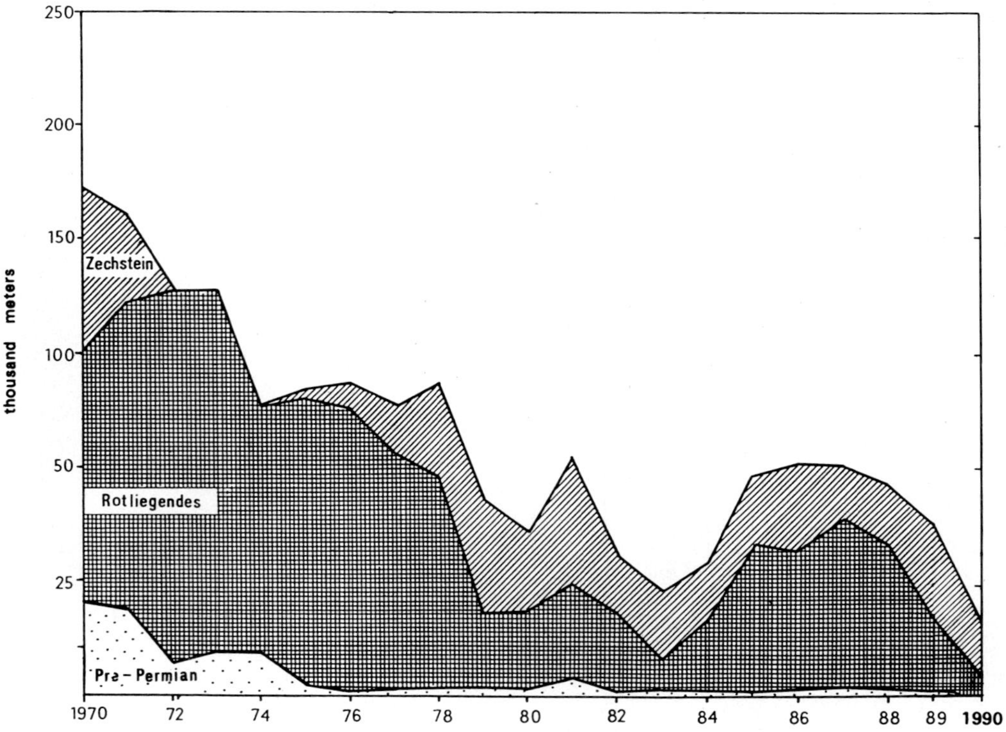

Fig. 7. Exploratory drilling meterage per target (1970–1990)

Drilling with standard equipment available in most cases required special hole program design and technical measures e.g.:

- Installation and cementation of heavy casings with installation depths of:

18 5/8″	deeper than 1310 m
13 3/8″	deeper than 3670 m
9 5/8″	deeper than 5030 m
7″	deeper than 6890 m
5″	deeper than 8000 m.

- Successful deviations at depths of more than 6000 m.
- Drillstem tests at depths exceeding 6900 m with temperatures of 198 °C.
- Utilization of oil-based muds up to depths of 8000 m.

4 Exploration Results of Zechstein

By the end of 1990, a total of 1084 wells had been drilled to the Stassfurt Carbonate. Exploration has been performed in the platform slope, carbonate sand barrier and lagoon facies of the northern and southern basin areas.

The Stassfurt Carbonate is a facies-zoned formation (Fig. 2). Dark laminated carbonates (Stinkschiefer) dominate in the central area of the basin and reach the platform slope. They consist of mudstones and wackestones with thicknesses of 2 to 10 m. Shallow marine oolitic and algal carbonates 10 to 100 m thick develop on the shelf edges. Barrier facies, consisting of skeletal grainstones and packstones, occur in the near shore areas. The carbonate sands overlie the thickness anomalies made up of the A_1 Werra Anhydrite. They were subjected to transgressions and regressions on an eustatically controlled platform. Lagoonal facies developed between barrier ridges and shoreline. The lagoonal micrites are dolomitic-calcitic mixtures.

On the northern edge of the Zechstein basin, as well as in the centre of the basin, the Stassfurt Carbonate has a patchy development within the coeval evaporite facies. Highly concentrated formation waters, relict solutions of evaporite sequences characterize stagnant hydrodynamic conditions and petrostatical pressure conditions (1.8–2 at/

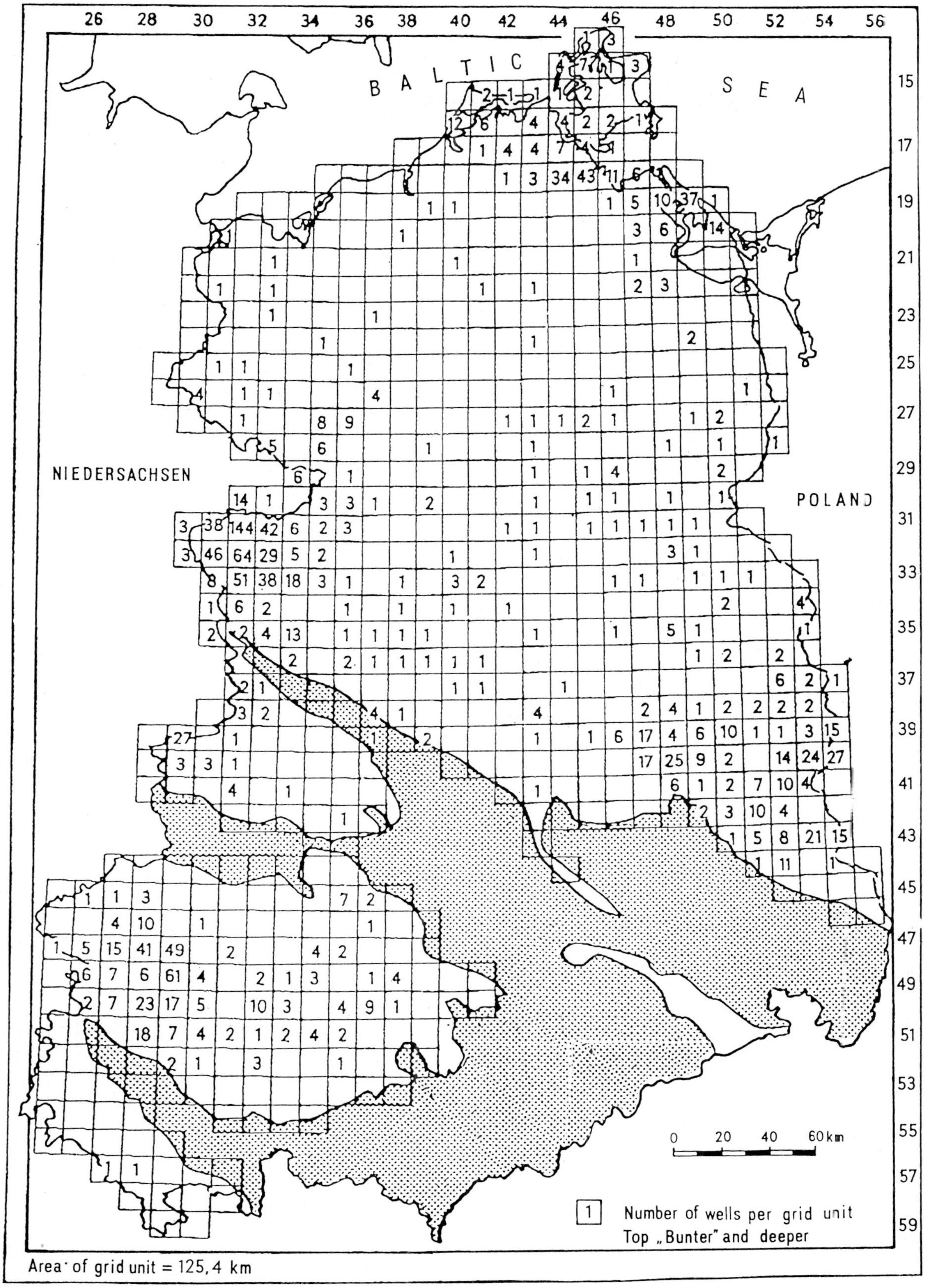

Fig. 8. Drilling density of pre Mid-Triassic objectives

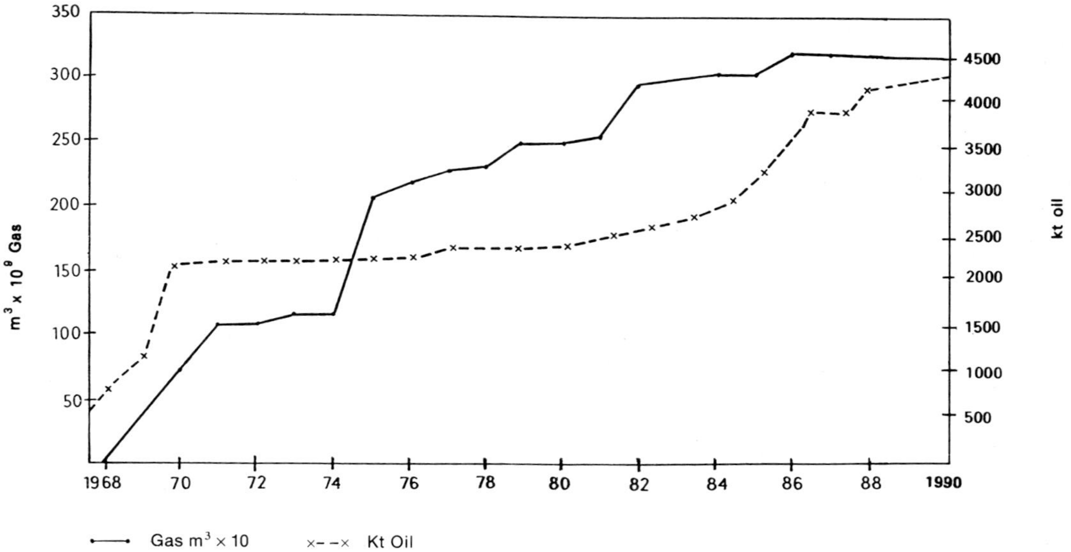

Fig. 9. Recoverable oil and gas cumulative reserves

Table 1. Production and reserves as of 1st January 1991

Area	Production	Number of producing fields (1990)	Proved reserves
1. Gas (in 10^6 m^3)			
Altmark	6,584.4	8	45,178
Thuringia	131.9	5	1,390
Subhercynian	7.9	1	362
Lausitz	21.5	13	50
NE-Mecklenburg	5.1	5	
	6,750.8	32	46,980
2. Oil (in 10^3 t)			
NE-Mecklenburg	20	7	296
Lausitz	33.4	13	560
Thuringia	0.6	5	3
	54	25	859

10 m), so that a migration of hydrocarbons from the overlying and underlying strata was less probable. On the southern edge, in the South Brandenburg-Lausitz region and in the Thuringia Basin, changing hydrogeological and hydrodynamic conditions are characteristic.

The hydrocarbon fields in Stassfurt Carbonate reservoirs are mainly controlled by structural factors. Combined lithologic-structural traps are characteristic and were found using the analysis of wave images between reflection horizons Z_1, at the basal anhydrite surface and Z_3, at Zechstein's base (Fig. 10). Fields were found in fractured and fractured-porous reservoirs. Porous reservoirs, with porosities of 15–20%, dominate in the platform slope and the barrier ridge zones interfingered in mudstones.

The source rock potential (Fig. 11) is determined by the lithologically controlled distribution of organic substance and its thermal history. They have: 0.2–0.7% TOC and 0.5–2.5% R_o.

In the *Mecklenburg-Vorpommern* area, 12 hydrocarbon fields have so far been discovered (Fig. 12; Table 2):

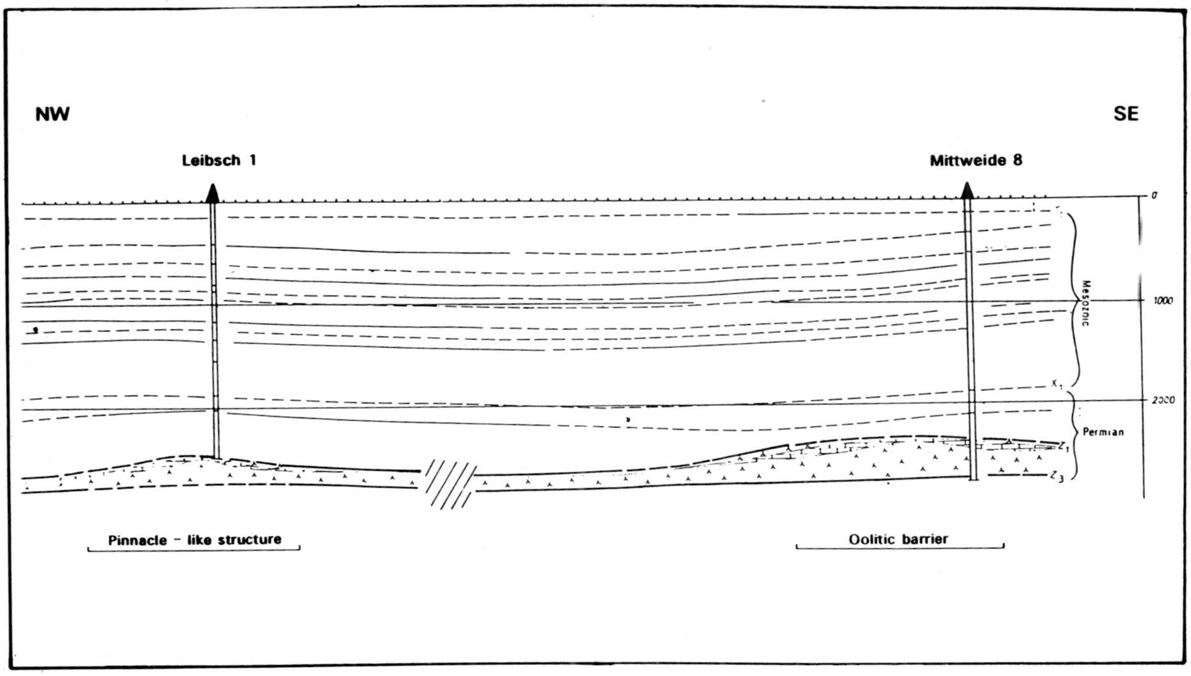

Fig. 10. Definition of play concepts in Stassfurt Carbonate using digital recording survey interpretation

Field name	Fluid type	Discovery year
Reinkenhagen	Oil	1961
Grimmen	Oil	1963
Richtenberg N	Oil	1964
Barth	Oil	1966
Luetow	Oil	1965
Grimmen SW	Oil	1970
Wustrow	Oil	1977
Papenhagen	Oil	1980
Heringsdorf	Gas condensate	1981
Kirchdorf	Oil	1963

In the Stassfurt Carbonate of the *Brandenburg and Lausitz* areas, 26 hydrocarbon reservoirs have been found (Fig. 13; Table 2):

Field name	Fluid type	Discovery year
Staakow	Natural gas	1960
Doebern	Oil	1962
Guben I	Oil	1962
Burg	Oil	1964
Drebkau	Natural gas	1964
Luebben	Natural gas	1965
Raden	Natural gas	1965
Tauer	Oil	1966

Field name	Fluid type	Discovery year
Atterwasch	Oil	1968
Guben II	Oil	1968
Drewitz	Oil	1971
Lakoma	Oil	1977
Leibsch	Gas condensate	1980
Wellmitz	Oil	1980
Wellmitz NW	Oil	1983
Steinsdorf	Oil	1986
Maerkisch-Buchholz	Gas condensate	1986
Kietz	Oil	1987
Ratzdorf	Oil	1988
Wellmitz SE	Oil	1989

In the *Thuringia Basin*, eight hydrocarbon reservoirs have been found (Fig. 14, Table 2).

Field name	Fluid type	Discovery year
Kirchenheiligen	Natural gas	1958
Fahner Hoehe	Natural gas	1960
Behringen	Natural gas	1962
Krahnberg	Natural gas	1963
Mehrstedt	Natural gas	1970
Lauchstedt	Natural gas	1970

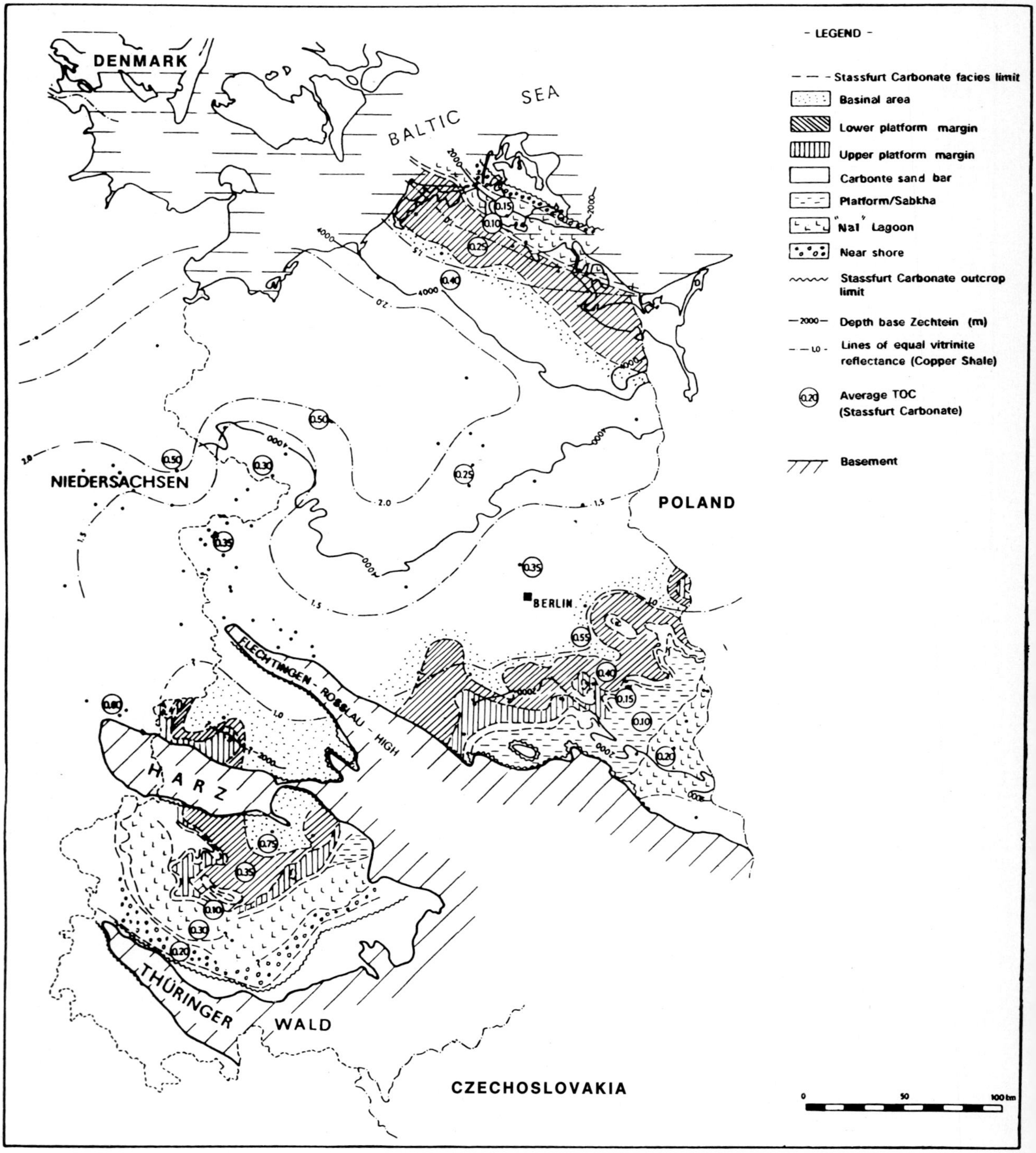

Fig. 11. Source rock characteristic of Stassfurt Carbonate depositional area

The crude oil and condensate from the Stassfurt Carbonate are of paraffinic-naphtenic type with densities of 0.700 to 0.880 g/cm^3 and have $\delta^{13}C$ values of -23.3 to -28.5‰. Associated gas and free gas show a high variability of composition: the hydrocarbon content with a dominance of CH_4 can range from 90–10% vol.; the second main element is nitrogen; H_2S content may vary from 0.1–7%.

The autochthonous oil, condensate and natural gas fields in the Stassfurt Carbonate originated by

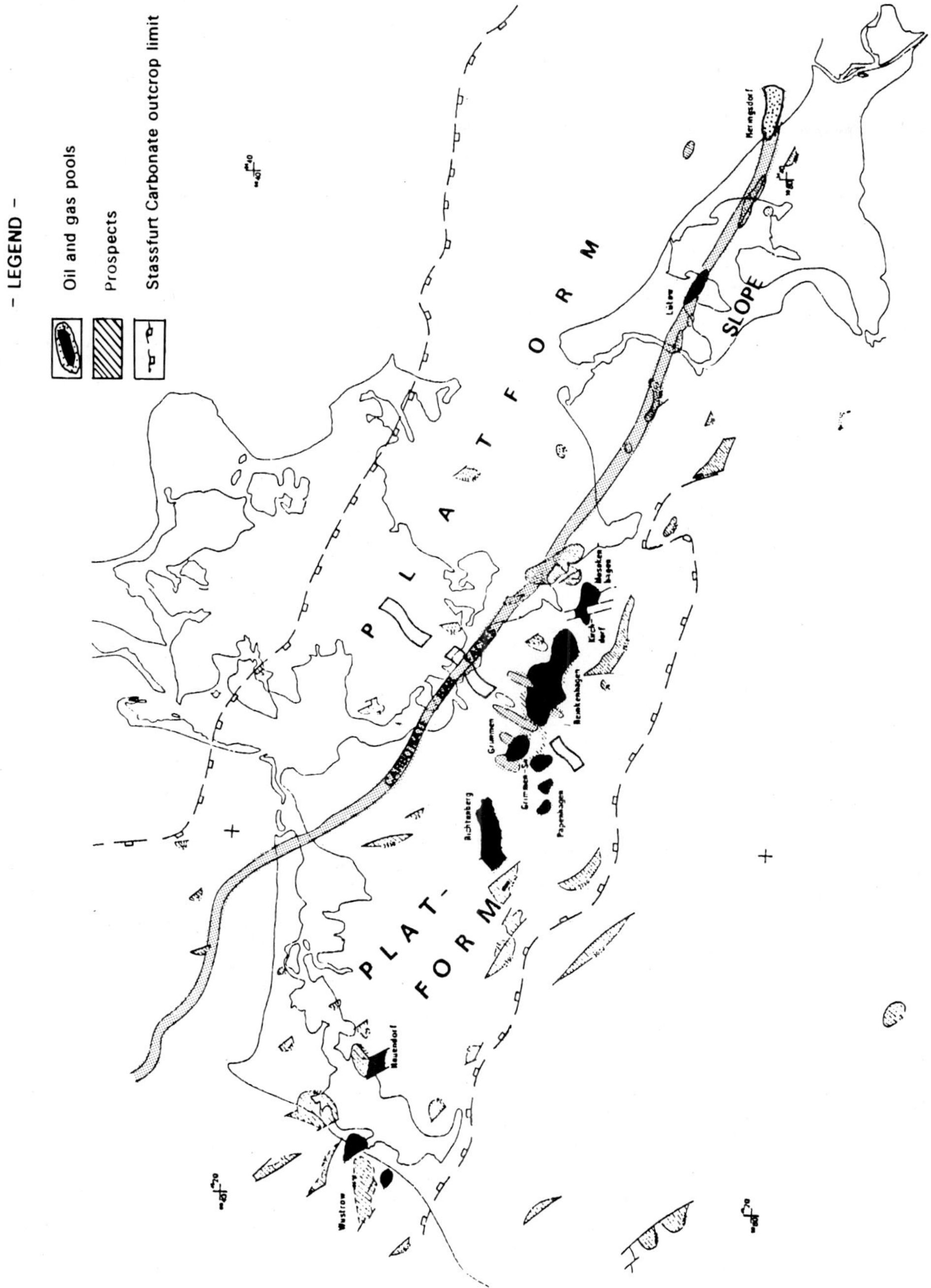

Fig. 12. Oil and gas fields in Stassfurt Carbonate, Mecklenburg-Vorpommern

Table 2. Status of oil and gas fields as on 1 January 1991

Area	Field name	Trias, Buntsandstein, Upper	Trias, Buntsandstein, Middle	Trias, Buntsandstein, Lower	Perm, Zechstein, Staßfurtkarbonat	Perm, Rotliegendes, Upper	Perm, Rotliegendes, Lower	Carbon, Upper, Stefan	Carbon, Upper, Westfal	Carbon, Lower, Namur	Carbon, Lower, Dinant	Devon, Upper	Devon, Middle	Devon, Lower
Thuringian and Subhercynian basins	Allmenhausen		○											
	Langensalza		○											
	Fahner Höhe		○											
	Volkenroda				●									
	Mühlhausen				○									
	Kirchheilingen				◒									
	Langensalza				○									
	Fahner Höhe				○									
	Behringen				○									
	Krahnberg				○									
	Mehrstedt				●									
	Holzsußra				● ⚒									
	Rockensußra				● ⚒									
	Holzthalleben				● ⚒									
	Fallstein		○		●									
	Lauchstedt					○ ⚒								
Mecklenburg/ Vorpommern Area	Reinkenhagen				●									
	Grimmen				●									
	Barth				●									
	Richtenberg-N				●									
	Grimmen-SW				●									
	Lütow				●									
	Papenhagen				●									
	Wustrow				● ⚒									
	Bansin				● ⚒									
	Heringsdorf				◒									
	Kirchdorf				●									
	Krummin					○								
	Mesekenhagen				● ⚒									
Altmark Area	Salzwedel-Peckensen (GDE)					○								
	Winkelstedt					○								
	Mellin-Süd					○								
	Wenze					○								
	Sanne					○								
Brandenburg/Lausitz Area	Rüdersdorf					○								
	Staakow				○ ⚒									
	Guben I				●									
	Döbern				● ⚒									
	Drebkau				○									
	Burg				● ⚒									
	Lübben				● ⚒									
	Tauer				● ⚒									
	Guben II				●									
	Atterwasch				●									
	Komptendorf				○ ⚒									
	Raden				○									
	Schenkendöbern				○									
	Branitz				● ⚒									
	Fürstenwalde				●									
	Drewitz				●									
	Lakoma				● ⚒									
	Schlagsdorf				● ⚒									
	Mittweide				●									
	Leibsch				◒									
	Wellmitz				●									
	Wellmitz-NW				●									
	Wellmitz-SE				● ⚒									
	Steinsdorf				●									
	Steinsdorf-N2				● ⚒									
	Ratzdorf				●									
	Kietz				●									
	Märk. Buchholz				◒									
	Breslack				● ⚒									
	Breslack-NE				● ⚒									

○ Gas field
◒ Gas - condensate field
● Oil field
⚒ Depleted field

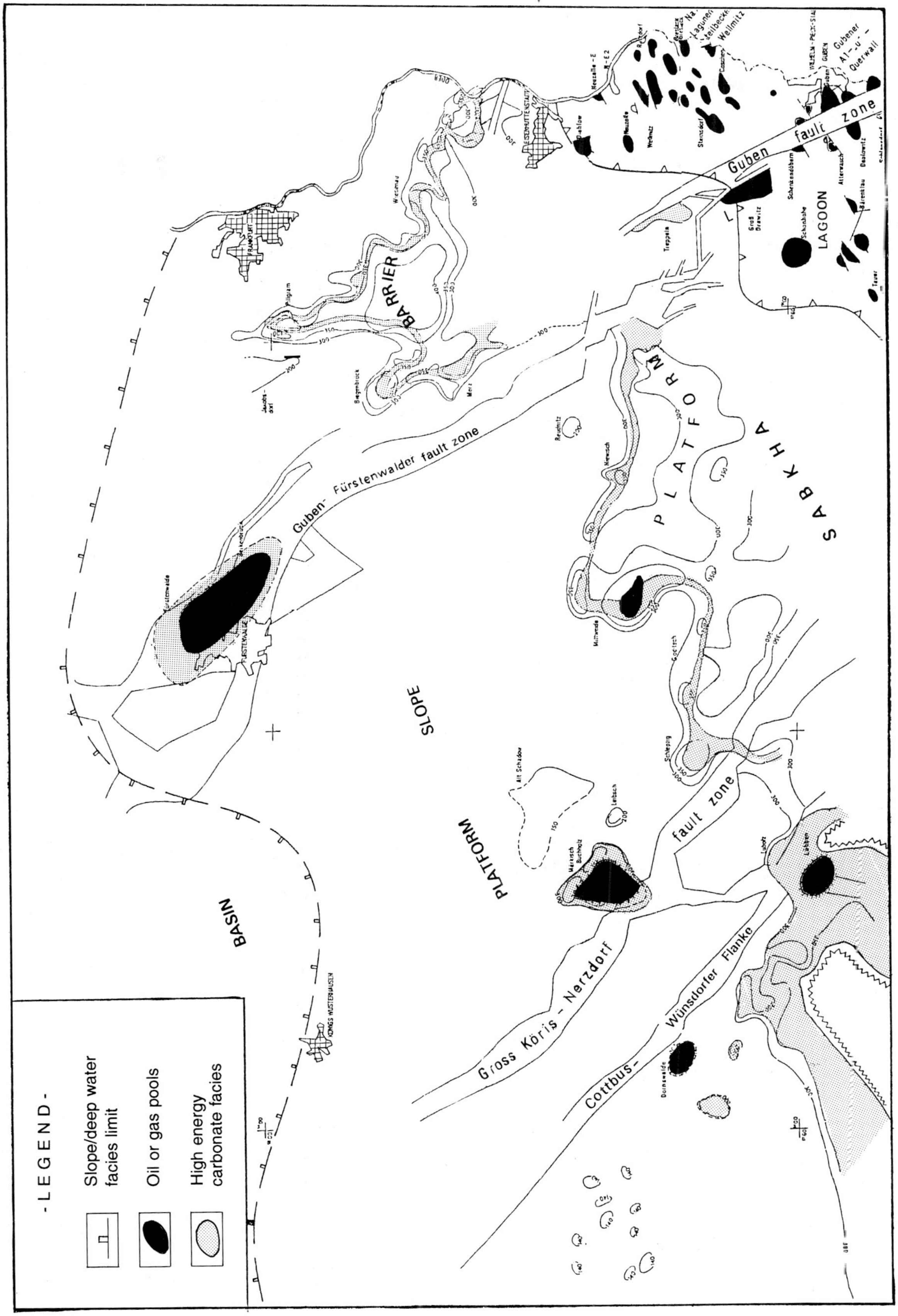

Fig. 13. Oil and gas fields in Stassfurt Carbonate, Brandenburg-Lausitz

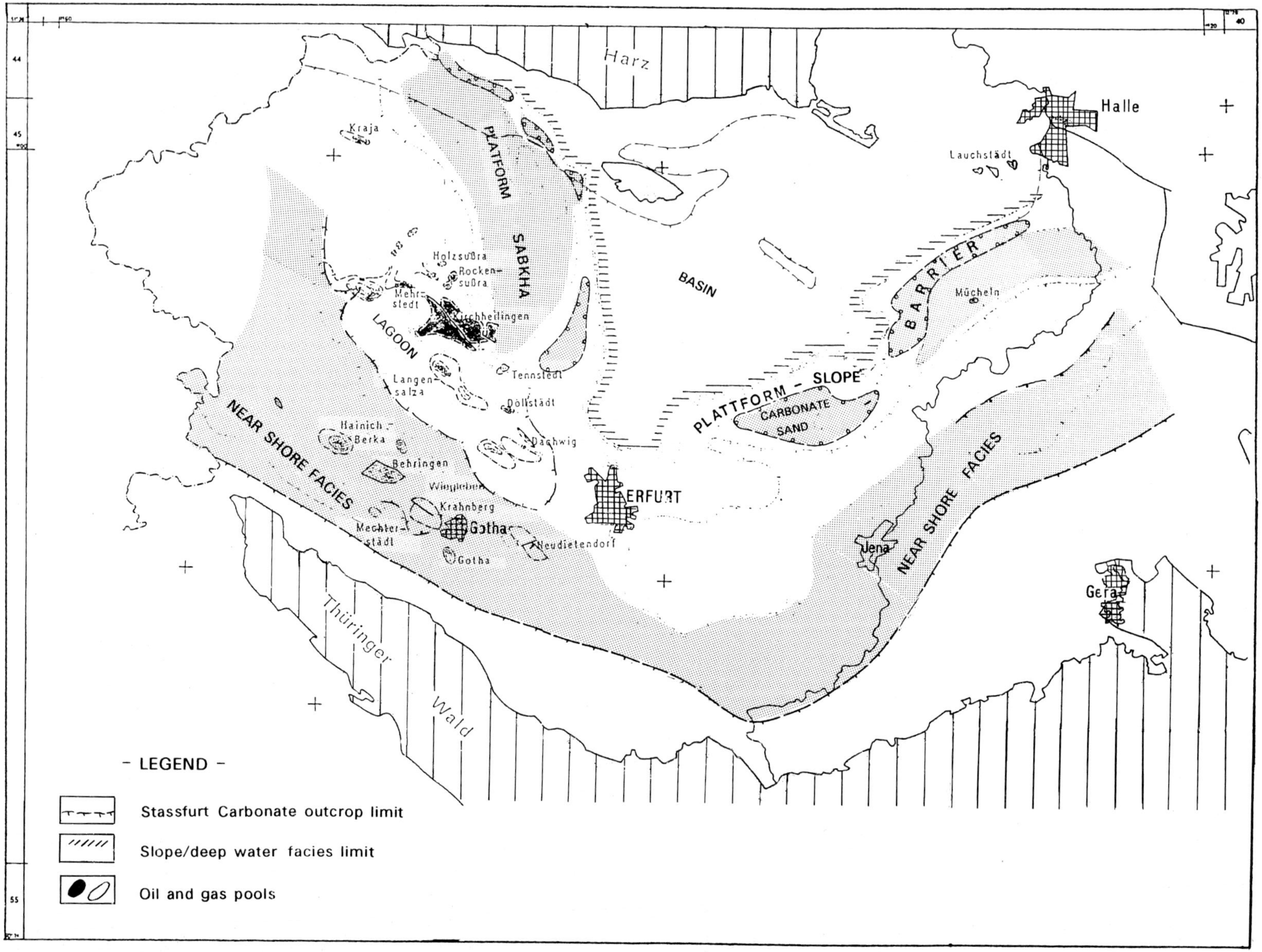

Fig. 14. Oil and gas fields in Stassfurt Carbonate, Thuringian Basin

multiple-phased processes dependent on thermal evolution of the laminated carbonate mudstone source rocks. The lateral migration paths of the fluids were short. The bulk of migration was towards the carbonate sand barrier ridge zone or towards fractured reservoirs of the carbonate buildups. The gaseous phase, with a subsequent isotopic and chemical composition differentiation, migrated the opposite way, towards the lagoonal mudstones.

The genesis of oil and natural gas deposits in an intra-evaporite carbonate can be the result of a multiple-phased process:

1. The synsedimentary, early diagenetical, formation of hydrocarbons by microbiologic processes had no significant influence on accumulation of hydrocarbons in the Stassfurt Carbonate since they have $\delta^{13}C$ values which do not correspond to the microbial interval (oil: $\delta^{13}C$, -23.3 to -28.5‰; natural gas: -35 to -52‰).

2. The formation of "paleoaccumulations" in the Stassfurt Carbonate took place during the Upper Permian by the thermochemolysis which occurred in the Zechstein Evaporite. The thermochemical formation of oil was mainly controlled by the crystallization energy of carbonates and salt minerals. The maturation of organic substances occurred with participation of the clayey component of the clastic-evaporite subsequent sedimentation. For the alginitic kerogen types in question, this was already protocatagenetically possible with $R_o < 0.3\%$.

3. The expulsion of hydrocarbons and of nitrogen from the source rocks occurred in the Permian on very short migration paths to the interwedging area of laminated carbonates and oncolites. The migration was related to the presence of structural and lithofacies traps with a chemical and isotope differentiation of fluids in hydrodynamically closed systems in the gaseous phase.

Post-Permian subsidence and tectogenetic processes led to the catagenetic formation of hydrocarbons during the Austrian phase (example: the Luetow oil field). The oil generation window in the Stassfurt Carbonate is at about 4000 m, a value inferred from the results of the coalification research on basin sediments.

4. The Laramian tectogenetic stress of Stassfurt Carbonate in the Thuringia Basin, and in the Brandenburg and Lausitz regions, led to a secondary migration of hydrocarbons redistribution of fluids and the mobilization of oils from the rock matrix was made through fracture systems with an enlargement of migration pathways and a change of the hydrodynamic systems.

Hence, the hydrocarbon accumulation was intimately related to the geologic evolution of the Middle European Depression and was primarily related to the paleofacies distribution and to subsequent structural and thermal-catagenetic evolution.

5 Oil and Gas Production from Zechstein Stassfurt Carbonate

In 1949, 7.1 million m^3 (only 687 MCFD) of natural gas were produced for local utilization from the natural gas fields of Stassfurt Carbonates of the Thuringia Depression, the only producing area at that time. Oil production was also low: in 1952, output from the Fallstein field was at an annual rate of 82 metric tonnes (575 bbl). A few years later, in 1959, and annual output of 1400 tonnes of oil (9800 bbl) an 22 million m^3 (2730 MCFD) of natural gas was produced from the Fallstein oil field and from the Langensalza and Muehlhausen natural gas fields respectively.

As a result of exploration activities during the 1950s, production from the Reinkenhagen oil field started in 1961. Two years later, the Doebern, Guben and Grimmen fields were put onstream. From 1964 to 1969 oil and natural gas production from ten additional fields, partly very small occurrences with reserves of less than 20,000 tonnes of oil and one billion m^3 of natural gas, was established in the Lausitz, Mecklenburg-Vorpommern and Thuringia production areas.

The highest oil production was reached in 1969 (Fig. 15) with an annual production of 350,000 tonnes (some 2.45 million bbl/6700 BOPD) of oil. In the same year, 450 million m^3 (43, 540 MCFD) of natural gas were produced, of which the Salzwedel-Peckensen natural gas field in the Altmark region, contributed 116 million m^3 (11,223 MCFD).

The drop in oil production after 1969 was, on the one hand, caused by the fact that the fractured Zechstein reservoirs were rapidly depleted. On the other hand, since exploration activities had been concentrated only in the Altmark region, new fields such as Drewitz and Lakoma were only put onstream after 1976.

The natural gas yield from the deposits of the Stassfurt Carbonate in Mecklenburg/Vorpommern, Brandenburg/Lausitz, Subhercynian and

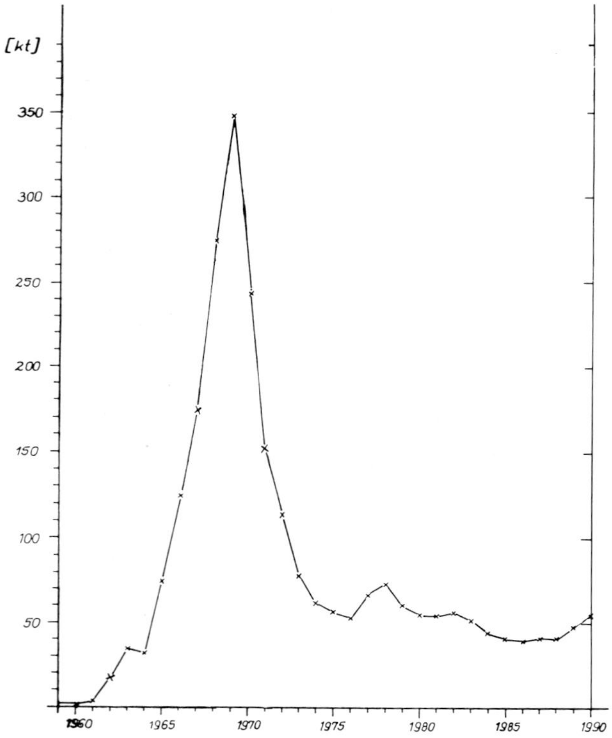

Fig. 15. Oil production (1959–1990)

Thuringia areas was supplied to the local consumer networks only.

6 Exploration Results from the Rotliegendes

The main exploration areas of the southern and northern margins of the Saxonian depocenter were the Altmark Swell, Brandenburg Ridge and partly the central basin areas (Fig. 16). Until the beginning of 1991, 406 wells had been drilled to the Rotliegendes. They led to the discovery of the Gas Dynamical Unit (GDU) and of four natural gas deposits in the Altmark region related to brachyanticlines or combined traps in the Elbe sequence (Fig. 17, Table 2):

- Salzwedel-Peckensen (gas dynamical unit) 1968
- Winkelstedt 1971
- Mellin-Sued 1971
- Wenze 1975
- Sanne 1981

The Saxonian is divided into two groups (Fig. 18): the Havel and the Elbe groups which are lithostratigraphically subdivided into Parchim, Mirow, Rambow, Eldena, Peckensen and Mellin formations. The Saxonian transgressively overlays the Autunian and Carboniferous effusives. Progressively younger Saxonian sedimentary formations set in towards the basin margin (Fig. 19). The following Saxonian facies were identified in the study area: aeolian, fluvial, continental sabka or playa, and evaporitic marine.

The natural gas fields in the Saxonian of the Middle European Depression, as in the Altmark region, are mainly related to structural traps especially in the Elbe Group. They are sealed by Zechstein evaporites. The natural gas fields in the Altmark region are located on the Altmark Swell, with traps formed by brachyanticlines or combined type whose average gas/water contact average depth is 3442 m.

In the northern part of eastern Germany, the Rotliegendes depocentre was morphologically divided into shallow sags, grabens and uplifted faulted blocks. The basal sandstones of the Parchim and Mirow formations are potential natural gas reservoirs in the central basin area. Cap rocks may be Saxonian clayey-silty and partly, halitic rocks. Lithologic-stratigraphic and combined traps have to be expected.

7 Natural Gas Production from the Rotliegendes

The exploration for Rotliegendes natural gas has been considerably intensified since 1967. This led to the discovery, development and exploitation of five large fields in the Altmark region.

The production from the Altmark group of fields is fed into a supra-regional distribution network, that supplies consumers in Rostock, Schwerin, Brandenburg, in the districts of Magdeburg, Halle and Leipzig, as well as in the southern part of eastern Germany. The consumers are adapted to a minimal calorific value of 3 kWh/m^3 due to the low content of methane and high content of nitrogen supplied by the Altmark fields.

In Salzwedel-Peckensen fields, methane content averages 35.3% with a deviation ranging from 17.9 to 57.2%. In the Winkelstedt field, the average methane content is 22% (12.6 to 27.6% range).

Present and future challenges of the natural gas exploitation are mainly determined by production conditions of natural gas fields in the Altmark area.

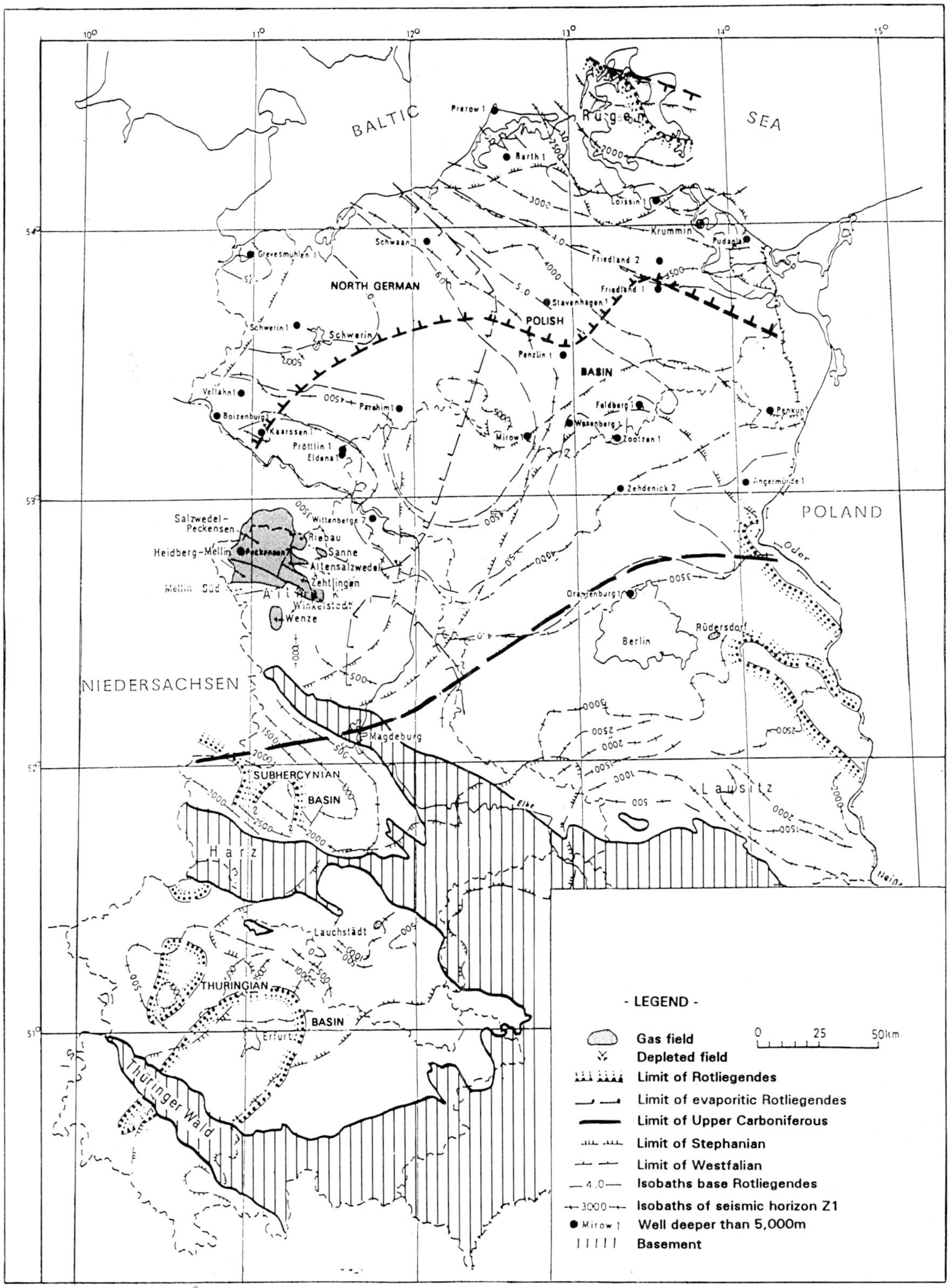

Fig. 16. Map of Rotliegendes gas fields

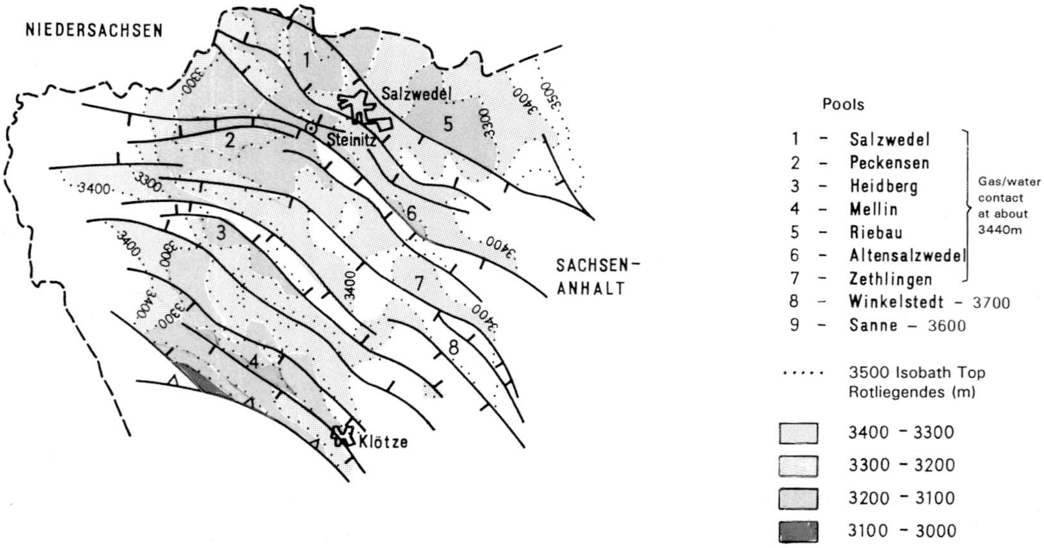

Fig. 17. Western Altmark, gas dynamic unit

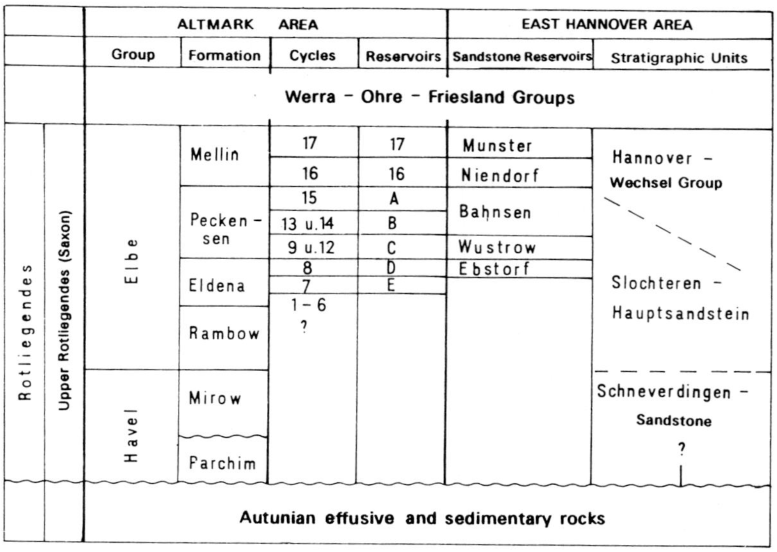

Fig. 18. Simplified stratigraphic chart of northeastern Germany Rotliegendes

One of the principal geologic and engineering-technical problems is and will be related to the production of natural gas from inhomogeneous, thick clayey-silty sandstones overlying and underlying higher permeable natural gas reservoirs.

These tight gas sands are characterized by the following parameters: porosity 8%; permeabilities less than 10 md; thickness 5–20 m. An undiscovered potential of about 100 billion m^3 of natural gas can be estimated for the Altmark region.

8 The Distribution and Chemical Composition of Natural Gas Fields of the Altmark Region

The chemical composition of gas changes horizontally and vertically within the reservoirs of the group of Altmark fields, i.e. in the Rotliegendes reservoirs 17, 16, A, B, C, D, E and the lower sequence (Fig. 20):

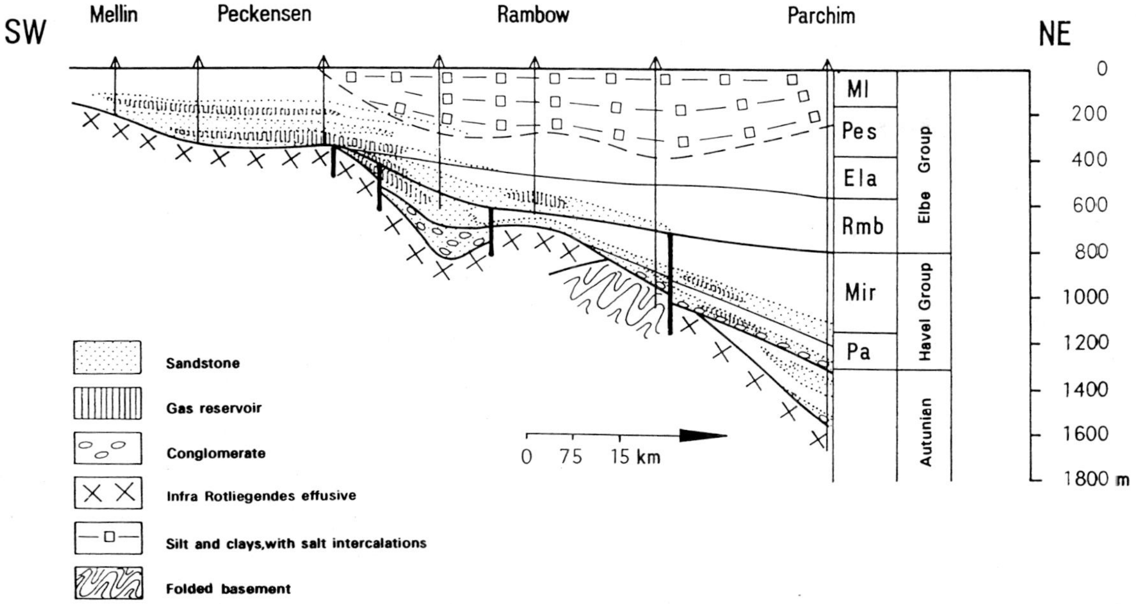

Fig. 19. Schematical cross-section, Altmark (SW) – central basin (NE)

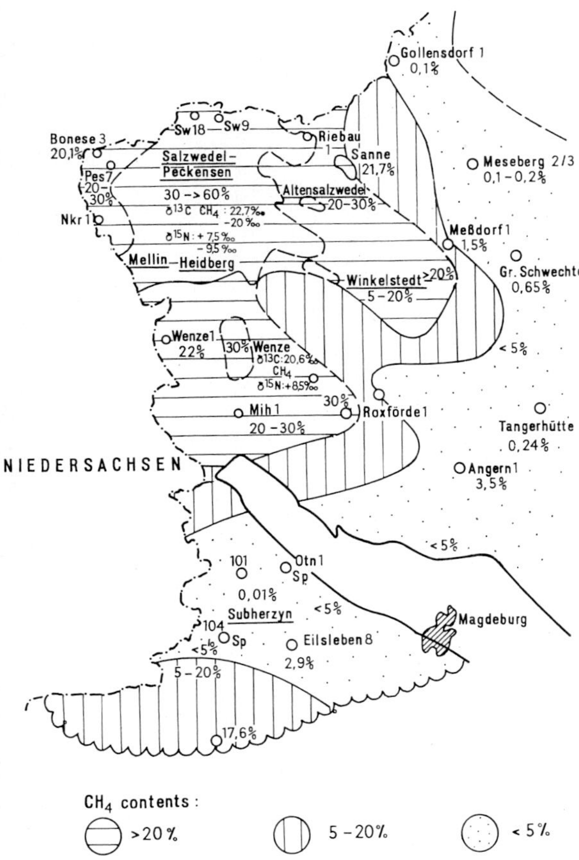

Fig. 20. Chemical composition of natural gases within Rotliegendes, Altmark area

- The highest methane content with more than 40% is restricted to the NW part of the group of fields.
- With ± 30% of methane, the gas composition is relatively homogeneous in the central part of the group of fields including the fields Heidberg, Pueggen, Mellin up to the block of Calvoerde (Wenze-Dannefeld).
- The eastern flank is characterized by average methane contents of 20% with a drastic drop to 5% in the Gollendorf 1 and Angern 1 wells.
- A differentiation has also been noticed on the western flank of the Altmark group of fields.

The variations in the methane content depend on the lithological composition of the reservoir and local tectonic conditions. They are also the result of migration processes. Parallel to the changes in methane content (from 20 to 60%), the following parameters vary:

N2 content	80–40%
heavy hydrocarbon content	0.002–0.04%
n-C_4/i-C_4	0.4–1
CO_2	0.2–0.5%
He	0.15–0.24%
Hg	3–5 mg/m^3
$\delta^{13}C$ (CH_4)	– 19.4 to – 22.4‰
$\delta^{13}C$ (C2H_6)	– 19.1 to – 24.7‰
$\delta^{15}N$ (N_2)	+ 5.3 to + 9.5‰
δD (CH_4)	– 98 to – 111‰

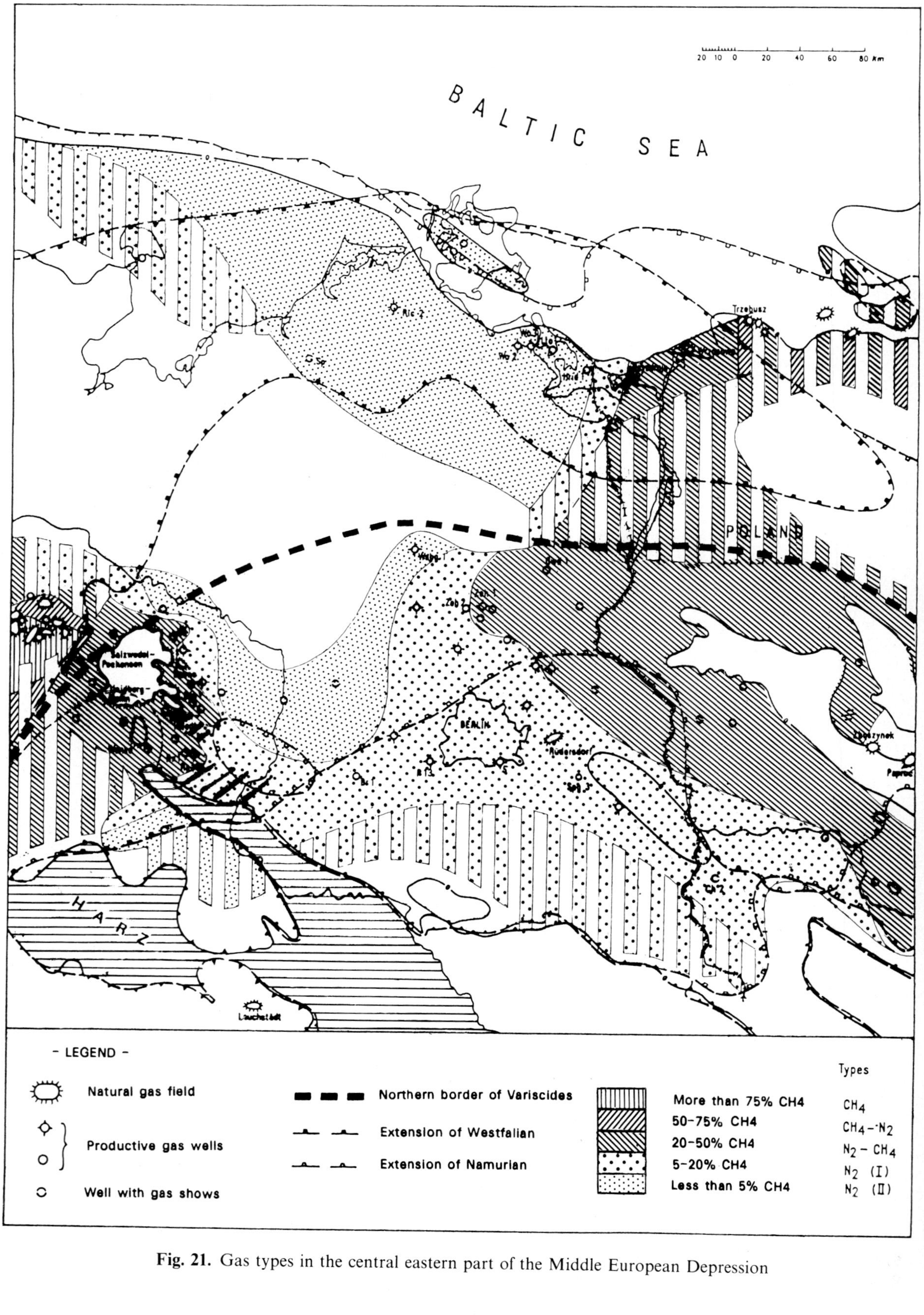

Fig. 21. Gas types in the central eastern part of the Middle European Depression

Vertical changes in chemical composition are strongly pronounced in the Elbe Group (Salzwedel-Peckensen field) on the western flank of the Altmark Swell (Figs. 21, 23). They consist of a prevailing upwards increase in CH_4 content. On the eastern flank of the Altmark Swell, a similar phenomena occurs in the Havel Group.

The high nitrogen content is a special feature of the Rotliegendes reservoirs in the eastern part of the Middle European Depression. This is expressed not only in the high nitrogen content of the Salzwedel-Peckensen fields, but also in numerous pure nitrogen occurrences as well as in the formation waters containing dissolved nitrogen up to the saturation point. For instance, Ruedersdorf (1964) and Krummin (1986) are typical nitrogen fields

The natural gas of Rotliegendes reservoirs was generated by coalification of coal seams or dispersed organic matter from the Carboniferous source rocks (Fig. 22). In the eastern part of the

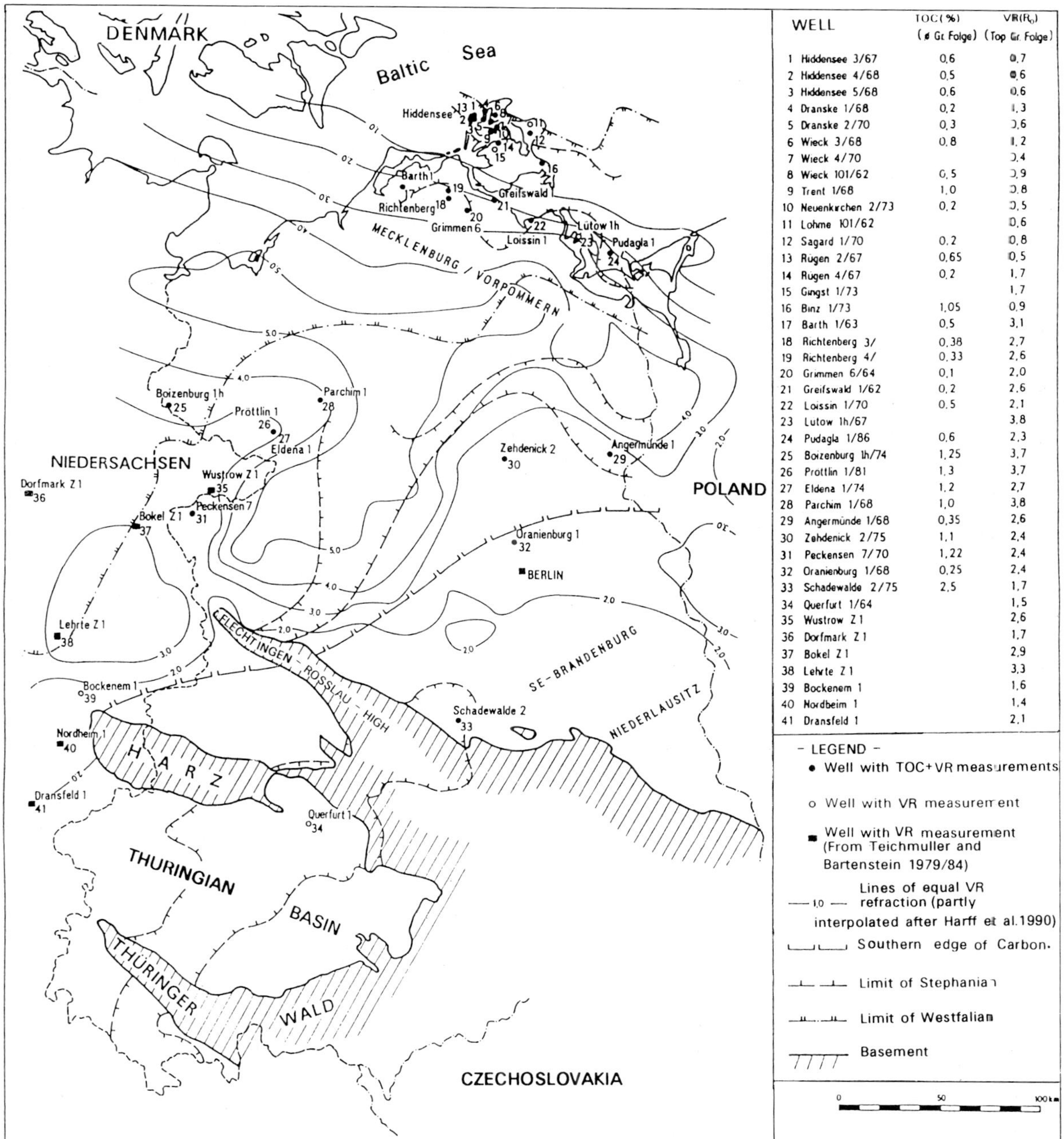

WELL	TOC(%) (ø Gr Folge)	VR(R_o) (Top Gr. Folge)
1 Hiddensee 3/67	0,6	0,7
2 Hiddensee 4/68	0,5	0,6
3 Hiddensee 5/68	0,6	0,6
4 Dranske 1/68	0,2	1,3
5 Dranske 2/70	0,3	0,6
6 Wieck 3/68	0,8	1,2
7 Wieck 4/70		0,4
8 Wieck 101/62	0,5	0,9
9 Trent 1/68	1,0	0,8
10 Neuenkirchen 2/73	0,2	0,5
11 Lohme 101/62		0,6
12 Sagard 1/70	0,2	0,8
13 Rügen 2/67	0,65	0,5
14 Rügen 4/67	0,2	1,7
15 Gingst 1/73		1,7
16 Binz 1/73	1,05	0,9
17 Barth 1/63	0,5	3,1
18 Richtenberg 3/	0,38	2,7
19 Richtenberg 4/	0,33	2,6
20 Grimmen 6/64	0,1	2,0
21 Greifswald 1/62	0,2	2,6
22 Loissin 1/70	0,5	2,1
23 Lutow 1h/67		3,8
24 Pudagla 1/86	0,6	2,3
25 Boizenburg 1h/74	1,25	3,7
26 Prottlin 1/81	1,3	3,7
27 Eldena 1/74	1,2	2,7
28 Parchim 1/68	1,0	3,8
29 Angermünde 1/68	0,35	2,6
30 Zehdenick 2/75	1,1	2,4
31 Peckensen 7/70	1,22	2,4
32 Oranienburg 1/68	0,25	2,4
33 Schadewalde 2/75	2,5	1,7
34 Querfurt 1/64		1,5
35 Wustrow Z 1		2,6
36 Dorfmark Z 1		1,7
37 Bokel Z 1		2,9
38 Lehrte Z 1		3,3
39 Bockenem 1		1,6
40 Nordbeim 1		1,4
41 Dransfeld 1		2,1

Fig. 22. Source rock potential of Carboniferous formations

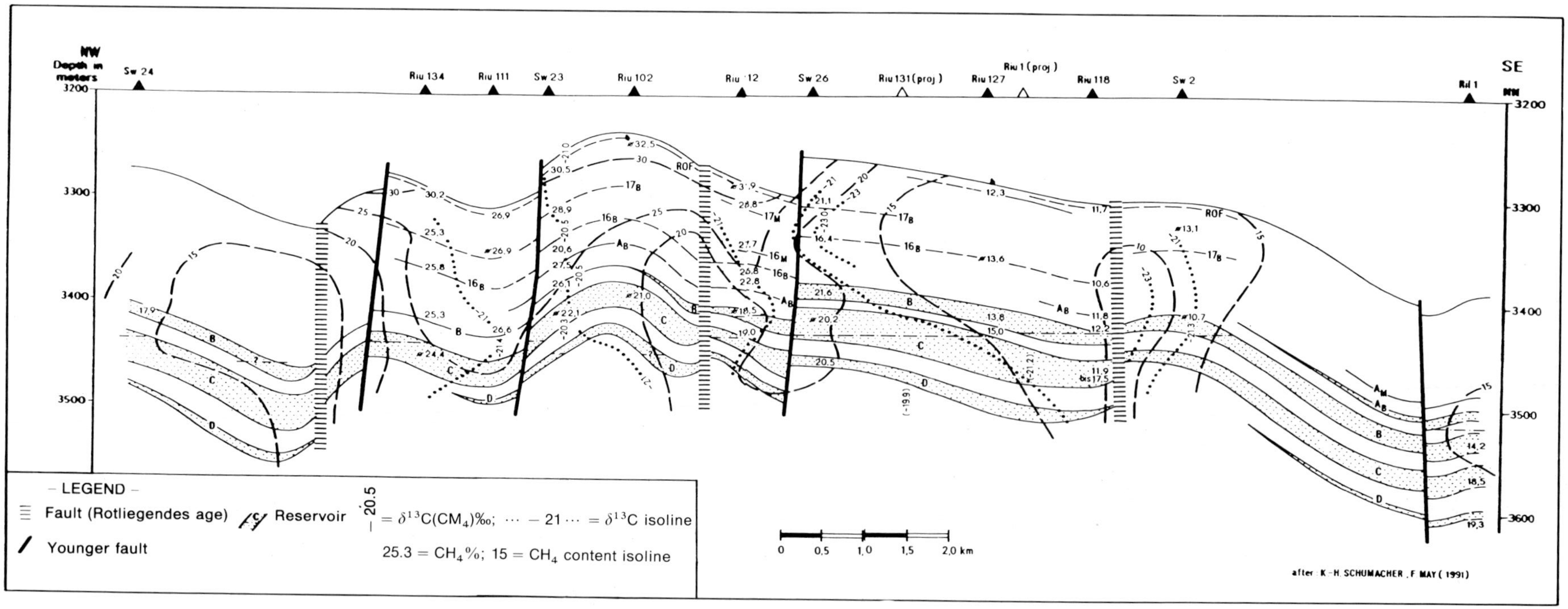

Fig. 23. Changes in chemical and isotopic composition of natural gas in West Altmark Swell

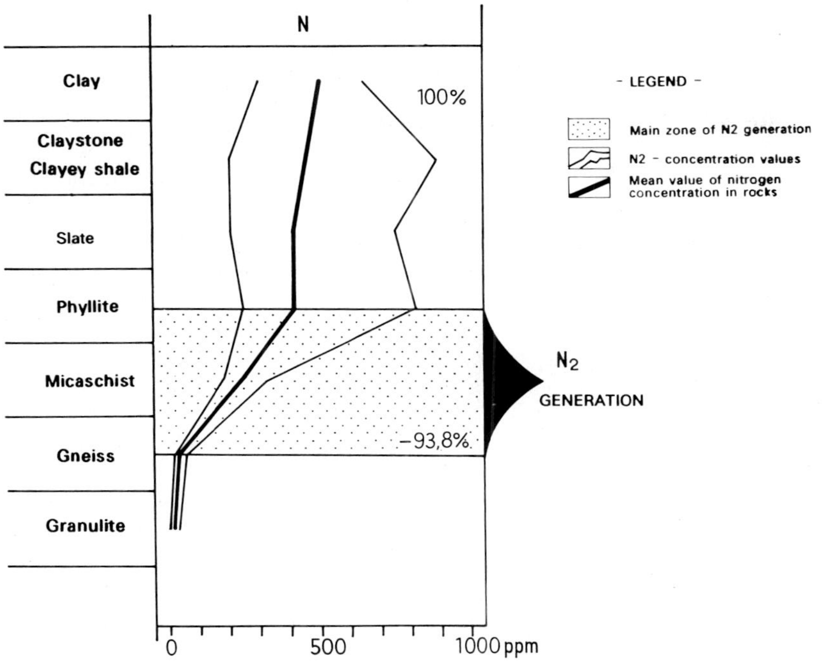

Fig. 24. N_2 generation from rocks by catagenesis and metamorphosis

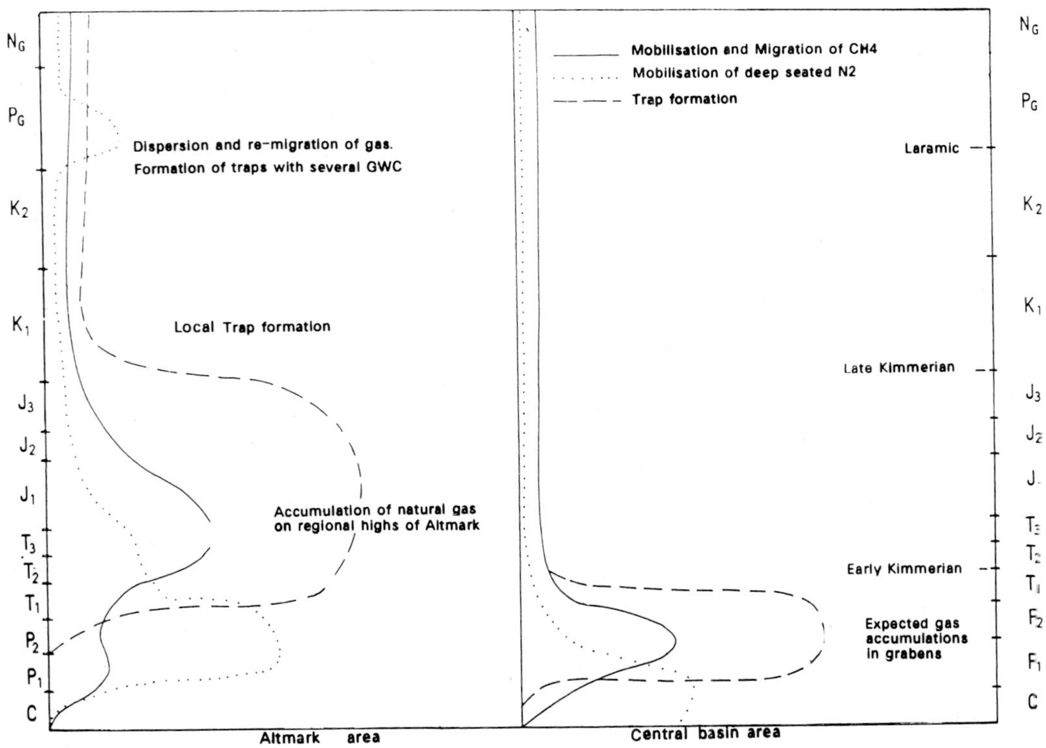

Fig. 25. Genetic scheme for accumulation of natural gas in Rotliegendes

Middle European Depression (Fig. 23), five gas types have been described: CH_4; CH_4–N_2; N_2–CH_4; N_2(I) and N_2(II). The nitrogen of the CH_4 and CH_4–N_2 types is attributed to processes of coalification, while the nitrogen of the N_2–CH_4 and N_2 types has a "deep-seated" origin. Nitrogen is generated by combined action of catagenesis, metamorphosis, tectonic activity and permanent desorption of the earth crust and resulting mobilization of nitrogen compounds from rocks.

The following multiple phased process is proposed for the genesis of gases in the Altmark region (Figs. 24 and 25):

1. Pre-Kimmerian accumulation of deep nitrogen.
2. Post-Kimmerian filling of the Altmark reservoir with natural gas of a normal chemical composition.
3. Austrian to Laramian redistribution and secondary migration of the initial homogenous "paleoaccumulation" with formation of local blocks having distinct GWC.
4. Cenozoic re-feeding of the field with HC and N_2.

In the central basin area, natural gas may have accumulated in traps of the Havel Group by vertical migration from underlying Carboniferous source rocks during the Upper Permian. Redistribution processes were less probable, because of milder tectonic activity. In general it appears that here migration and trap formation took place earlier than in the Altmark area (Fig. 25).

9 Results of Exploration from the Deeper Rotliegendes and Pre-Permian

Besides numerous deep wells up to depths of 5000 m, 27 super-deep stratigraphic tests (deeper than 5000 m) were drilled from 1963 to 1989. Of these, 11 wells were drilled to the Carboniferous and the well Pudagla 1, to the Devonian (Fig. 26).

The Rotliegendes in the central basin region was explored by 16 super deep wells. Well Schwerin 1/87, with a total depth of 7343 m, had for objective potential reservoirs of the Havel Group. The well Wesenberg 1/72 with a total depth of 5160.1 m yielded significant natural gas with high N_2 content. Deeper Rotliegendes reservoirs of the central basin may contain potential natural gas deposits in combined traps, the investigation of which will be one of the future challenges.

Besides natural gas shows, the exploration of the Carboniferous yielded valuable results on the geological structure of the in pre-Permian formations as well as information on the source rock potential in the northern part of eastern Germany. Effective reservoir rocks in the Carboniferous have not been encountered yet.

In the Dinantian clastics, the contents of dispersed organic matter varies from 1.1–1.5% TOC, similar to those in the overlaying Silesian stage deposits which have 1.3–1.4% TOC. The organic matter is mainly humic-sapropelic (kerogene type III); the maturity degree achieved can lie in the Anthracite-Metaanthracite-stage, i.e. it varies from 2.5 to 5.0% R_o.

The Dinantian (Namurian) and the Lower Silesian are considered to be the source rocks of natural gases of the Rotliegendes. The well Proettlin 1 (Figs. 21, 26) penetrated Namurian and Dinantian source rocks with a thickness of more than 850 m in the central basin area.

Between the Harz Mountains, the Subhercynian Basin and the Ruegen Island, the potential of the Devonian has still not been fully evaluated. In the Subhercynian Basin, well Huy-Neinstedt 2, with a total depth of 2640 m, penetrated the Upper and Lower Devonian. In the Usedom Island, wildcat Pudagla 1 bottomed in Middle Devonian quarzites (Fig. 26). Natural gas shows from Devonian carbonates had been reported from a series of wells drilled in the Ruegen Island. Exploration of the Devonian concentrated on determining the existence of Upper Devonian reefs.

The Northern part of the study area and the largest part of the offshore area are a portion of a broad shelf region at the edge of the pre-Variscan platform which stretches from the Belgium to Poland. The geologic and depositional conditions of the Carboniferous and Devonian can be summarized as follows.

Upper Givetian to Upper Devonian is a marine, regionally developed facies-zoned carbonate. Shelf edge and internal shelf facies were dependent on tectonic-structural transverse elements which lead to development of local carbonate platforms with reef rims in the Upper Devonian. The distal and proximal shelf zone running from the Altmark region/SW Mecklenburg to N Brandenburg is a gas-prone area. According to drilling results from the Ruegen Island and in NW Poland, the gas-prone area also develops between the eastern part of the offshore area across the Usedom/Pasewalk Island to N. Brandenburg.

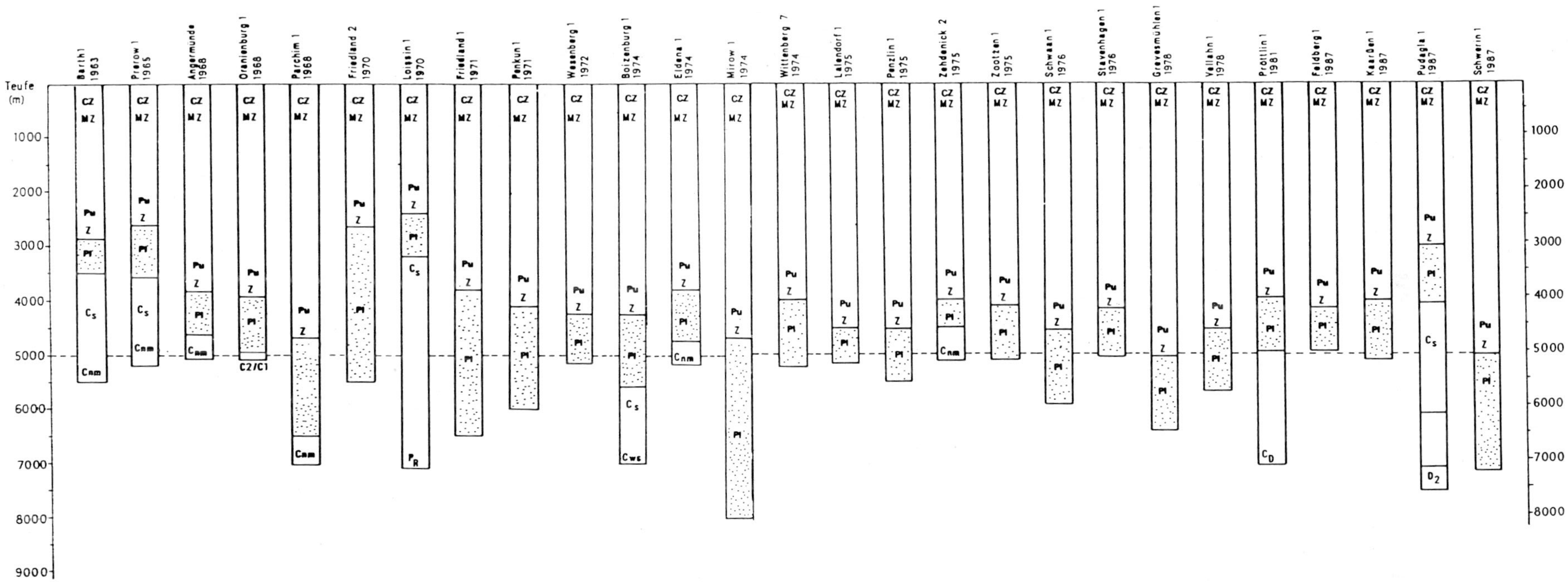

Fig. 26. Simplified stratigraphy of wells deeper than 5000 m

In contrast to the possible favourable conditions in the Devonian, the Carboniferous is less prospective. Above all, the Carboniferous has to be evaluated and investigated as the major source rock for the Rotliegendes gas reservoirs. It is mainly finely clastic with local development of substantial richness in organic matter.

Superdeep stratigraphic well Bonese 7, was planned to evaluate the oil and gas potential of Devonian buildups located on the Altmark regional high close to the area of the Salzwedel-Peckensen Rotliegendes field.

10 Conclusion

Exploration in northeastern Germany has reached a certain level of maturity and the easy-to-find fields have, perhaps, already been discovered. However, the geological conditions for the discovery of new fields do exist. Smaller, tectonically complex fields could be found in the traditional investigation areas of the Permian Basin. Deeper Rotliegendes reservoirs of the central basin at depths of more than 5000 m are also a reasonable target. Exploration concepts have also been developed for pre-Permian objectives: Carboniferous clastics and Devonian carbonate plays. Their exploration is still in a very early stage. The potential of the Triassic in the eastern German portion of the North German-Polish Basin is very low, due to the ubiquity of evaporitic impervious screen which prevented hydrocarbon migration from deeper source rocks.

Acknowledgements. The author would like to thank the Oil and Gas Company, Gommern, for permission to use unpublished reports on its activities over the last 40 years. Discussions with numerous colleagues were of great help. Special thanks are due to W. Eiserbeck, H. Dubslaff, S. Schretzenmayr. B. M. Popescu of Petroconsultants, Geneva, kindly reviewed the manuscript and made suggestions for improvement.

Suggested Reading

Blechert KH, Eiserbeck W, Piske J (1991) Hydrocarbon prospecting methods in Stassfurt Carbonate (Permian). EAPG Conference, Florence, 72 pp

Deubel F (1954) Betrachtungen über das Auftreten von Erdöl und Erdgas im Zechstein des Thüringer Beckens. Geologie 6/7, pp 804–831

Katzung G (1988) Tectonics and sedimentations of Variscan molasses. Z Geol Wiss 16: 823–84

Müller EP (1984) Zur Genese von Erdölen in Karbonaten am Beispiel der Lagerstätten im Oberen Perm des Territoriums der DDR. Z Angew Geol 30(5): 214–218

Müller EP (1990) Genetische Modelle der Bildung von Erdgaslagerstätten im Rotliegenden. Nds Akademie Geowiss Veröff 4, Hannover, pp 77–90

Müller EP (1991) Genetic models of natural gas deposits in the "Rotliegendes" in eastern Germany. Abstract EAPG Conf, Florence, pp 127

Müller EP, Dübslaff H, Eiserbeck W, Sallüm R (1992) Die Entwicklung der Kohlenwasserstoff-exploration zwischen Ostsee und Thüringer Wald. Geol Jahrb (in press)

Schretzenmayr S (1989) Grundzüge für die tektonisch-genetische Modellierung der Fallenbildung im Saxon der DDR. Z Angew Geol 35: 225–230

Schwab G (1985) Paläomobilität der Norddeutsch-Polnischen Senke. Diss B, ZIPE, Potsdam

5 Exploration History and Future Possibilities in Hungary

Janos Kokai[1]

CONTENTS

1 Hungary: Key Data

(Situation as of January 1, 1991)

Sedimentary basin surface area: 75,000 km^2
Total seismic lines: 110,000 km
Refraction = 8000 km
Analog = 12,000 km
Conventional = 29,000 km
Digital = 61,000 km

Exploratory wells: about 5200
Development wells: about 2500
Average depth of wells: 2150 m
Wells deeper than 4500 m: 17
E&D targets and reservoir rocks: Pliocene, Miocene and Oligocene clastics/Cretaceous, Jurassic and Triassic carbonates, Paleozoic metamorphics.
Cumulative discovered "geological" reserves (1935–1990):
Oil: 263 million tons
Gas: 319 billion m^3
CO_2: 46.2 billion m^3
Remaining recoverable (proven) reserves:
Oil: 21.1 million tons
Gas: 101.6 billion m^3
Cumulative production:
Oil: 70.1 million tons
Gas: 139.3 billion m^3
CO_2: 5.5 billion m^3

[1]Ministry of Industry and Trade, P.O.B. 22, Budapest H-1518, Hungary

2 Exploration History

Hungary is a relatively young hydrocarbon producer: the first oil and gas fields were discovered between 1937 and 1940 at Budafa, Lovászi and Bükkszék. They were the result of some 30 years of geophysical exploration and drilling by the Hungarian state and by subsidiaries of the Standard and Anglo-Persian oil companies. After the Second World War, tenements held by foreign companies were nationalized. A Soviet-Hungarian venture operated during the late 1940s and early 1950s. The bulk of discoveries were, however, made after the mid-1950s by the state oil company OKGT, now MOL.

2.1 Geophysical Exploration

Hungary was among the first countries in the world to apply systematic geophysical exploration for natural resources. Before the First World War, a national gravity programme was implemented using the Eötvos torsion balance. Not surprisingly, the first wells in the Great Hungarian Plain were drilled on gravity anomalies. Hungary has a fairly good coverage of gravity, magnetic and geoelectric measurements; most of these data are available on published maps.

The first seismic survey was carried out in 1937. Between 1937 and 1991 seismic acquisition was

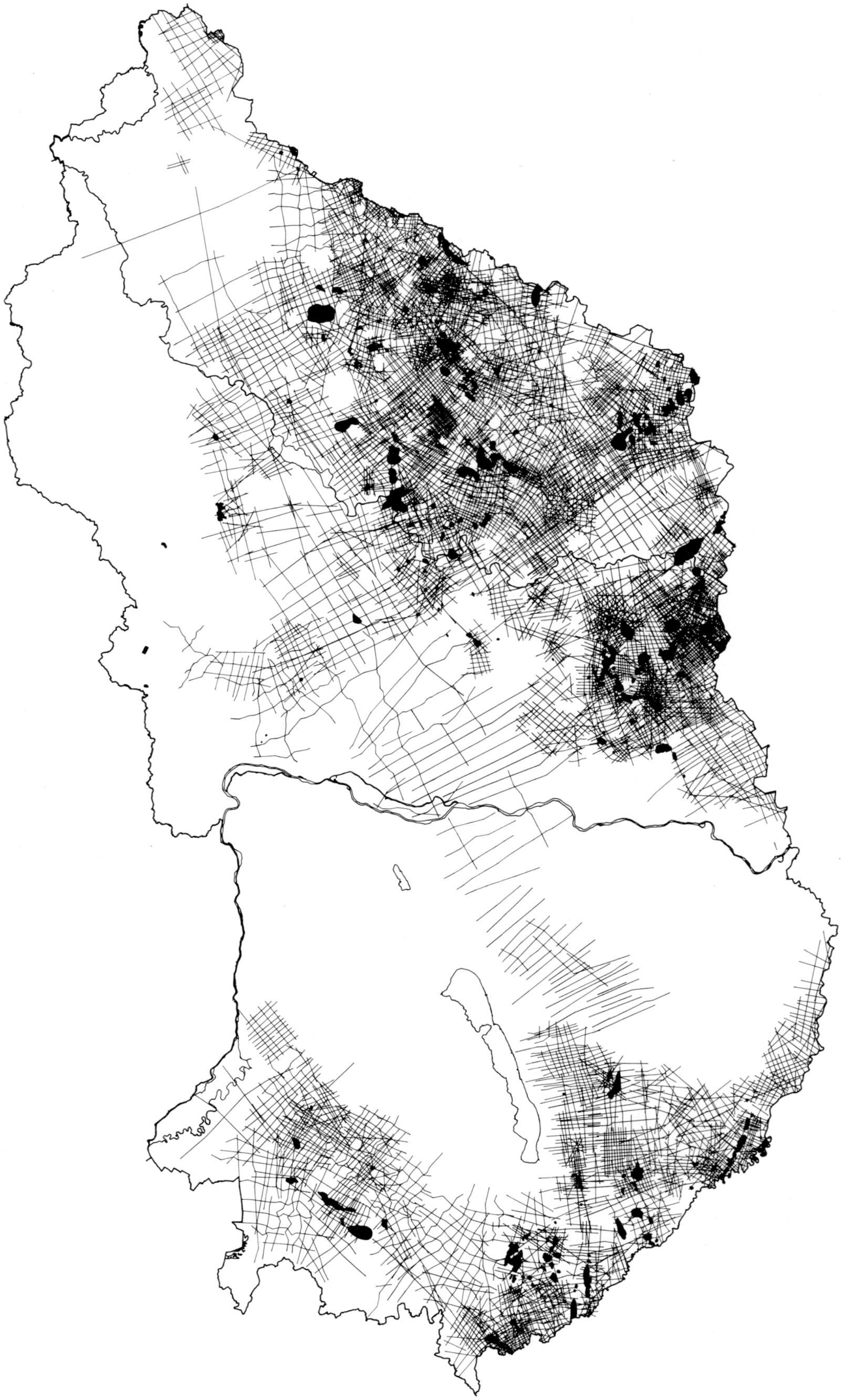

Fig. 1. Map showing the seismic grid and location of hydrocarbon fields

conducted mainly by the national oil company and in much smaller volume by L. Eötvos Geophysical Institute (MAELGI).

Some 100,000 km of seismic lines have been acquired in Hungary. Their quality varies from poor to good, depending on the method used for recording: conventional, analogue or digital.

Over the last 15 years, OKGT subsidiary GKV and ELGI (a subsidiary of the Hungarian Central Geological Institute), recorded altogether some 40 million seismic traces. They cover various subbasins with rather different line-densities (Fig. 1). The coverage varies from 6- to 48-fold. The most common CDP interval is 15 to 25 m. The first 3-D surveys were run in 1985 and in 1991 by CGG and GKV.

The recorded line-km versus new seismically defined leads during the period 1974–1991 is shown in Fig. 2. Geophysical methods used so far identified 781 leads, many of which became drilling prospects. In 186 of them, various kinds of oil/gas accumulations were observed, while 301 were dry.

It is noteworthy to mention that in the last few years, while drilling and production declined continuously, the volume of seismic acquisition was maintained above the 4000 line-km/year mark.

Use of advanced, modern seismic acquisition and seismic stratigraphy has shown that the potential of important thick Neogene formations was not yet fully evaluated. This is even more true for the Mesozoic and Palaeozoic "basement" formations whose pre-Neogene tectonic evolution created a complex structural framework.

2.2 Exploratory and Development Drilling

About 7700 exploration, development and production wells, with an average depth of 2150 m, have been drilled so far in Hungary.

Reliable statistics are available from 1935 onwards (Table 1). After a peak during World War II (in 1943) and a decline in 1944 and 1945, activity reached the pre-war levels in 1949. A record high was reached in 1962 (Fig. 3) when a very high success ratio (57%) was also achieved. A sharp decline in activity began in 1988, and in 1990 drilling volume was only marginally higher than in 1950.

The distribution versus depth of the drilled structures is as follows:

< 1500 m depth: 30%
1500–3000 m depth: 58%
3000–4500 m depth: 10%
> 4500 m depth: 2%

It is also noteworthy that, in spite of mature exploration of the "shallow" mainly Tertiary deposits, only 17 wells deeper than 4500 m (Table 2)

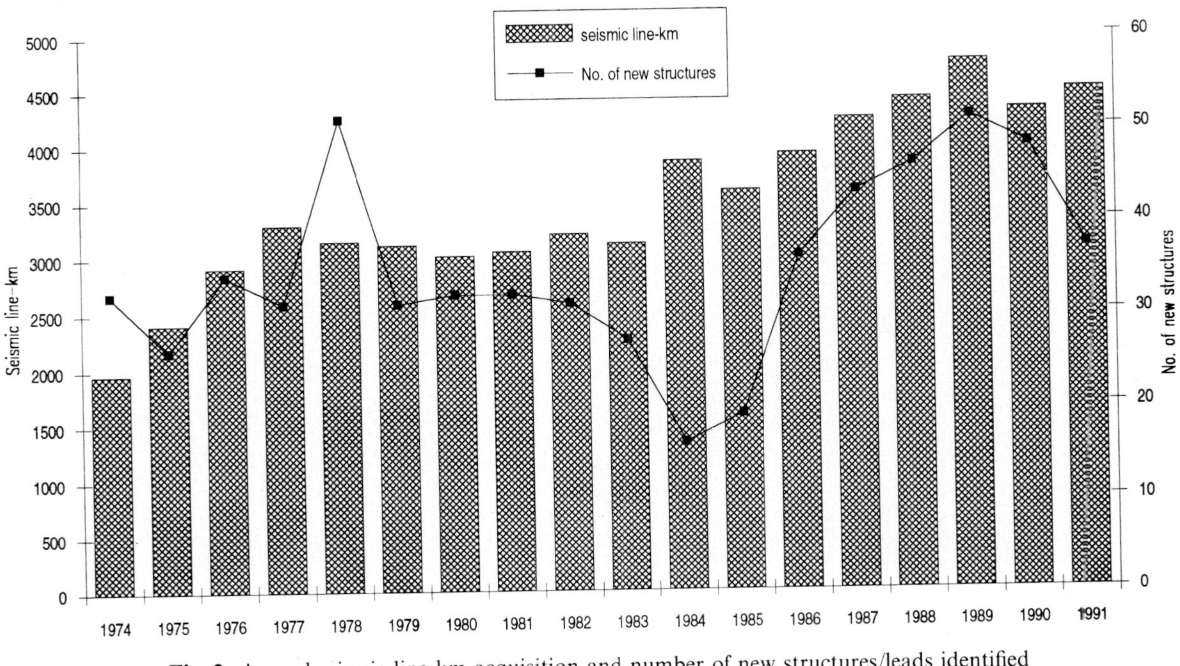

Fig. 2. Annual seismic line-km acquisition and number of new structures/leads identified

Table 1. Results of drilling for hydrocarbons in Hungary during the 1935–1990 period

	Exploratory wells						Development wells				
Year	Metrage	Oil	Gas	Dry	Total	Year	Metrage	Oil	Gas	Dry	Total
1935	2695		1		1	1935					
1936	4872		2	1	3	1936					
1937	4507	1			1	1937					
1938	10,064	8		1	9	1938					
1939	29,872	18	2	3	23	1939					
1940	54,838	35	2	4	41	1940					
1941	64,502	30	4	12	46	1941					
1942	90,186	41	9	9	59	1942					
1943	97,649	39	7	26	72	1943					
1944	74,403	32	4	16	52	1944					
1945	17,250	5	2	3	10	1945					
1946	26,626	10	3	4	17	1946					
1947	52,981	17	8	10	35	1947					
1948	63,034	17	7	18	42	1948					
1949	61,326	24		22	46	1949	27,369	18	1		19
1950	105,758	41	6	25	72	1950	36,900	28	1		29
1951	150,333	56	8	41	105	1951	68,909	49	3	2	54
1952	211,810	57	15	75	147	1952	78,851	56	1		57
1953	252,823	61	14	118	193	1953	40,889	22	3	2	27
1954	168,554	35	14	93	142	1954	28,788	14	7	2	23
1955	145,902	26	12	69	107	1955	27,399	16	1	1	18
1956	128,360	38	13	62	113	1956	12,750	6			6
1957	151,316	30	10	49	89	1957	9968	12			12
1958	250,103	36	18	91	145	1958	7419	3		1	4
1959	314,466	61	45	95	200	1959	66,061	42		5	47
1960	405,262	96	45	107	248	1960	22,329	11	1		12
1961	375,577	67	39	111	217	1961	46,938	27	1	3	31
1962	388,338	44	37	137	218	1962	87,996	30	13	16	59
1963	311,684	50	34	128	212	1963	80,174	40	10	9	59
1964	320,998	30	39	116	185	1964	54,803	25	10	1	36
1965	280,007	37	36	80	153	1965	74,877	35	7	6	48
1966	295,919	55	19	80	154	1966	66,601	23	12	7	42
1967	266,987	48	26	54	128	1967	144,142	54	7	19	80
1968	188,612	37	13	41	91	1968	188,748	79	14	9	102
1969	205,459	25	14	55	94	1969	151,746	49	17	10	76
1970	204,874	33	15	47	95	1970	171,203	34	37	22	93
1971	164,826	29	20	47	96	1971	160,889	32	13	23	68
1972	142,459	13	15	37	65	1972	94,653	26	23	5	54
1973	165,387	25	10	41	76	1973	79,899	15	18	10	43
1974	186,604	21	10	41	72	1974	92,692	13	22	16	51
1975	189,971	16	19	46	81	1975	114,673	24	15	23	62
1976	204,363	19	22	51	92	1976	132,529	31	9	20	60
1977	206,017	8	25	40	73	1977	163,942	46	19	24	89
1978	202,108	15	23	42	80	1978	166,643	48	28	13	89
1979	211,770	18	30	35	83	1979	143,975	54	12	13	79
1980	209,706	12	25	48	85	1980	141,597	58	14	14	86
1981	199,882	25	21	37	83	1981	144,827	37	24	22	83
1982	206,077	18	29	46	93	1982	152,228	49	21	13	83
1983	208,265	16	24	54	94	1983	178,432	44	30	14	88
1984	202,093	13	32	50	95	1984	184,345	69	24	13	106
1985	192,970	12	21	48	81	1985	195,490	53	25	23	101
1986	205,687	17	19	45	81	1986	177,068	48	33	24	105
1987	205,327	22	25	43	90	1987	177,625	40	37	22	99
1988	175,935	23	13	62	98	1988	157,025	33	38	4	75
1989	149,303	13	19	52	84	1989	135,579	38	32	7	77
1990	118,416	4	12	34	50	1990	53,975	24	7	10	41
Total	9,525,113	1579	937	2702	5217	Total	4,342,946	1455	590	428	2473

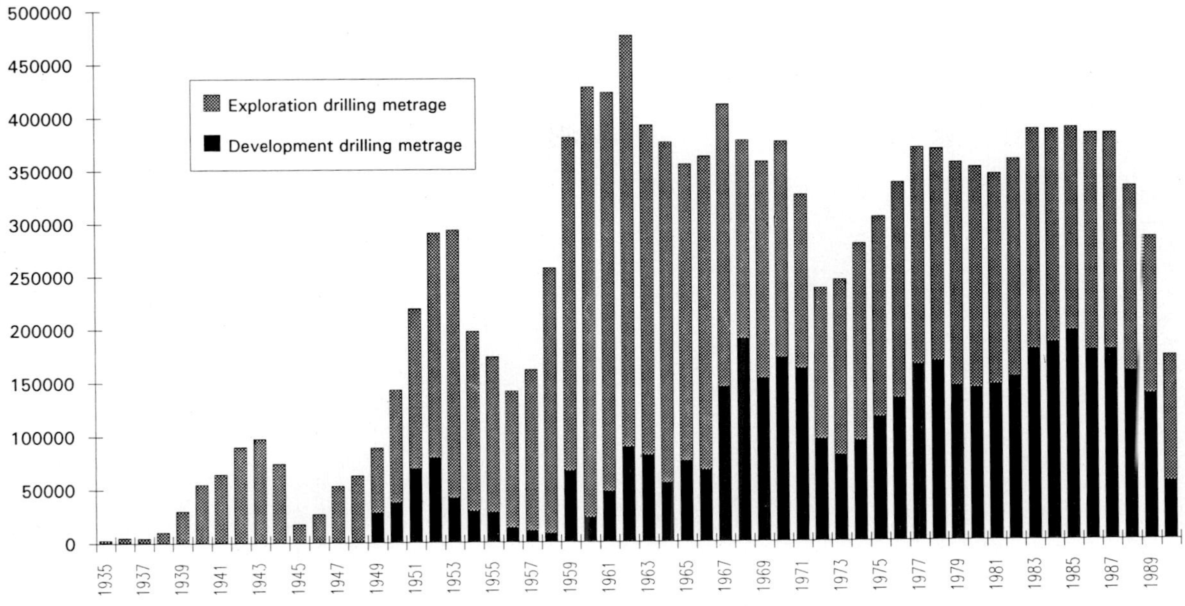

Fig. 3. Metres drilled during the 1935–1990 period

Table 2. List of wells deeper than 4500 m

Year	Name of Well	T.D. (m)
1967	Kerkaskapolna-1	4510.00
1968	Bosarkany-1	4517.00
1969	Lovaszi-II	5400.00
1971	Hodmezovasarhely-1	5842.50
1972	Mako-2	5060.00
1974	Barszentmihalyfa-1	5075.50
1975	Sarkadkeresztur-1	4841.00
1975	Budafa-IV	5265.50
1977	Derecske-1	5205.00
1978	Gyekenyes-1	4675.00
1979	Doboz-1	4656.00
1980	Inke-1	5000.00
1982	Sarand-1	4800.00
1985	Alpar-1	5305.00
1986	Bekes-2	5500.00
1987	Kiskunhalas-1	4500.00
1988	Gater-Mely-1	4800.00

have been drilled, some with assistance from the US Geological Survey. All were abandoned dry.

Exploratory drilling peaked in the early 1960s (Figs. 3,4) with a maximum of some 245 completions and 275,000 m drilled. In 1986 and 1987 meterage was still in the range of 200,000 m. Activity declined to 150,000 m drilled in 1989 and to some 118,500 m in 1990.

Development drilling had three peaks: in the early 1950s, late 1960s and mid-1980s (Figs. 3,5). As for the exploratory drilling, activity declined sharply to only 65,000 m in 1990.

The reason for the exploratory and development drilling meterage decline has been the financial shortage in the whole Hungarian economy. As a result, over the last decade the amount of discovered hydrocarbons was not significant.

2.3 Production and Reserves

After three decades of exploration in Hungary, commercial crude oil and natural gas production started in 1937. Over 53 years the cumulative oil production was: 70 million tons (some 527 million bbl), and for gas: 139.3 billion m^3 (5295 bcf).

Sixty-two percent of Hungary's oil production is from sediments of the Tertiary Pannonian basin's wrench sub-basins and interior sags. Approximately 24% of the oil production is from Mesozoic carbonates underlying the Pannonian Basin. Production has also been established from fractured and fissured Paleozoic reservoirs. Seventy percent of gas production is from Tertiary sediments.

Oil production was 1.95 million tons (some 40,000 bopd) in 1990 (Table 3) covering only a quarter of Hungary's demand. The gas production was 6.3 billion m^3 in 1989, decreasing to 5 billion m^3 in 1990 (Table 4). The Algyö giant field still produces half of the nation's hydrocarbon output. Hungary's demand is 12.0–12.5 billion m^3 of gas and 7.0–7.5 million tons of oil per year.

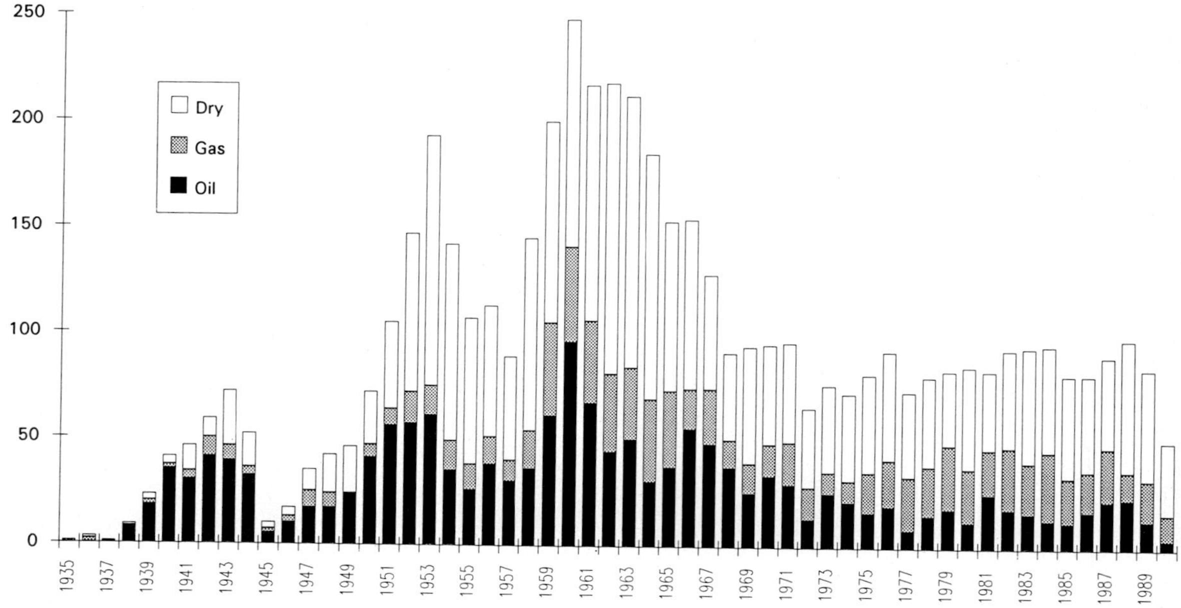

Fig. 4. Exploratory, drilling completions (1935–1990)

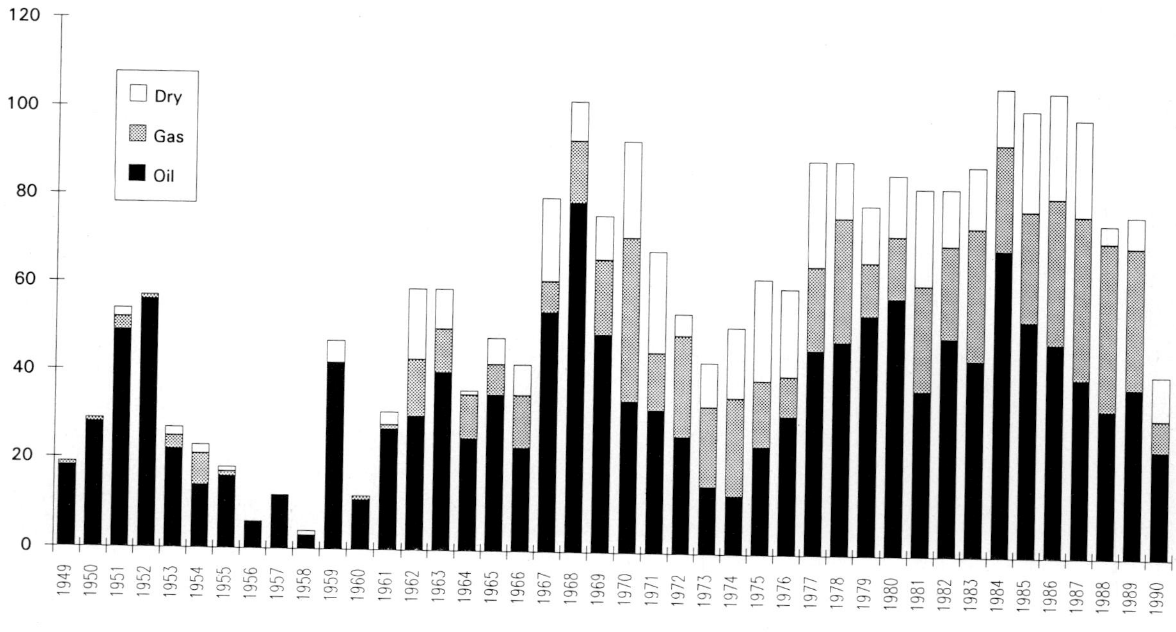

Fig. 5. Development drilling completions (1949–1990)

Up to 01.01.91 there were 126 oil and gas fields discovered, but multiple oil and gas pools may exist in the same field. They contained initial geological (in-place) reserves of 263.3 million tons oil and 314.8 billion m^3 of natural gas (Figs. 6, 7). Initial recoverable reserves were estimated at 93.5 million tons of oil and 245.8 billion m^3 of gas. Estimated proven remaining reserves of the same date were 21.1 million tons of oil and 101.6 billion m^3 of gas. About 28% of the total gas reserves contain non-combustible components, mainly CO_2. It is noteworthy that most of the discovered reserves are located at relatively shallow depths (Fig. 8) having a high geothermal gradient (Fig. 9).

Among them are the following large fields (50 to 190 million toe recoverable reserves each): Algyö,

Table 3. Oil field main characteristics and 1990 production

Field name	Discovery year	Depth (feet)	Degree API	Producing wells	Total	Average BOPD	Cumulative 01.01.91
Szolnok district							
Algyö	1965	5320–8317	41.00	723	1210	21,593	174,197,625
Szeged	1971	8629	41.70	18	25	2220	23,841,785
Kiskundorozsma	1964	4846–9711	41.90	26	36	1861	4,093,093
Szeghalom	1980	6545–6608	44.50	42	55	1152	1,300,200
Asotthalom	1967	3199	32.10	15	20	961	10,202,320
Battonya East	1970	2913	45.40	114	126	938	3,848,833
Ferencszallas	1969	7001–7674	41.10	29	52	463	3,054,192
Pusztaföldvár	1958	5341–5538	30.20	69	107	427	15,411,992
Ruzsa	1979	6445–9403	45.40	3	3	307	135,788
Demjen	1954	725–3937	38.20	193	224	267	9,239,855
Kelebia Sud	1970	2428	17.40	6	8	257	3,236,223
Kaba Sud	1978	6548	34.10	3	4	206	709,467
Mezosas	1978	8432	40.00	3	3	201	79,837
Csanadapaca	1967	6063	37.00	6	6	181	143,849
Ferencszallas E + Kiszombor	1973	7283	41.10	13	16	77	3,155,712
Battonya	1959	3041	39.20	17	28	74	1,404,763
Martfu Sud	1979	6519	40.00	8	10	73	26,762
Kismarja	1979	2178–2536	25.10	27	36	71	1,156,456
Mezohegyes	1960	3504–3589	38.90	6	9	64	1,324,674
Kelebia NE	1968	3113–3258	37.00	4	4	38	1,083,245
Kaszaper Sud	1971	5056	32.80	14	14	27	217,316
Totkomlos SW	1966	5991	37.00	4	6	6	378,379
Ulles	1962	5848–7464	36.00	35	39	5	68,599
Szolnok	1953	5994	30.00	3	6	2	1,118,000
Fuzesgyarmat	1974	5753–5802	22.90	1	1	2	4805
Almosd	1977	8143	33.80	2	2	1	367
Others						0	2,895,544
Nagykanisza district							
Nagylengyel	1951	5545–8235	16.00	71	460	5004	128,534,063
Zalakaros + Savoly	1979	5463–5791	20.70	6	34	1112	2,885,329
Kiskunhalas NE (Pz)	1974	6119–6594	27.50	22	74	864	9,086,012
Budafa-Kiscsehi	1936	2713–4167	41.10	123	514	597	41,570,828
Szank	1964	5827–5932	38.70	22	110	507	16,482,527
Lovaszi	1940	3510–4301	41.30	75	479	443	49,828,477
Kiskunhalas NE (Mz)	1974	5905	31.20	10	34	204	1,794,381
Ortahaza	1970	4580–6089	35.20	10	36	117	4,127,418
Tazlar Nord	1986	7129–7415	32.50	1	7	90	43,741
Ortahaza East	1986	5413	31.10	4	12	57	62,901
Szank NW	1977	5495	41.10			48	913,598
Pusztaapati	1973	8300	22.30	2	11	32	1,036,499
Tazlar	1966	6365–6890	24.00	2	23	21	393,304
Mezocsokonya West	1987	5938	27.50			21	9,895
Belezna	1963	8176	41.10	5	31	8	283,054
Szank West	1966	6106–6201	28.70	1	10	8	19,234
Buzsak	1954	1683	17.40		17	4	234,389
Bugac	1970	4924	38.20	1	5	2	82,963
Others						0	7,339,704
			Total	1739	3907	40,613	527,057,998

Nagylengyel and Hajduszoboszlo, some medium size (15–50 million toe): Ulles, Lovászi, Budafa, Szank, Szeghalom and Pusztaföldvár. Most of the remaining discoveries were small (Fig. 10).

Table 5 shows the number of producing wells as of 01.01.91 and their productivity. Tables 6 and 7 show the production record of the country's major fields to 01.01.91.

Hungarian oil and gas production has been declining over the last 10 years, when more reserves were produced than found. In spite of intensive, improved recovery methods it is expected that the

Table 4. Gas production by field in 1990

Field name	Discovery year	Million m^3	1000 m^3/day
Algyö	1965	2075.33	5685
Ulles deep	1962	655.27	1795
Hadjuszoboszlo	1958	311.38	853
Kisujszallas West	1969	309.46	847
Szeghalom	1980	223.47	612
Endrod-III	1976	220.70	604
Pusztaföldvár	1958	163.49	448
Zsana	1978	133.50	365
Szank West	1964	104.50	286
Ferencszallas	1969	84.55	231
Martfu Sud	1979	77.07	211
Sarkadkeresztur	1976	71.96	197
Endrod-I	1971	65.72	180
Fegyvernek + Fegyvernek East	1969	59.10	162
Gorgeteg-Babocsa	1954	48.50	132
Szank West	1966	43.40	119
Szeged	1971	36.67	100
Nagykoru + Tiszapuspoki + Surjan	1964	36.39	99
Tazlar	1966	36.20	99
Tatarules-Kunmadaras	1956	34.67	95
Ereszto	1976	28.60	78
Barca West	1979	26.90	73
Ebes	1960	21.89	60
Endrod Nord	1978	20.68	56
Kiskundorozsma	1964	20.67	56
Battonya East	1970	15.97	43
Kiskunhalas NE	1974	15.90	43
Battonya	1959	14.18	38
Kiskunhalas NE	1974	10.80	30
Mezocsokonya	1964	9.00	24
Ferencszallas East + Kiszombor	1973	8.94	24
Kaszaper Sud	1971	8.40	23
Szarvas	1961	7.97	22
Vizvar Nord	1982	7.70	21
Harka	1975	6.50	18
Zalakaros + Savoly	1979	6.40	17
Kiskunhalas	1967	5.30	14
Kismarja	1979	4.90	13
Uraiujfalu	1945	4.30	12
Ortahaza	1970	4.20	11
Mezohegyes	1960	3.94	10
Csanadapaca	1967	3.93	10
Bajcsa	1955	3.50	9
Belezna	1963	2.30	6
Kaba Sud	1978	2.16	6
Peneszlek	1982	1.66	4
Mezosas (Komadi)	1978	1.64	4
Ruzsa	1979	1.62	4
Battonya Nord	1982	1.59	4
Asotthalom	1967	1.44	4
Kismarja Sud	1984	1.10	3
Totkomlos SW	1966	0.42	1
Demjen (East Punkosdhegy)	1954	0.32	neg.[a]
Tazlar Nord	1986	0.30	neg.
Fuzesgyarmat	1974	0.27	neg.
Kelebia Sud	1970	0.20	neg.
Szank NW	1977	0.20	neg.
Totkomlos	1941	0.11	neg.
Kelebia NE	1968	0.10	neg.
Ortahaza East	1986	0.10	neg.
Szolnok	1953	0.01	neg.
Almosd	1977	0.00	0
	Total	5067.44	13,861

[a]neg. = negligible.

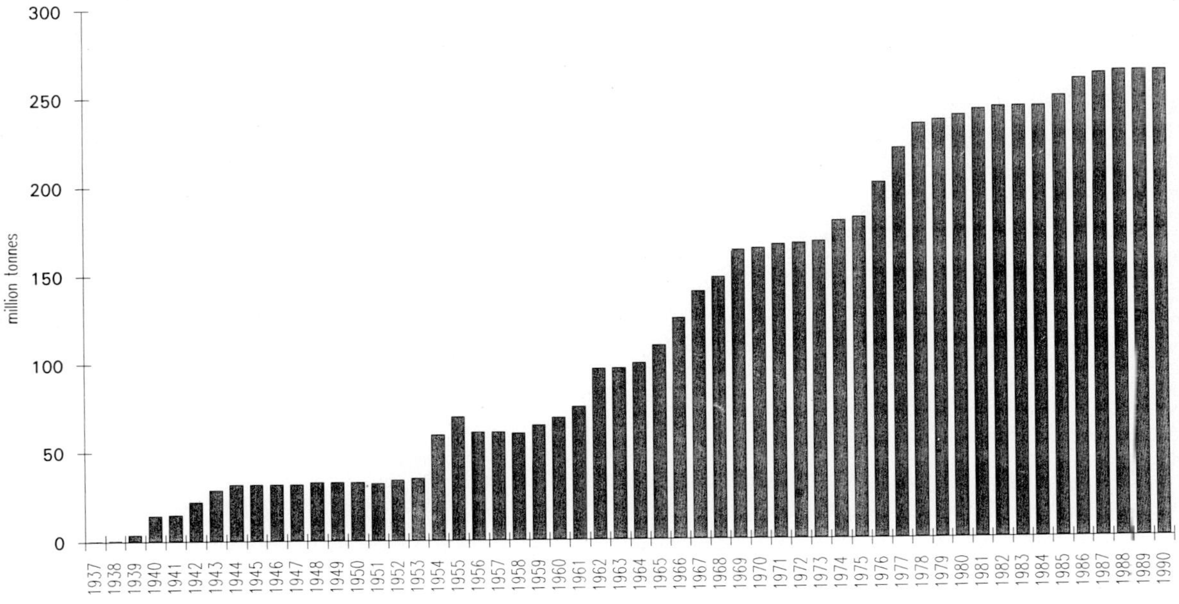

Fig. 6. Cumulative evolution of initial in-place oil reserves

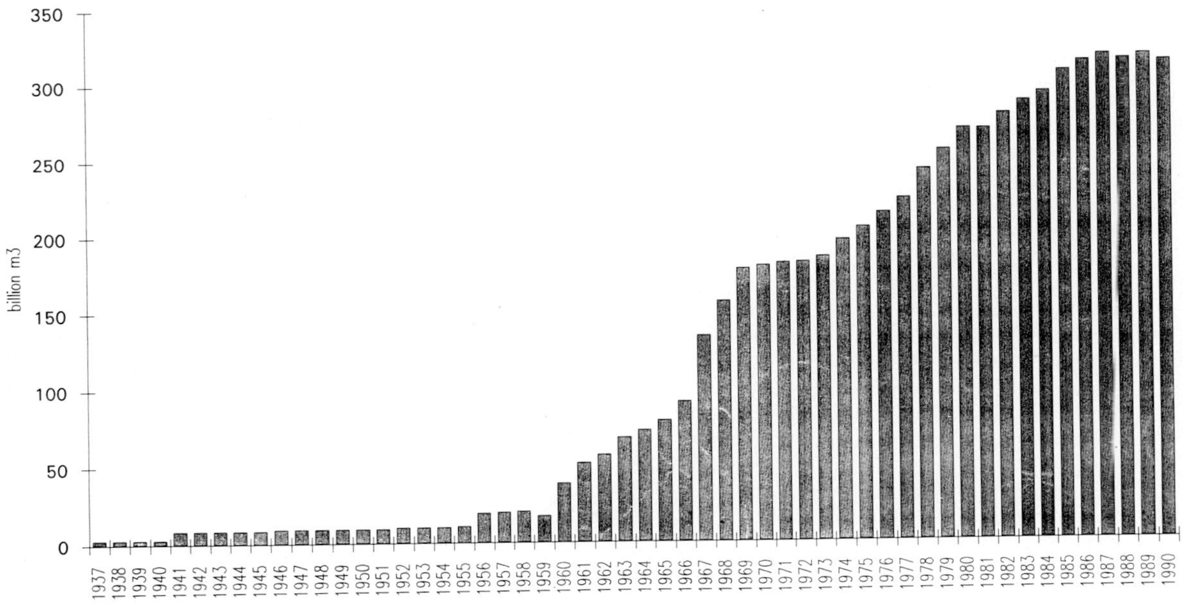

Fig. 7. Cumulative evolution of initial in-place gas reserves

decline will continue at least over the next few years.

A bidding round open to foreign companies was announced in 1991, but has not yet been formally launched. It is expected that foreign investment and renewed exploration efforts by the national oil company will increase reserves and stabilize production. Hungary has a dense pipeline network (Fig. 11), therefore only 10–20 km of new pipelines would be needed to connect with a new oil or gas field.

3 Petroleum Geology

On Hungarian territory, the Tertiary Pannonian basin (Fig. 12) covers some 75,000 km^2 and is the country's main petroliferous basin. It lies over a

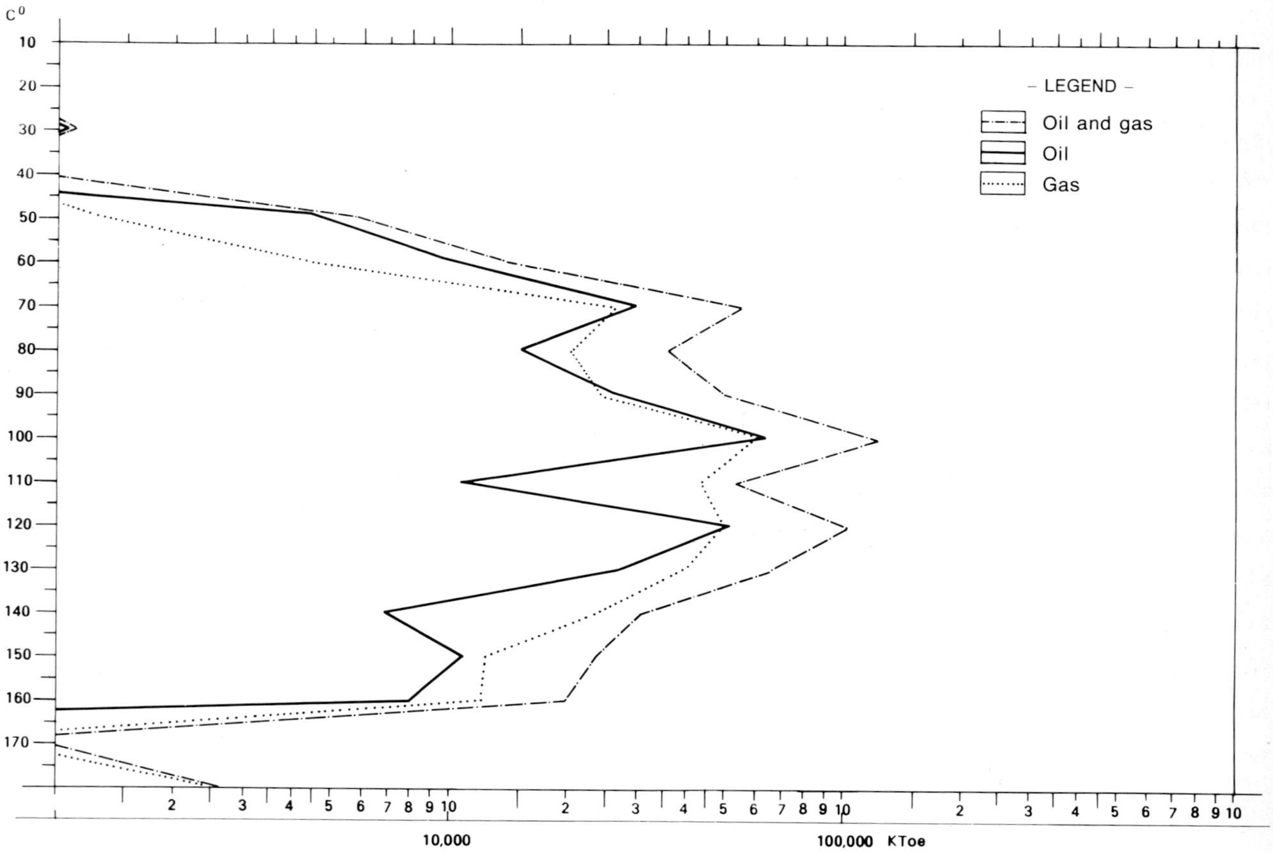

Fig. 8. Initial in-place reserve/temperature relation

"basement" which consists of two main Paleozoic-Mesozoic terranes: Pelso and Tisza. They also proved to be hydrocarbon-bearing.

3.1 Basin Structure

The Pannonian basin is a Neogene back-arc, transtensional basin associated with crustal extension which culminated in the Middle Miocene and locally has continued with reduced intensity until recent times. The Pannonian Basin lies over an Eocene-Middle Miocene foredeep compressional basin. These Tertiary basins are superimposed over older sedimentary and crystalline series which were considered for many years as the economic basement.

The pre-Tertiary "basement" consists of Paleozoic and Mesozoic limestones and shales overlaying a metamorphic rock basement of a Paleozoic-Precambrian age. They belong to two main units, or terrains, named Pelso and Tisza (Fig. 12) by Fülöp et al. (1987).

The Pelso and Tisza units separated in the Late Permian break-up of Tethys. Their paleogeography was controlled by various rifting phases and plate movements during the Mesozoic Alpine evolution of the Tethys. The Austrian and Laramic orogenic phases induced their present thrust sheet and nappe structure.

The Eocene-Middle Miocene foredeep basin installed after a long period of erosion of the Pelso and Tisza uplifted blocks, first on Pelso then on Tisza. The Paleogene of the Pelso block was folded and thrusted in the Oligocene. From Oligocene onwards the Pannonian area is part of the Paratethys, a distinct tectonic setting and faunal realm developed north and northeast of the Tethys.

The rapid subsidence of the Szolnok-Maramures flysch trough situated between Pelso and Tisza blocks was controlled by transform faults and back

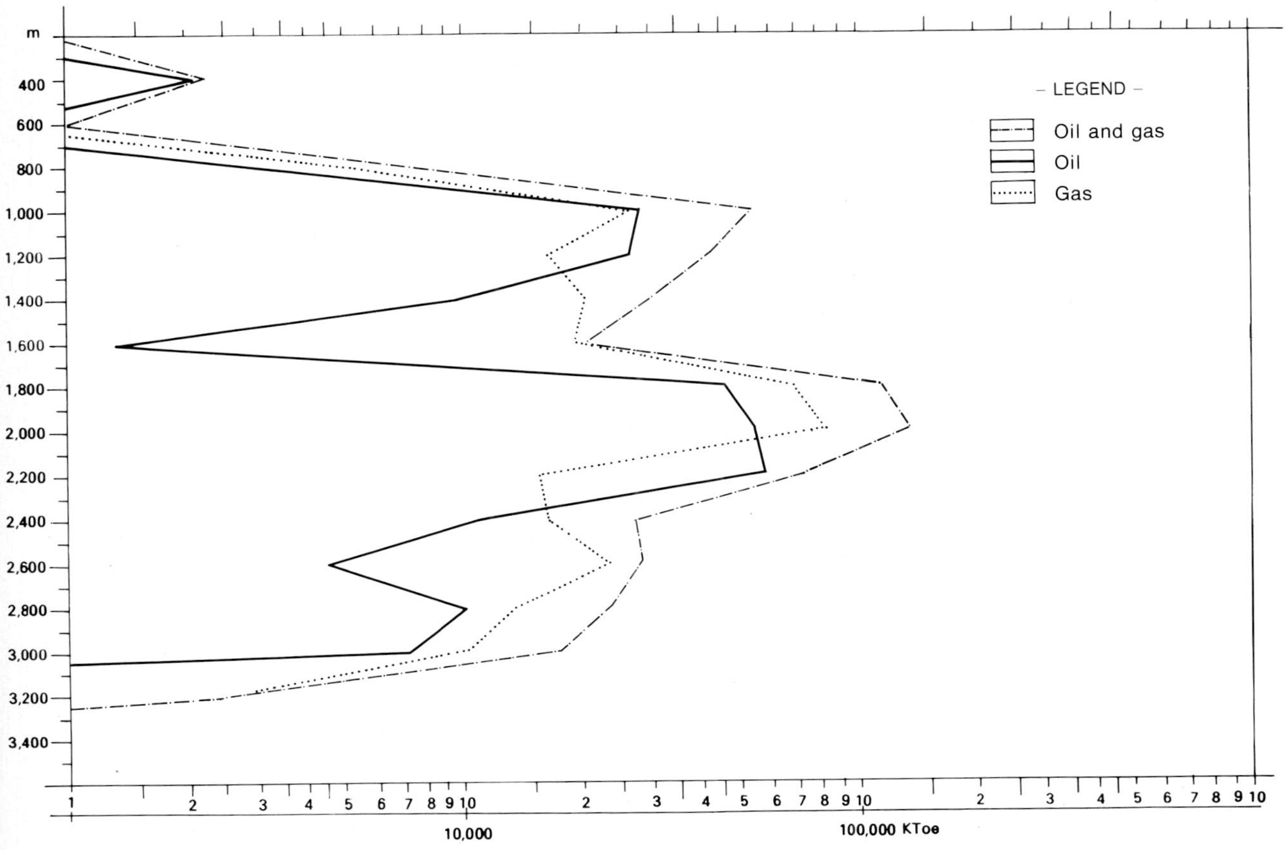

Fig. 9. Depth distribution of in-place reserves

thrusting of the western Carpathians. Wrenching and fault movements created local depocentres in the central part of the Hungary. In the Lower Miocene uplifting movements and subaerial exposure were widespread.

The Pannonian Basin (Late Miocene-Quaternary) is a result of rapid, thermally controlled post-rift subsidence. It was totally isolated from the Tethyan sea in Sarmatian (Late Seravallian), some 10.5 million years ago. On the Hungarian territory, some very deep (up to 8000 m), medium deep (2000–3500 m) and relatively shallow (1000–2000 m) sags were defined (Fig. 13) in recent years.

3.2 Reservoirs and Source Rocks

Reservoir rocks were reported in the whole sequence ranging from the Precambrian basement to Pliocene (Figs. 14, 15). They are weathered and faulted metamorphics, carbonates and clastics. Source rocks with proven potential are in the Mesozoic and Oligo-Pliocene.

The various hydrocarbon plays (Fig. 16) in Hungary belong to three depositional megasequences: Permian-Mesozoic, Paleogene-Middle Miocene and Upper Miocene-Quaternary.

3.2.1 Permian-Mesozoic Megasequence

The Lower Paleozoic shallow water sediments were folded during the Hercynian orogeny then subaerially exposed until the Permian. They were partly metamorphosed and represent the basement of

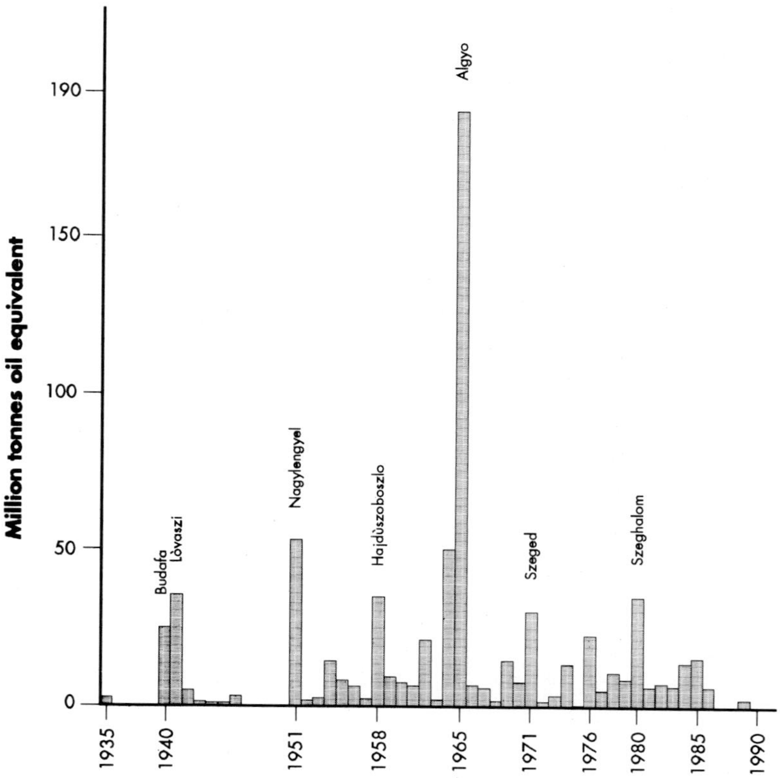

Fig. 10. Evolution of in-place oil and gas reserves discovered during the period 1935–1990

Table 5. Producing wells, 1990

	Well number	Oil produced (1000 tons)
Spontaneous	262	1271
Gas-lift	334	475
Pumped	472	218
Others	4	2
Total	1072	1966[a]
	Well number	Gas produced (billion m^3)
Free gas + gas cap	255	6
Associated Gas	965	1
Total	1220	7

[a]Oil produced using EOR methods are included in the above figures.

Hungary's sedimentary basins. Highly fractured and weathered, Pre-Cambrian and Lower-Paleozoic metamorphics are reservoirs of a number of producing fields: Sarkadkeresztur, Szeghalom, Dorozsma, Barcs and Kismarja.

During the Permian continental evaporites and volcanics of early Tethys rifting phase accumulated in the southern part of the Tisza unit. The Triassic to Early Cretaceous sedimentation was mainly carbonate with shales and pelagic carbonates deposited in deeper parts of troughs.

The stratigraphic evolution of the Pelso and Tisza units was markedly different (Fig. 14). During the Austrian movements the Mesozoic sediments were folded and uplifted. Erosion, karstification and bauxite formation characterized this period. In the Upper Cretaceous a shallow to deeper marine regime installed over both the Pelso and Tisza terranes (Debrecen Sandstone Fm.) and in the Szolnok-Maramures flysch trough. The Early Tertiary movements resulted in a new uplift and erosion of the Mesozoic, especially of the Upper Cretaceous.

Good quality *source rocks* are in the Upper Triassic of the Pelso unit, (Veszprem and Kössen formations) and in the Late Triassic-Late Cretaceous of the Tisza unit's Mecsek Coal and Gosau facies formations (Fig. 14).

Table 6. Main field oil production 1937–1990 (1000 tons)

Year/ field	Budafa	Lovászi	Nagylengyel	Algyö	Szeged	Kiskunhalas-NE
1937	1					
1938	37					
1939	142					
1940	247	3				
1941	283	138				
1942	298	340				
1943	263	497				
1944	223	493				
1945	201	405				
1946	205	393				
1947	164	341				
1948	149	286				
1949	174	284				
1950	208	260				
1951	189	266	3			
1952	187	271	90			
1953	158	253	378			
1954	139	235	788			
1955	125	210	1219			
1956	106	162	891			
1957	102	134	391			
1958	101	125	545			
1959	109	119	727			
1960	106	128	862			
1961	109	126	1053			
1962	114	118	1159			
1963	100	89	1258			
1964	82	79	1285			
1965	66	69	1303	21		
1966	60	55	1223	62		
1967	54	42	1081	198		
1968	50	35	847	544		
1969	45	38	698	585		
1970	47	36	524	896		
1971	46	30	410	919		
1972	36	30	348	973	3	
1973	36	26	289	1000	3	
1974	36	22	234	1000	1	1
1975	29	18	183	1046	4	0
1976	41	15	156	1103	95	1
1977	48	14	133	1053	171	4
1978	45	14	120	1041	234	27
1979	35	15	110	1023	268	46
1980	30	18	112	1031	291	92
1981	29	22	104	1042	248	164
1982	37	25	103	1031	234	191
1983	40	27	88	988	233	199
1984	45	30	87	968	221	181
1985	48	29	80	1002	216	176
1986	49	28	78	1074	205	129
1987	44	26	68	1011	205	109
1988	40	25	72	1074	198	89
1989	32	22	217	1016	161	69
1990	28	21	275	1010	105	55
Total	5418	6487	19,592	22,711	3096	1533

Table 7. Main field gas production 1937–1990 (million m^3)

Year/ field	Budafa	Lovászi	Puszta- foldvar	Hajdu- szoboszlo	Szank	Algyö	Ulles	Sarkad	Zsana
1937	2								
1938	15								
1939	46								
1940	74	2							
1941	98	39							
1942	107	105							
1943	113	151							
1944	108	200							
1945	114	248							
1946	128	282							
1947	105	268							
1948	94	225							
1949	117	265							
1950	151	311							
1951	149	376							
1952	153	397							
1953	165	341							
1954	182	292							
1955	178	279							
1956	175	234							
1957	167	207							
1958	149	200							
1959	124	179	9						
1960	120	197	27						
1961	97	200	52	3					
1962	89	198	101	6					
1963	92	201	148	278					
1964	80	166	187	438	2				
1965	94	123	190	708	5	1	2		
1966	77	77	141	1160	13	10	12		
1967	46	62	186	1315	32	48	2		
1968	33	57	272	1415	156	78	0		
1969	25	46	549	1560	443	102	0		
1970	23	39	609	1585	480	157	8		
1971	28	31	638	1456	765	345	37		
1972	22	21	658	1522	753	754	36		
1973	16	19	627	1538	585	1734	35		
1974			409	1454	563	2298	32		
1975			472	1092	543	2707	20		
1976			472	886	742	3539	1		
1977			477	1032	644	3874	0		
1978			480	1217	593	4197	0	73	
1979			220	971	330	3800	4	201	
1980			183	816	240	3520	33	407	52
1981			239	788	227	3187	31	603	248
1982			244	906	298	3164	58	544	478
1983			325	787	273	2880	123	575	414
1984			381	755	237	2769	493	421	532
1985			297	641	302	2915	809	353	487
1986			234	533	280	2489	1069	314	400
1987			224	513	209	2458	1260	208	352
1988			151	417	137	2429	927	160	271
1989			229	461	131	2376	832	119	257
1990			163	311	105	2075	655	72	134
Total	3556	6038	9594	26,564	9088	53,906	6479	4050	3625

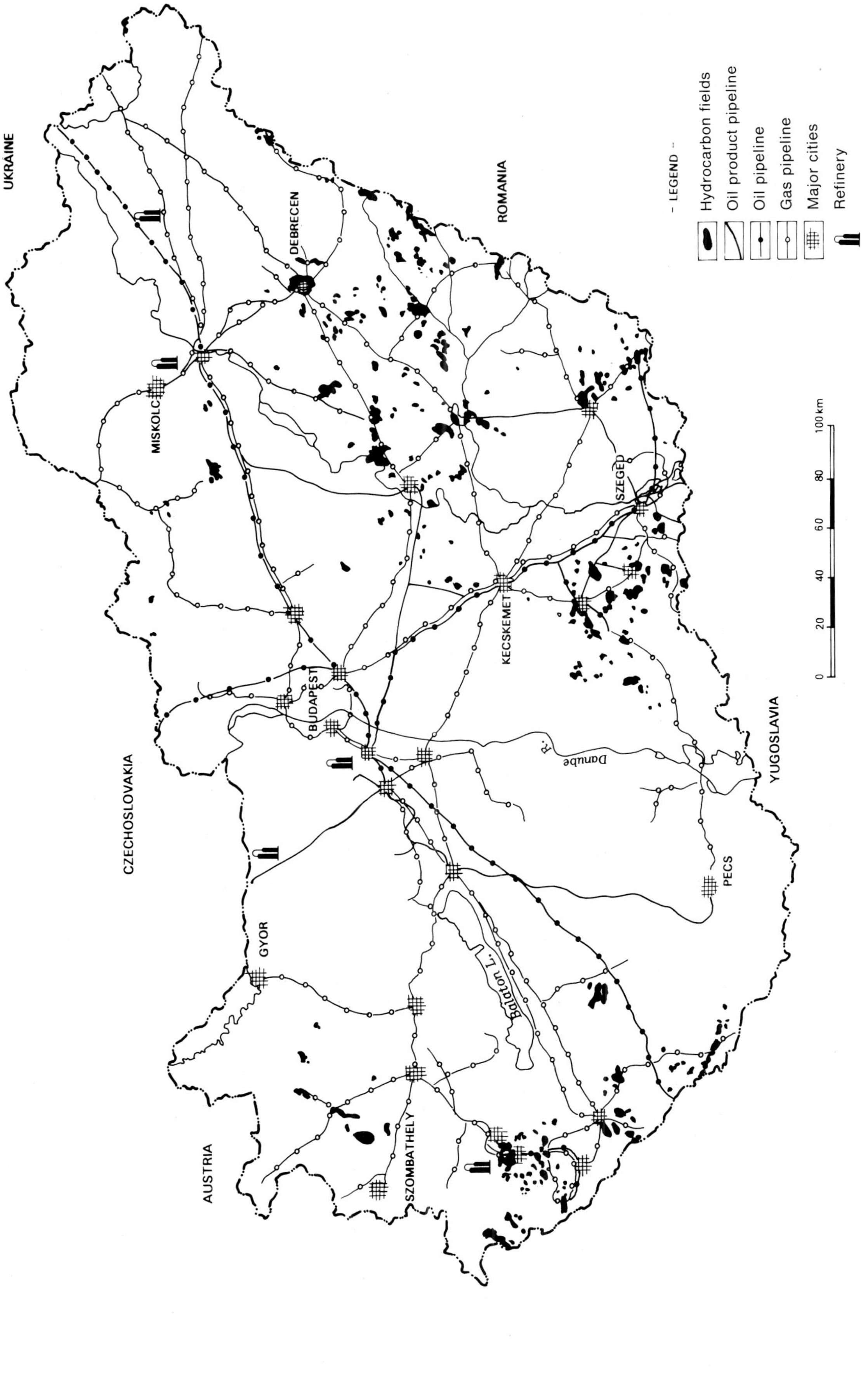

Fig. 11. Pipeline and refinery map

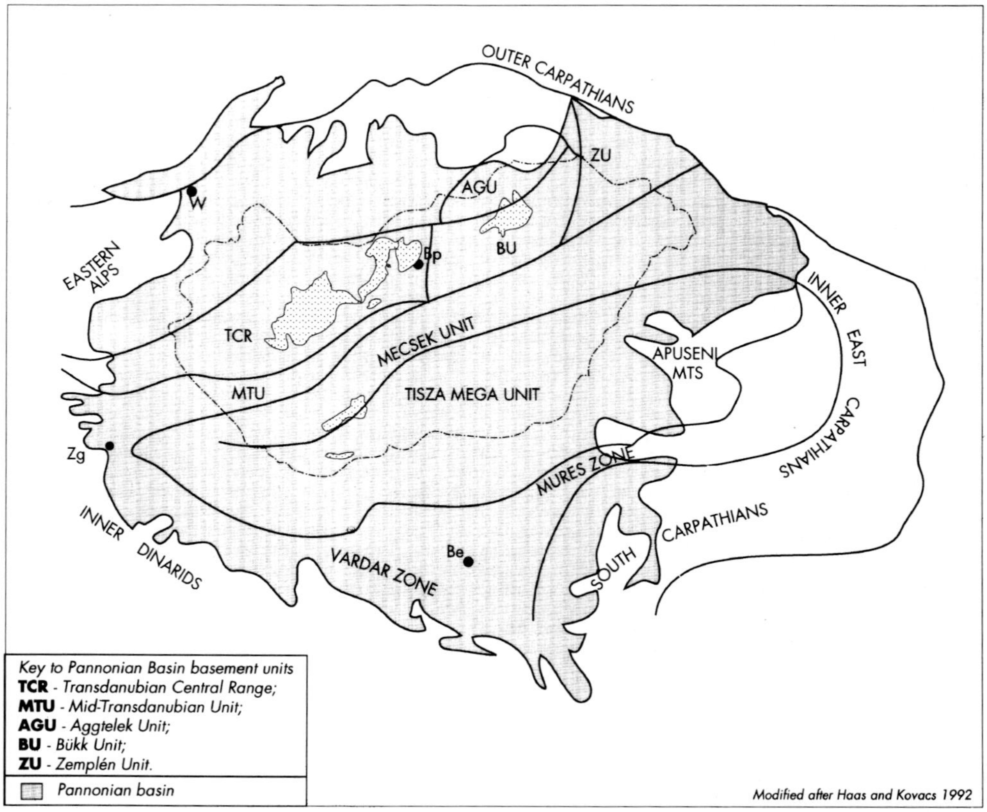

Fig. 12. Simplified location map of the Pannonian Basin and its pre-Neogene basement. (After Haas and Kovacs 1992)

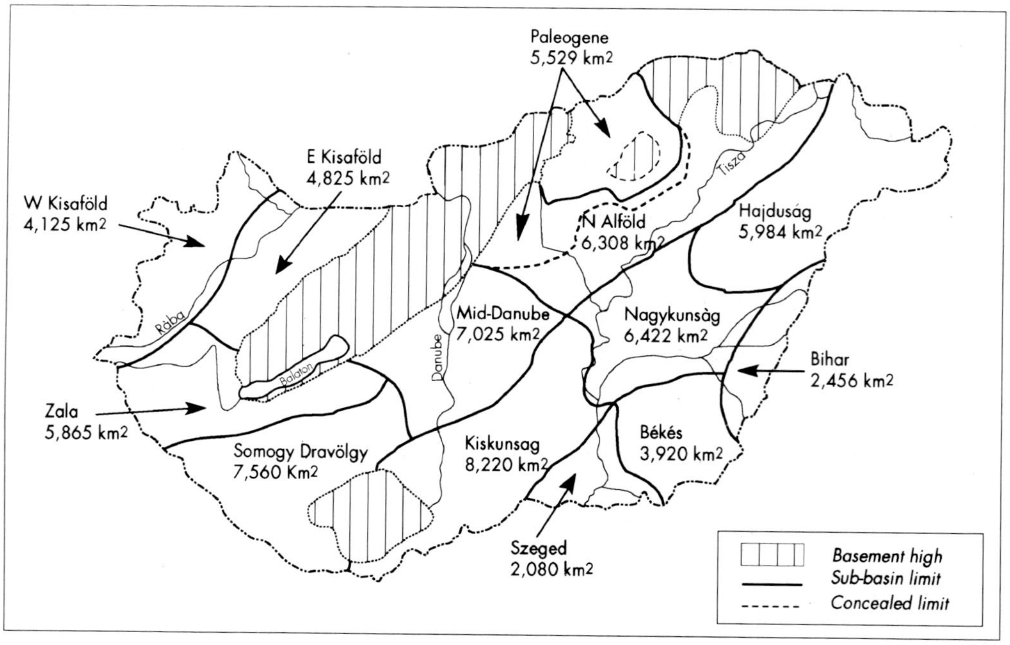

Fig. 13. Sub-basin map of the Pannonian Basin

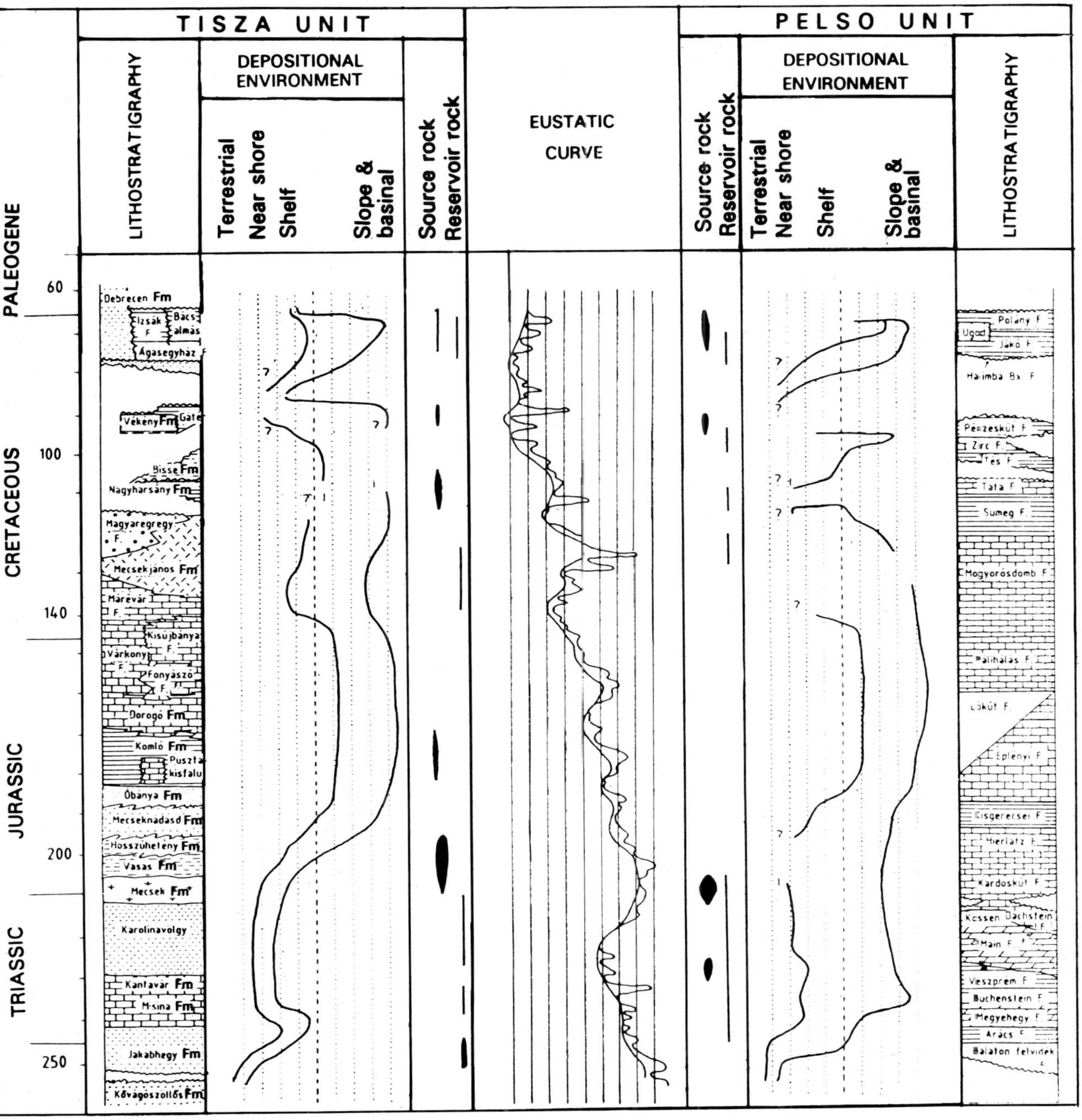

Fig. 14. Mesozoic stratigraphic chart. (Kokai and Pogacsas 1991a)

The Jurassic sediments have favourable hydrocarbon-generating qualities. The organic carbon content TOC is 0.31–0.53% by weight in clayey rocks. The Upper Liassic and Dogger shales have been found to contain autochthonous bitumen. In addition, some of the Liassic and Dogger carbonates can be considered moderate to good source rocks. Recent geochemical studies (Kokai and Pogacsas 1991b; Szalay and Koncz 1991) show that 7% of the Upper Cretaceous volume of sediments could be source rocks. Gosau facies formations contain the best source rocks. The kerogen is mainly type III while mixed type II-III kerogen is rare.

On the basis of vitrinite reflectance, the position of the oil-generation zone ($R_0 = 0.6$–1.3%) is in the 2200–2500 m depth interval in the Mesozoic sequence. The bottom of the wet-gas generating

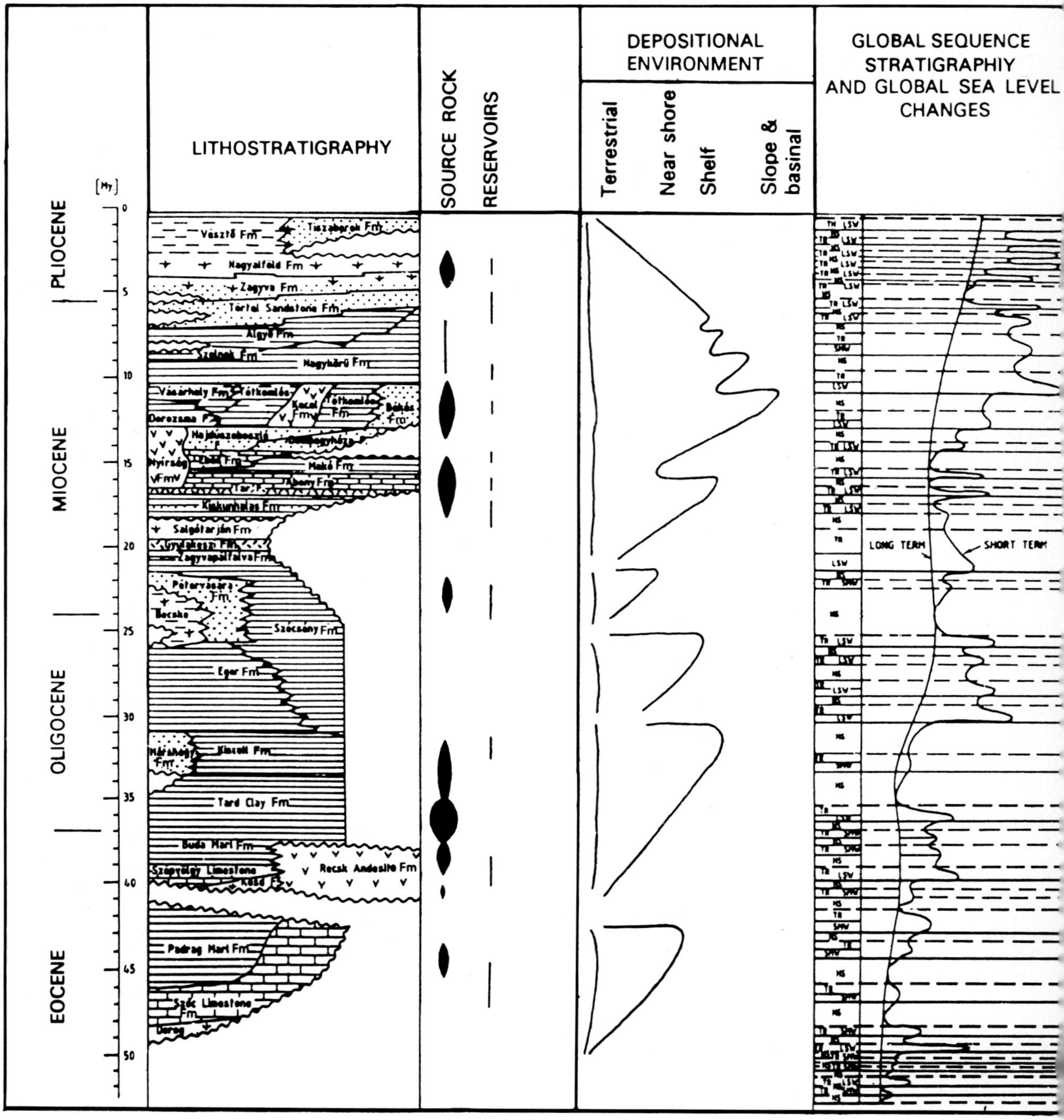

Fig. 15. Tertiary stratigraphic chart. (Kokai and Pogacsas 1991a)

zone ($R_0 = 2.1\%$) is around 4000 m and that for dry gas ($R_0 = 3.2\%$) is placed at 5000 m depth.

As a result of the high rate of subsidence in Neogene times, Mesozoic rocks rapidly passed through the wet-gas generation zone and generated dry gas. Analysis of the subsidence and thermal history indicate that the Liassic and Dogger source rocks started hydrocarbon generation in the Late Senonian and lasted until Late Miocene. Hydrocarbon generation from the Upper Cretaceous sediments also occurred during the Neogene time.

Significant *reservoirs* of the Permian-Mesozoic megasequence are the Triassic Dachstein Limestone and Main (Haupt) Dolomite as well as the Cretaceous Ugod Limestone (Fig. 14).

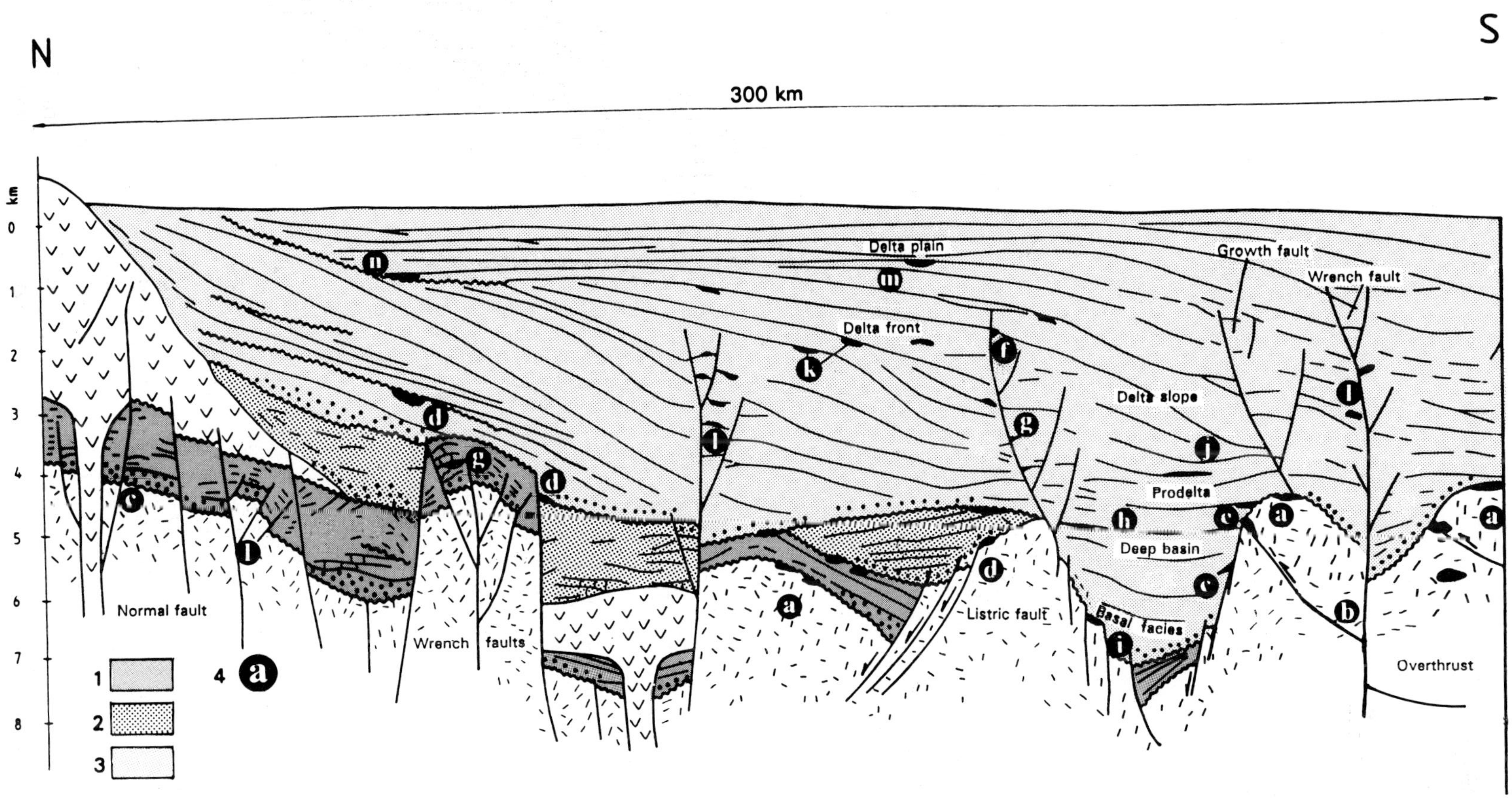

Fig. 16. Theoretical cross-section showing the principal plays of the Pannonian Basin and underlying units. Sedimentary cycles: *1* Paleogene; *2* Lower Miocene; *3* Pannonian-Quaternary; *4* Trap location and type; *a* buried hills; *b* thrust faults; *c* normal fault block; *d* unconformity; *e* drape-over; *f* growth-fault; *g* roll-over anticline; *h* turbidite pinch-out; *i* basal transgressive sand conglomerate; *j* pro-delta-delta slope fan, slump, canyon fill; *k* delta-front sand; *l* wrench fault; *m* delta-plain bar; *n* onlapping sandstone. (After Kokai and Pogacsas 1991b)

3.2.2 *Paleogene-Middle Miocene Megasequence*

This megasequence is made up of three cycles separated by regional unconformities: Eocene, Oligocene and Lower/Middle Miocene (Figs. 14, 15).

The Eocene-Oligocene sediments accumulated in half-grabens mainly on the Pelso unit while the Lower-Middle Miocene cycle covered the Tisza unit as well. They are continental to shallow marine clastics and carbonates bearing a limited number of hydrocarbon accumulations mainly related to unconformity traps (Fig. 16).

The main *source rocks* of this megasequence are Tard and Kiscell anoxic clay formations. In North Hungary they have an average TOC by weight of 0.5–1.0%; locally, however, concentrations can reach 0.8–1.8%. The bulk of the kerogen is type I-II but type III kerogen is present in the upper part of the Paleogene cycle. At greater depths it is in the oil window.

Reservoirs were encountered in the sands of the Rupelian Kiscell Formation and in the Eocene Szöc Limestone (Fig. 15).

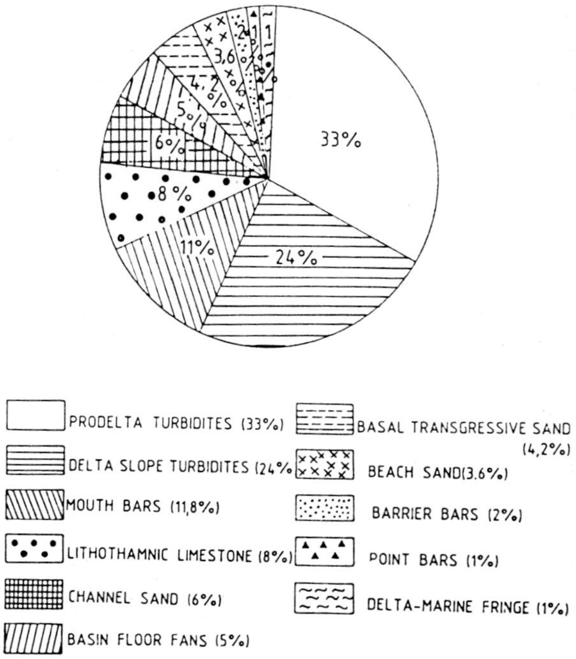

Fig. 17. Genetic types of the reservoirs (Neogene plays, West Hungary). (After Kozma 1990)

3.2.3 *Upper Miocene-Quaternary Megasequence*

Sediments of this megasequence accumulated in the Pannonian area's interior sags. Facies are marine, lacustrine and fluviatile. The typical sedimentary development of this megasequence consists, from bottom to top, of basal conglomerates passing into shallow marine sediments. This cycle is followed by deep water deposits, usually black shales with turbiditic intercalations.

These are overlain by a complex of Pannonian age prodelta, to delta slope and delta-plain deposits containing good source rocks and reservoirs (Figs. 15, 16). Coal-bearing and recent fluvial deposits end the megasequence.

On the basis of the organic carbon content, more than 60% of the Upper Miocene-Pliocene siltstones and mudstones in deltaic facies could be hydrocarbon-prone *source rocks*, but only 1% of their volume is richer than the average.

The average organic carbon content of the Neogene clayey sediments is 0.85% and the average bitumen content is about 0.07% (Clayton et al. 1990). Higher organic carbon contents can be found only in the calcareous marls. The bitumen content of the calcareous marls can reach 0.12%. Most of the producing fields in Hungary are located in these megasequence's plays (Fig. 16). Figure 17 shows the main facies of the Neogene hydrocarbon reservoirs.

TOC values of more than 2500 Neogene samples, show higher than 1.0% content in 20% of the samples and over 2% only in 9% of the samples. The discovered gas reserves in most cases are of good quality, but in 28% of the total reserves there are inverted components, primarily CO_2.

Reservoirs are in basal transgressive prodelta turbidites and delta plain plays (Figs. 16,17) of the Dorozsma, Bekes and Tortel sandstones and marlstones (Fig. 15).

Over recent years, seismic stratigraphy evaluations and basin modelling studies have been applied. The Miocene sediments are characterized by high amplitude, high continuity, parallel to slightly divergent seismic events (Fig. 18). The undergoing work has already resulted in the definition of many new leads and prospects.

3.3 Hydrocarbon Traps

The overwhelming majority of the country's hydrocarbon reserves were found in structural traps associated with Pliocene sandstone reservoirs of a delta plain and delta front facies. The Upper Miocene turbiditic sandstone reservoirs contain mainly gas, while the basal transgressive clastics have oil pools with gas caps. The weathered and fractured upper zone of the metamorphic basement also contains oil and gas.

The Precambrian, Late-Paleozoic and Middle-Triassic formations are thermally over-mature and the generated hydrocarbons were probably dispersed by the strong alpine tectonic events. Where the thermal conditions were favourable, hydrocarbons could have been generated from the Upper Triassic, Jurassic, Upper-Lower Cretaceous and Paleogene source rocks.

The largest part of the hydrocarbon reserves has been generated by the Miocene shales. In the sub-basin areas where the Lower Pannonian (Pliocene) is over 3500 m thick (Fig. 19), hydrocarbons were expelled not only from the Miocene but also from the Pliocene shales.

The oil generation zone in the Neogene sediments is between 2.1 and 3.5 km depth while in some cooler areas it is up to 4.5 km deep. The temperature limit for hydrocarbon generation was no higher than 200–255°C. Thousands of thermal data measured in wells were used to establish maps showing the isotherms for the whole area of the Hungarian territory (Fig. 20). They demonstrate the high heat flow typical to the Pannonian Basin.

The oil and gas fields of Hungary are commonly associated with major basement structures. The most common traps are structural but stratigraphic and combination traps were also recognized in a number of fields (Fig. 16).

Structural Traps:

The structural traps are associated with:

Wrench Faults. Wrench-faulted structures are the most widespread throughout the Pannonian Basin.

In recent years, several Pliocene-Quaternary strike-slip fault zones have been seismically identified in the Pannonian Basin. These played an important role in the migration and accumulation of hydrocarbons in uplifted en echelon folds, faulted en echelon folds and compressive uplifted blocks.

Gas fields associated with typical flower structures have been discovered in the region of Kengyel-North, Földes, Sáránd and Kokad. They are associated with good reservoirs of delta slope, prodelta, delta front and delta plain facies.

Anticlinal Folds. This structural style is represented by a large anticlinal belt in the westernmost part of the Pannonian basin tending from west to east. Typical anticlinal trapped fields are Budafa and Lovaszi. It is assumed that anticlinals are compression-related inversion features which post-dated the deposition of the Pliocene deltaic sequence. The tectonic process may still be active today, although with lesser intensity.

In certain sub-basins, e.g. Zala and Kiskunsag, the structural inversion could have been caused by transpressional movement on earlier normal faults and resulted in prolific oil and gas anticline traps.

Normal Faulted Blocks. Extensional block faulting in the Miocene syn-rift period created conditions for significant accumulations of hydrocarbons. While the post-rift (Upper Miocene to Quaternary) sediments are only slightly deformed, the syn-rift Miocene and the pre-rift Mesozoic sediments are strongly faulted and rotated. A few fields have been discovered in this type of trap in the Mid Danube-Tisza River area and in the eastern part of the Pannonian Basin.

Thrust Faults. Compressive thrust blocks have only recently been identified. They are uncommon features and their development was probably also accompanied by significant strike-slip movements. Basement thrusts of Cretaceous age are present in the whole Pannonian Basin but this style of deformation represents a little-tested play, hence no discoveries have been made yet in this type of structure.

Stratigraphic Traps:

These can be found in the fractured-fissured metamorphic basement and in sandstone bodies associated with the overlying Mesozoic and Cenozoic sedimentary section. These simple, large closures are the focal points for oil migration from the surrounding deep sub-basins. They have provided the trap for Algyö, the largest Hungarian hydrocarbon field. The uplift of basement block at Algyö was probably in the Middle Miocene and was bordered by NW-SW hinge lines. Post-rift Neogene sediments were deposited unconformably on the strongly eroded metamorphic basement.

In addition to the pools discovered in or above the basement highs, some pools have also been discovered in much deeper positions along the main basement warp. These reservoirs were basal transgressive sands and turbiditic sandstones. Using seismic inversion techniques, the thickness and

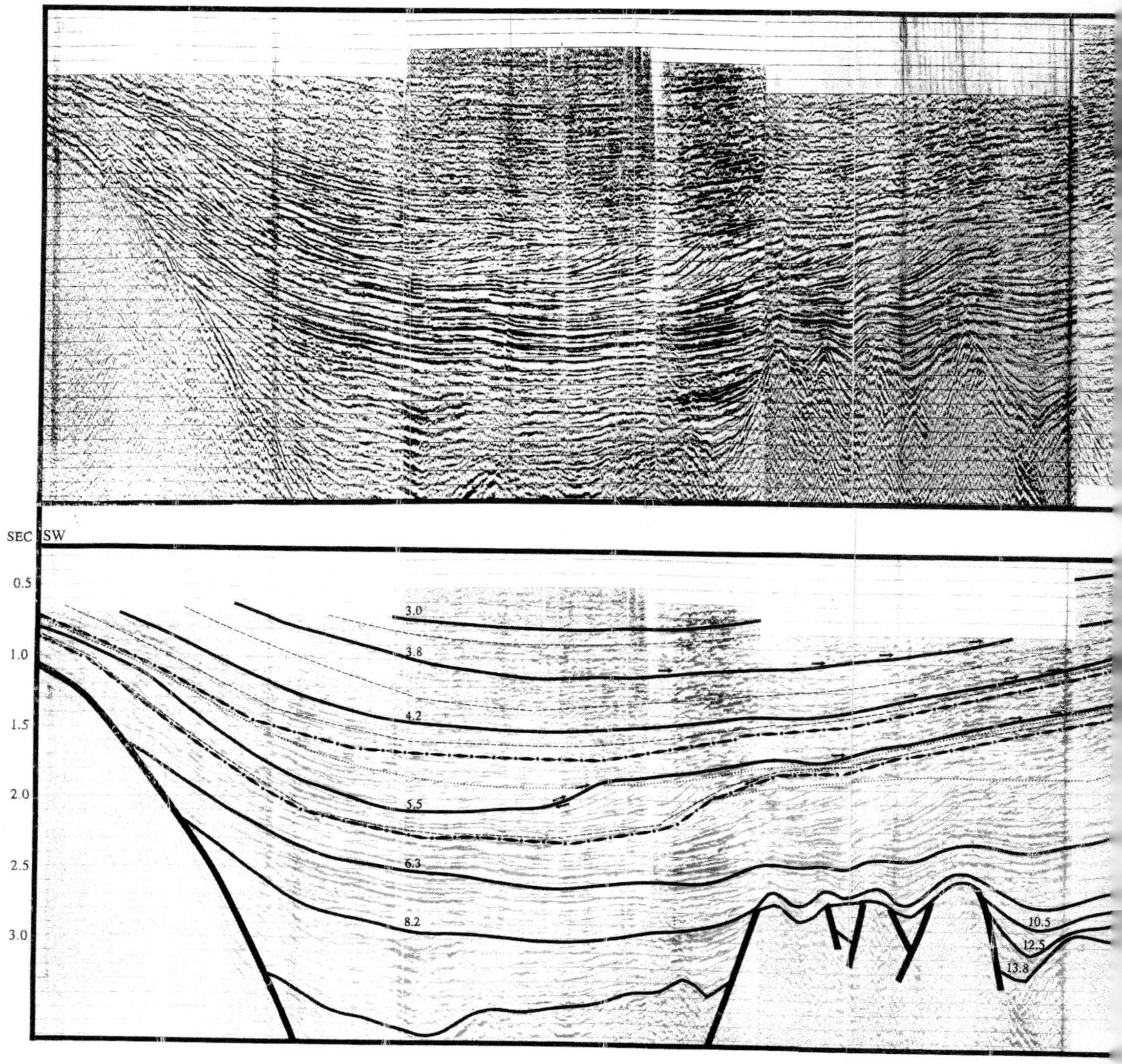

Fig. 18. Seismic cross-section through Hajdusag, Bihar and Bekes sub-basins. (After Vakarcs et al. 1992)

lateral extent of these reservoirs could be identified at depths down to 4 km.

4 Future Potential

A full assessment of the hydrocarbon potential of Hungary still requires refined geochemical and basin modelling studies as well as new seismic data acquisition or reprocessing (Szeghalom field was discovered this way) of existing data. Recent development in western technology and new exploration concepts should be integrated into current Hungarian oil industry organisation. The main future exploration targets are:

- Neogene stratigraphic traps using state-of-the-art seismic investigation.
- Seismic stratigraphically determined subtle traps in the thick Neogene and Mesozoic formations.
- Biogenic gas accumulations in areas where the Neogene thickness is less than 1000–1200 m.

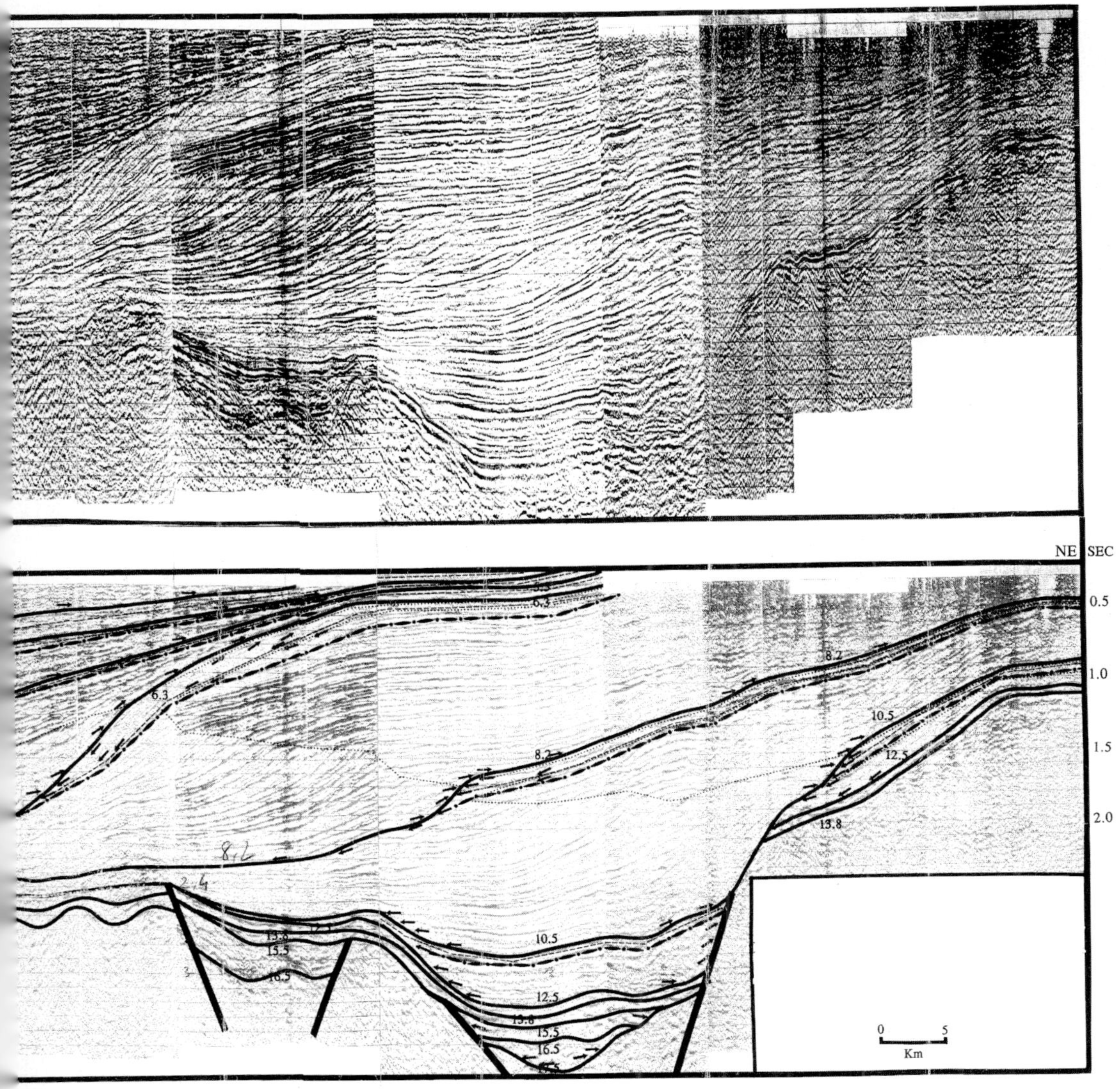

– LEGEND –

——	Sequence boundary (3rd order)
– –	Top of the transgressive systems tract
–•–	Top of lowstand systems tract
.....	Offlap break point
5.5	Age in million years
←	Reflectiion terminations

- Various structural traps: wrench fault, roll-over anticlines, and growth faults could be among the most prospective structural traps not fully explored as yet.
- Basement thrusts due to Alpine deformation may also provide significant prospects. The pre-deformation, immature Mesozoic strata may have become economically important source rocks during the rapid Miocene subsidence and burial.
- Fractured Mesozoic limestone and dolomite, especially in the areas of the basement highs covered by thick Neogene sediments.
- Fractured and fissured Paleozoic and Precambrian rocks which may have been sourced by Miocene shales of the deep wrench sub-basins.
- A better understanding of basin-forming tectonics and the paleogeography of the little explored Paleogene basins could result in new discoveries.

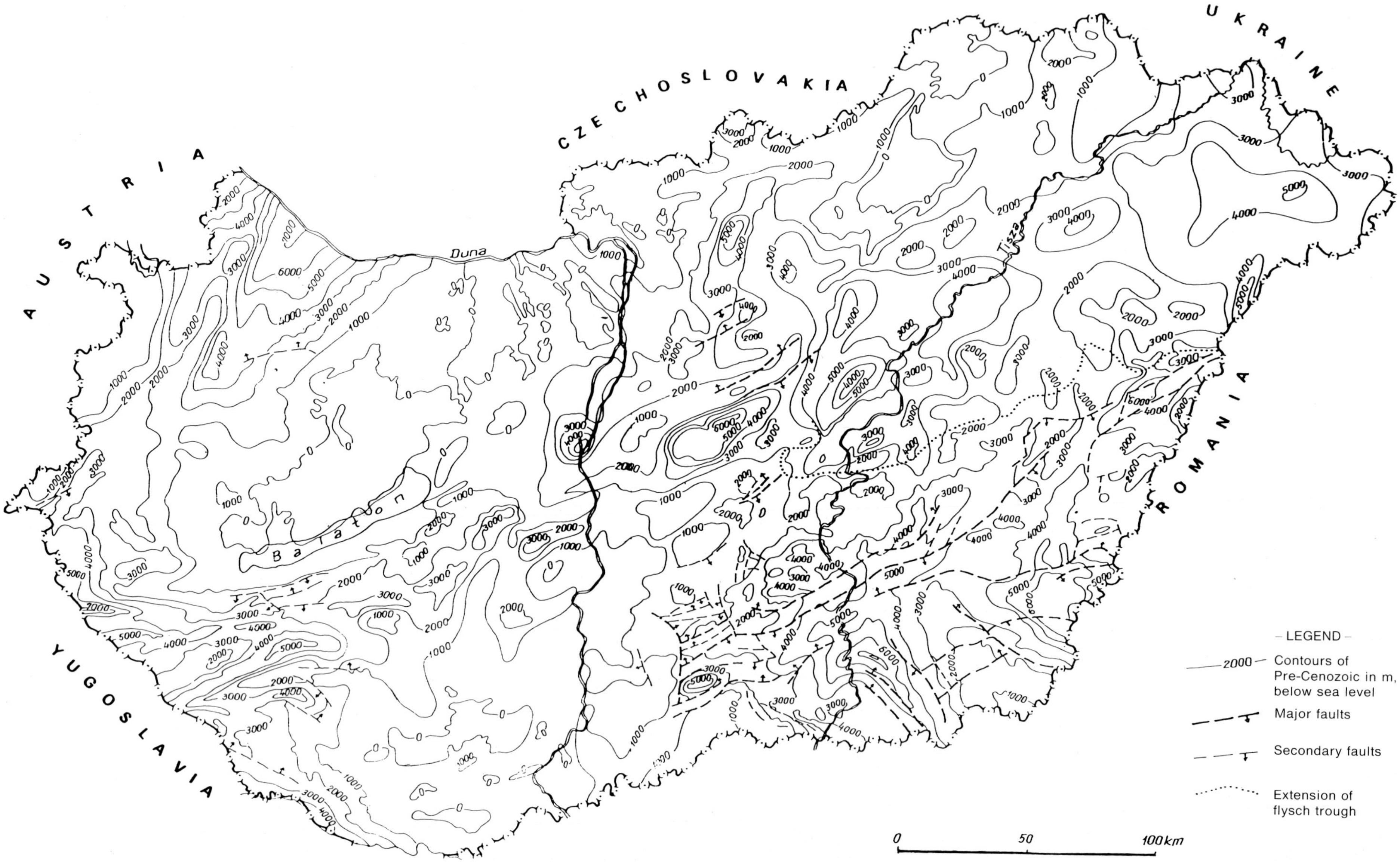

Fig. 19. Map of Cenozoic formations thickness

DEPTH CONTOUR MAP OF 100°C ISOTHERM SURFACE IN HUNGARY

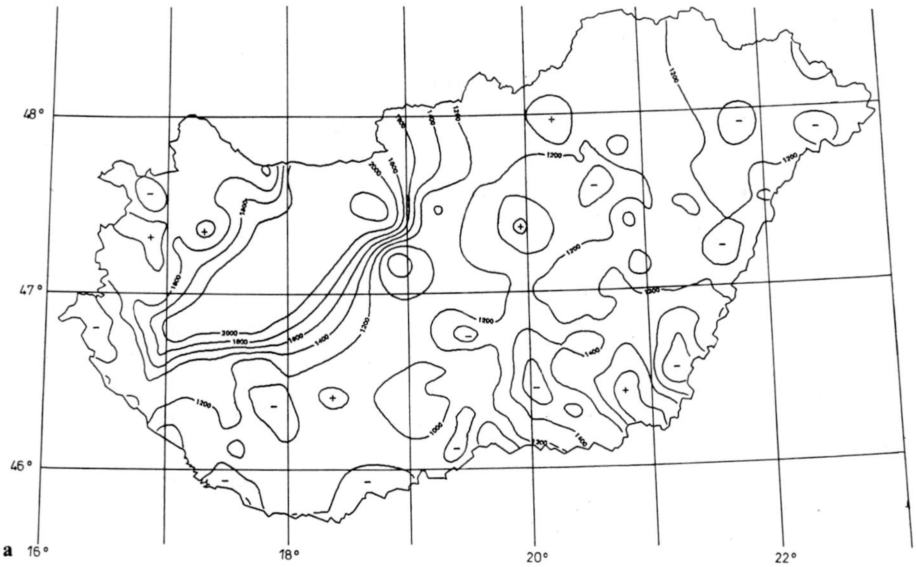

a

DEPTH CONTOUR MAP OF 120°C ISOTHERM SURFACE IN HUNGARY

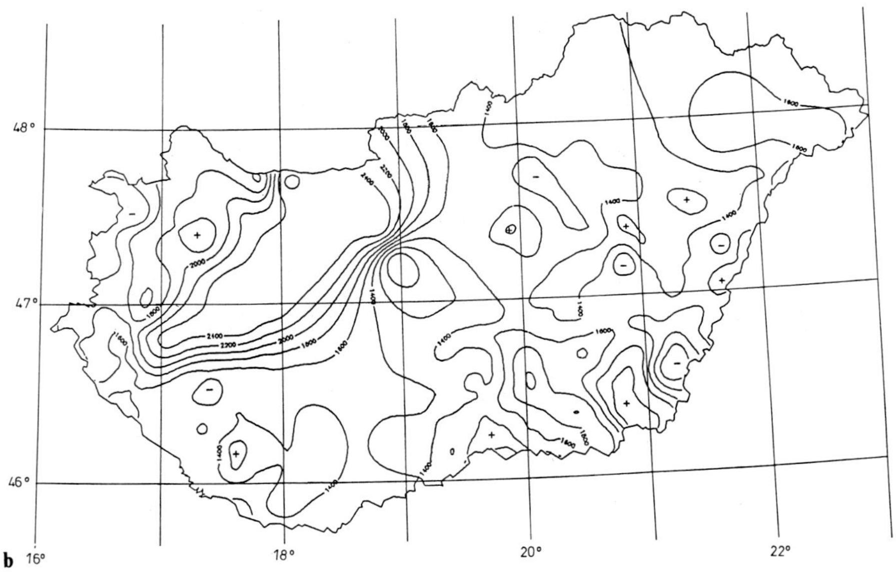

b

Fig. 20a–e (p. 171–173). Contour maps of 100 to 180 °C isotherms

DEPTH CONTOUR MAP OF 140 °C ISOTHERM SURFACE IN HUNGARY

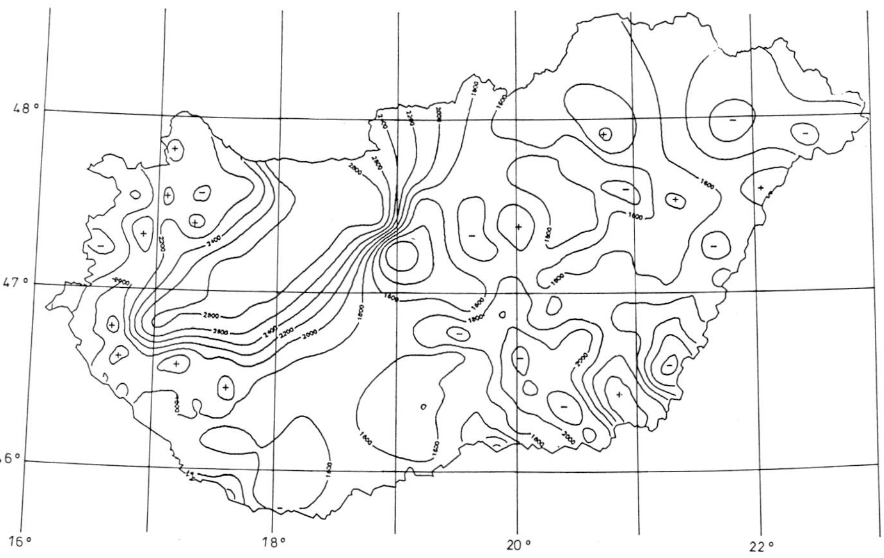

Fig. 20c

DEPTH CONTOUR MAP OF 160 °C ISOTHERM SURFACE IN HUNGARY

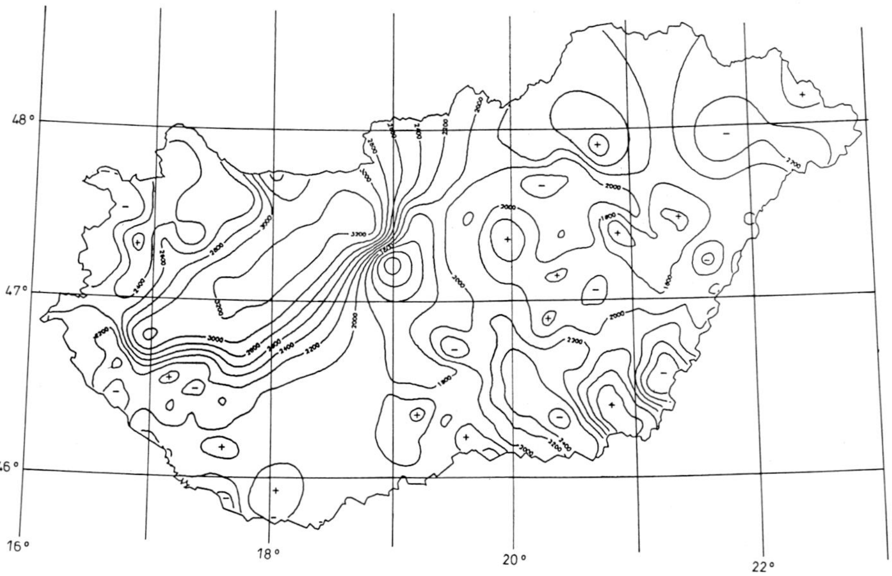

Fig. 20d

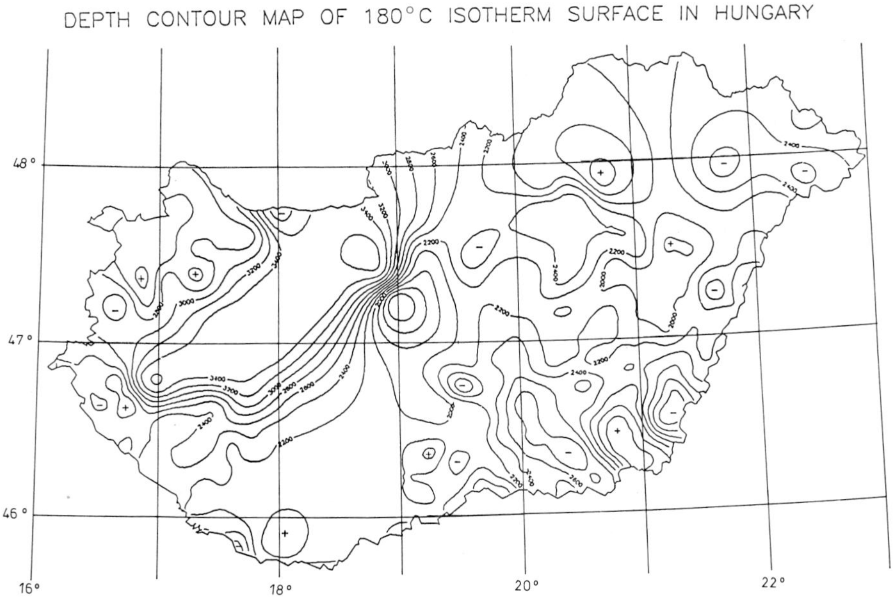

Fig. 20e

Acknowledgements. This work was fully supported by the management of the *Hungarian Oil Company*, who provided data from unpublished reports and gave their permission to publish this paper. Special thanks are due to my colleagues Gy. Pogacsas for his useful comments on the manuscript and A. Lelkes for skilful review and the assembling of various statistical data. I am grateful to B. Popescu for technical support, constructive criticism and improvement of the initial typescript.

References

Clayton J, Spencer CW, Koncz I, Szalay A (1990) Origin and migration of hydrocarbon gases and carbon dioxide, Bekes Basin, SE Hungary. Org Geochem 15: 233–247

Fülöp J, Baldi T, Baldi K, Balogh K, Barabas A, Csaszar G, Dudich E, Geczi B, Haas J, Hamor G, Jambor A, Jantsky B, Ronai A, Szederkenyi T (1983) Lithostratigraphy formations of Hungary. Spec Publ Geol Inst Budapest

Fülöp J, Brezsnyansky K, Haas J (1987) The new map of the basement of Hungary. Acta Geol Hung 30 (1–2): 3–20

Haas J, Kovacs S (1992) Tectonic versus eustatic control of the Tethyan continental margin evolution in Hungary. Sequence stratigraphy of Europe Basins. Abstract Volume, CNRS/IFP, Dijon, pp 246–247

Kokai J, Pogacsas Gy (1991a) Hydrocarbon plays in Mesozoic nappes, Tertiary wrench basins and interior sergs in the Pannonian Basin. First Break 9/7: 315–334

Kokai J, Pogacsas Gy (1991b) Tectono-Stratigraphical evolution and hydrocarbon habitat of the Pannonian basin. In: Spencer AM (ed) Generation, accumulation and production of Europe's hydrocarbons. EAPG, Oxford University Press, London, pp 307–316

Kozma T (1990) Seismic and electrofacies analyses of the Pannonian hydrocarbon reservoirs of Transdanubia. MSc Thesis, Budapest

Szalai A, Koncz I (1991) genetic relations of hydrocarbons in the Hungarian part of the Pannonian basin. In: Spencer AM (ed) Generation, accumulation and production of Europe's hydrocarbons. EAPG, Oxford Univesity Press, London, pp 317–322

Vakarcs G, Molnar K, Pogacsas Gy, Rumpler J, Lakatos L, Szabo A, Tari G, Vail P, Varkonyi L, Varnai P (1992) Third order Miocene-Pliocene depositional sequences in eastern Hungary, Pannonian Basin. Sequence stratigraphy of European basins. Abstract Volume, CRNS/IFP, Dijon, pp 260–261

Suggested Reading

Dank V, Kokai J (1969) Oil and gas exploration in Hungary. In: The exploration for petroleum in Europe and N. Africa. Proc Conf Brighton, The Institute of Petroleum London

Horvath F, Royden LH (1988) The Pannonian Basin. AAPG Memoir, Tulsa, 45 pp

Pogacsas Gy (1985) Seismic stratigraphic features Neogene sediments in the Pannonian Basin. Geoph Trans Eötvös Lorand Geoph Inst 30: 373–410

Teleki P, Kokai J Basin analysis for oil and gas exploration: a case history from Hungary. Graham and Trotman London (in press)

6 Oil and Gas Exploration in Poland

Jerzy Zagorski[1]

CONTENTS

1 Introduction

The aim of this chapter is to present the geologic framework of the oil and gas prospective zones and the hydrocarbon exploration history of Poland. In almost 100 years of exploration it has been proved that hydrocarbon prospective area covers roughly 82% of the whole territory, i.e. 257,000 km^2. This area falls into two major oil-and gas-bearing provinces (Fig. 1):

The Carpathian Province comprises folded Flysch (with an area of 19,000 km^2) and Foredeep (with an area of 17,000 km^2) tectonic zones. The Carpathian Foredeep has an external part situated north of the frontal overthrust fold of the Carpathian Flysch and an internal part, under the overthrust of the Carpathian Flysch.

Polish Lowlands cover the eastern part of the NE German-Polish basin and the western part of the E European Platform. It comprises platform areas of Poland and the Polish sector of the Baltic Sea with an area of 221,000 km^2.

[1]Geonafta, ul. Krucza 36, 00-921 Warsaw, Poland

2 Carpathian Province

This is the oldest oil-producing region in Poland. Exploration and production began around 1854 when the first oil field Bobrka was found. Up to now, 65 oil fields and 15 gas fields, most of them small in size, have been discovered in the Polish Carpathian province.

2.1 Polish Carpathian Flysch

The Flysch zone is divided into five main overthrust tectonic units. They are from the south to the north:

I. Magura Nappe
II. Dukla Folds
III. Silesian Nappe
IV. Sub-Silesian Nappe
V. Skole Nappe

Most oil and gas fields have been discovered in the so-called Central Carpathian Depression, the depressed part of the Silesian nappe between Sanok-Krosno-Jaslo.

The Carpathian Flysch deposits, of Cretaceous-Oligocene age, are up to 5000 m thick. The oil and gas reservoirs are Cretaceous, Paleocene, Eocene and Lower Oligocene sandstones. They have highly variable porosity, permeability and thickness. As a rule, reservoir properties are very poor.

The most important hydrocarbon bearing formations are the Upper Cretaceous Istebna Sandstone, the Paleocene-Lower Eocene Ciezkowice Sandstone and the Oligocene Kliwa Sandstone. More than 70% of recovered hydrocarbons come from these formations. The oil and gas accumulations in the Carpathian Flysch are in general structurally or sometimes structurally and lithologically trapped.

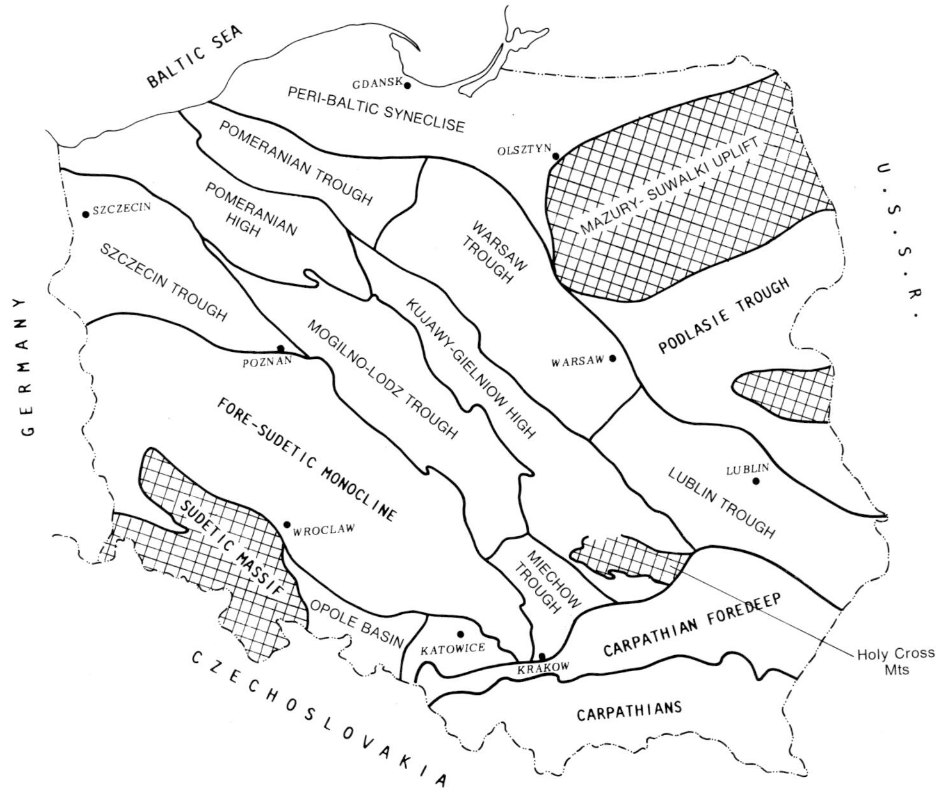

Fig. 1. Map of main tectonic units of Poland

The local oil- and gas-bearing structures have variable geometry: normal anticlines, asymmetrical folds, displaced central core folds, multiple folds, folds covered by overthrusts and various types of minor traps on folded limbs (Fig. 2).

The Carpathian Flysch oil and gas fields are classified as bedded deposits with edge waters. The initial reservoir pressures are close to hydrostatic pressure and it is only at greater depths (more than 5000 m) that the pressures are significantly greater than the hydrostatic pressure. The crude oil is methane-rich and sulphur-poor, its density varies from 0.820 to 0.850 g cm^{-3} (35 to 41°API). The main component of the gas is methane, and the content of heavy hydrocarbons does not exceed a few percent. There is no evidence of the presence of hydrogen sulphide gas. The source rocks are the organic matter-rich clays and argillaceous sediments which commonly occur in the whole sedimentary sequence of the Carpathian Flysch.

The tectonic structure, between depths of 200 and 2000 m, is fairly well understood, compared with a less well defined knowledge of the structural complexity, particularly below 3000 m. New and significant geological data have been supplied by deep wells such as Paszowa 1, Kuzmina 1, Lachowice 1 and 2, Zawoja 1, Czudec 1, Raclawowka 1 and Slopnice 1. Most of these have penetrated the flysch "basement", i.e. the NE German-Polish Platform deposits.

One of the most interesting discoveries has been well Nosowka 1 near Rzeszow (Fig. 3), where an oil flow with a specific gravity of 0.854 g cm^{-3} (34°API), and a stabilized production flow of about 30 tonnes/day was obtained from Visean limestones at the depth interval 3540–3465 m. The Lower Carboniferous deposits are covered by the Carpathian overthrust and a Miocene transgressive series about 200 m thick. Well Nosowka 1 bottomed at 3807 m in the Precambrian. At present, this field is being delineated by outpost wells, two of which have been successful.

Over the past few years, exploration activity was aimed at defining traps occurring at depths of 3000–5000 m and deeper. In the Polish Carpathians, deep exploration in conjunction with geological mapping focussed on confirming the presence of Boryslaw-Dolina type folds known from the eastern (Ukrainian) part of the Carpathians. Two deep wells, Paszowa 1 and Kuzmina 1, recently completed within this program, reached 7210 m and 7541 m respectively. Well Kuzmina 1 obtained the first significant gas shows from the Lower Cretaceous Spaskie Sandstone. However, these wells have yet to confirm the existence of deep-seated folds.

In the Carpathian Flysch province, there are a number of important horizons from the petroleum geology point of view:

Lgota Beds (Albian). Consisting mainly of glauconitic sandstones with 4–24.5% porosity and 0–974 md permeability in the Weglowka field.

Istebna and Inoceramus Beds (Lower Albian). Dominantly sandstones (up to 70%), with porosity rarely exceeding 15%. The permeability varies from 0 to 130 md.

Ciezkowice Sandstone (Eocene). With porosity 9.9–23.5% and permeability 60–500 md, this formation occurs in the main Carpathian fields: Burzyn, Szalowa, Kobylanka-Kryg, Biecz, Osobnica, Bobrka-Rogi, Sobniow-Jaszczew-Roztoki.

Krosno Beds (Oligocene). The lower part of these consists mainly of thick fine-grained sandstones with a thickness of 800 m in the north, and up to 300 m in the south. They occur in numerous fields such as Slopnice, Harklowa, Mokre, Tarnawa-Wielopole, Rajskie-Czarne, Zatwarnica and Dwernik. In the Gorlice area porosity is 10–22% and permeability 1.5 to 69.8 md. The middle part, which is 700–800 m thick, comprises a sandstone-shale series with erratic blocks of granites, porphyres and marbles. In the upper part, grey shales with sand and thin sandstone intercalations are predominant. In the Jaslo-Rymanow area the thickness of these beds is 700–800 m. The Krosno beds have an erosional top.

2.2 Carpathian Foredeep

The Carpathian Foredeep has been intensively explored, mostly after World War II. Numerous discoveries, mainly gas, have been made in the Miocene and the underlying Cretaceous, Jurassic, Triassic and Late Paleozoic (Cambrian, Ordovician, Devonian) "basement" formations.

The geological structure of the Pre-Miocene basement from Krakow to the eastern border is comparatively well known. A most interesting feature is the presence of paleo-valleys with a strongly marked NW-SE trend towards the Carpathians. Until recently, the course of these paleo-rivers was better recognized in the Miocene area located north of the Carpathian marginal overthrust, but presently their prolongation under the Carpathians has been outlined (Fig. 4). The clastic filling of the paleo-valleys is considered one of the most prospective plays of the Carpathian Foredeep "basement". The Dabrowka, Borek, Szczepanow and Kurylowka gas fields have been discovered in such features.

Generally, in the Carpathian Foredeep and its basement, three types of fields can be found:

1. Multilayered gas sands in Miocene molasse.
2. Fields in Jurassic carbonates underlying the Miocene.
3. Fields in paleo-relief onlapped by Miocene sandstones.

The largest fields existing in this area are: Przemysl, Mackowice, Jaroslaw, Lubaczow, Lancut and Krasne-Husow. It should be noted that the term "large" is relative to Polish conditions, and that according to international standards these are small fields. The biggest of the above-mentioned fields has 80 billion m^3 of gas reserves. The gas is predominantly methane type (up to 98% methane) with an admixture of higher hydrocarbon gases.

The maximum Miocene thickness has been reported from the eastern part of the basin in the Wielkie Oczy area, where it attains 3500 m. Generally, the thickness increases from west to east. The main portion of the Miocene sequence is represented by the Badenian and Sarmatian formations. Reservoir rocks occur throughout the autochtonous Miocene sequence from about 200 m to over 3000 m. Reservoir properties, again, vary greatly: porosity values are from 3 to 23%, most frequently 5–20%. Permeability is the most variable parameter, ranging from 0 to 4400 md, but is most frequently in the range of a few tens of md. The sand content is also very variable and ranges from 5 to 95% (Wielkie Oczy and Miekisz Nowy troughs), but 80% of the commercial hydrocarbon accumulations are in horizons containing 5–20% sand content. Porosity and permeability deterio rate with depth.

The optimum porosity and permeability values of 15–32% and several hundred md are in reservoir

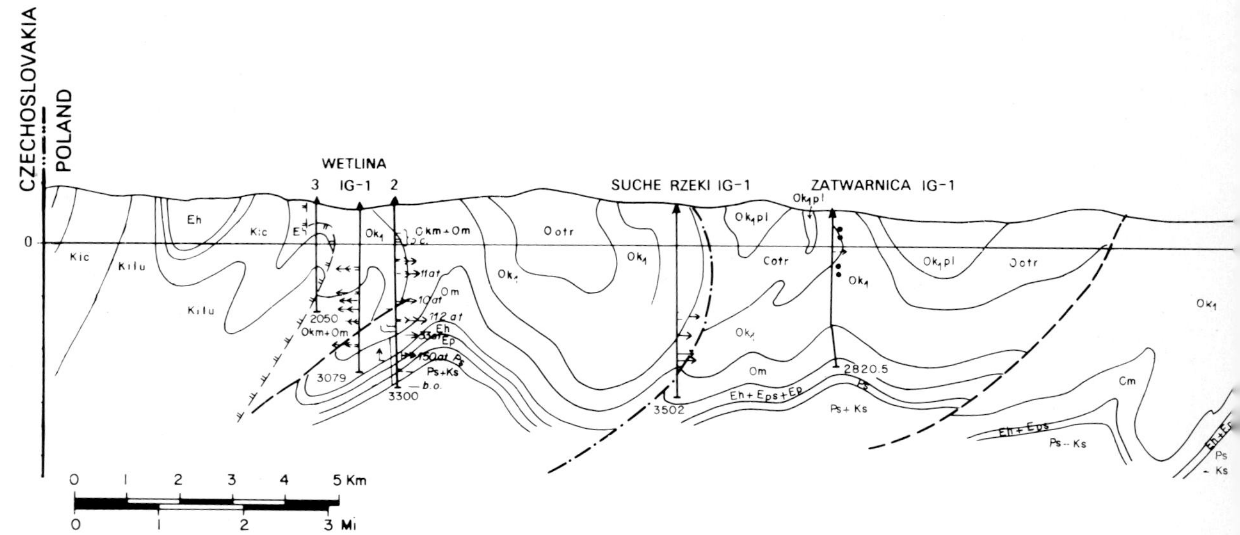

Fig. 2. East Carpathians – geological cross-section. (After Borys et al. 1989)

rocks occurring at depths of about 1000 m. In reservoirs occurring at depths down to about 2500 m, these parameters decrease to 5–20% porosity and to several tens md permeability. The poorest reservoir properties in the Miocene (5–12% porosity and 0.01 to 12 or so md permeability) occur below 2500 m and under the Carpathian overthrust. The reservoir rock thicknesses differ from about 1 m to more than 200 m, but are usually between 2–20 m. The seal for the Miocene traps are shale interbeds or massive shale and occasionally massive anhydrite.

The volume of reserves, composition of gas, and comparatively shallow depth of the Carpathian Foredeep fields are some of the reasons which will ensure that the foredeep remains one of the major production and prospective areas in coming years.

3 Polish Lowlands

This region comprises: (1) the Precambrian platform of the eastern and northeastern parts of the Polish Lowlands and (2) the Paleozoic platform with the Variscan and partly Caledonian elements overlain by the Mesozoic platform of the southwest of the Polish Lowlands. The western boundary of the East European Precambrian Platform is the Teisseyre-Tornquist tectonic zone. However, due to the 10–12 km thick overlaying sedimentary cover, the boundary between the Precambrian and Variscan/Caledonian elements is not always well defined.

The prospective gas-and oil-bearing areas of the Polish Lowlands belong to the eastern part of the Northeast German-Polish sedimentary province, the Paleozoic-Mesozoic platform of SW Poland.

The sedimentary cover includes all stratigraphic units from the Cambrian through to the Tertiary. Most of these units have prospective horizons (Figs.5–8). The main target for hydrocarbon exploration activity in the Lower Paleozoic members has been Lower and Middle Cambrian sandstones in the Peri-Baltic Syneclise, Warsaw Trough and Podlasie Depression. The reservoir rocks are quartzitic sandstones with intergranular and fracture porosity. The importance of the fracture porosity increases with depth. Traps are mainly structural-tectonic, influenced by lithological factors. Up to now, two small crude oil accumulations have been found in the Leba Uplift. In the Peri-Baltic Syneclise, promising shows of gas/condensate were found in well Malbork IG-1, in the Middle Cambrian sandstones at 3250 m depth. Overlying clayey and argillaceous sediments of Ordovician and Silurian age act as a regional cap rock.

The Middle Cambrian formations consist of fine- to medium-grained sandy sediments, with mudstone and claystone intercalations and frequent deformational structures (gravitational flows, bioturbations). In the eastern part of the Cambrian sediment areas the clastics are usually poorly compacted with a dominant clayey or clayey-ferruginous cement and have limestone interbeddings at the top. In the western part they are mostly clayey and have undergone a greater degree of diagenesis.

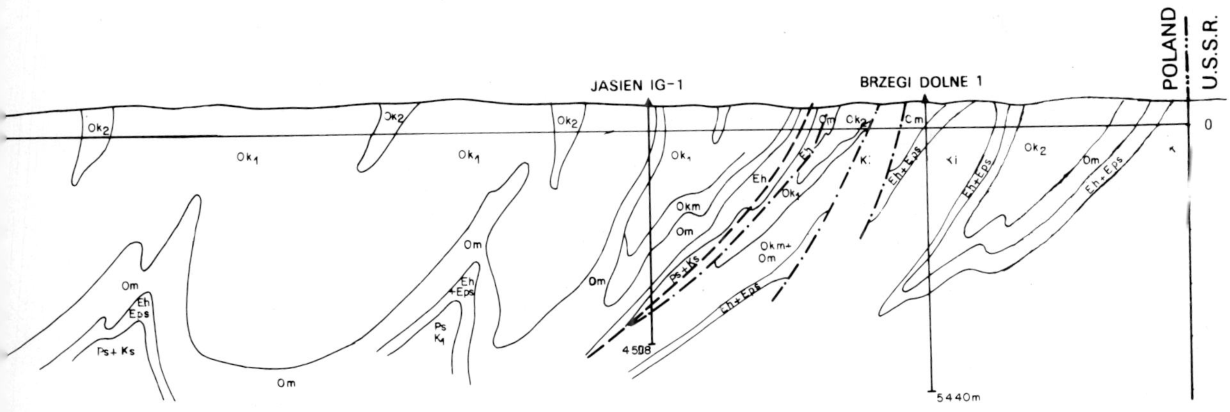

Mst	Stebnik Formation	Kic	Cisna Beds
Ma	Autochthonous Miocene	Kilu	Lupkow Beds
OK_2l	Upper Krosno Beds	Ks	Istebna Beds (non-divided)
Ok_2	Middle Krosno Beds	Ki	Inoceramus Beds (non-divided)
Ok_1	Lower Krosno Beds	Ki pst	Variegated Shales
Ok_1pl	Lower Krosno Beds-Supra Otryt Member	K ps	Variegated Marls
O otr	Lower Krosno Beds-Otryt Member	Kd	Lower Cretaceous (non-divided)
$O k_1$	Lower Krosno Beds-Infra Otryt Member	Pr	Precambrian
O km	Intermediate Beds	↑	Well location
O m	Menilite Series	—·—·—	Major revese fault
Eh+Eps	Hieroglyph Beds and Variegated Shales	— — —	Other fault
Ep	Ciezkowice Sandstones	⊥⊥ ⊥⊥ ⊥⊥	Overthrust Line
Ps	Istebna Shales	→>>>	Gas indication
Pma	Majdan Beds	•	Oil indication

The Middle Cambrian sediments were laid down in a regressive epicontinental basin under variable energy conditions, a higher energy environment in the east and a low-energy environment in the west. Total thickness of the Cambrian varies from 150 m (well Goldap IG-1) to over 700 m in the west (716 m in well Koscierzyna IG-1). So far, 93 wells have penetrated Cambrian sediments in the Peri-Baltic Syneclise. Laboratory analysis indicate that more favourable reservoir properties exist in the higher

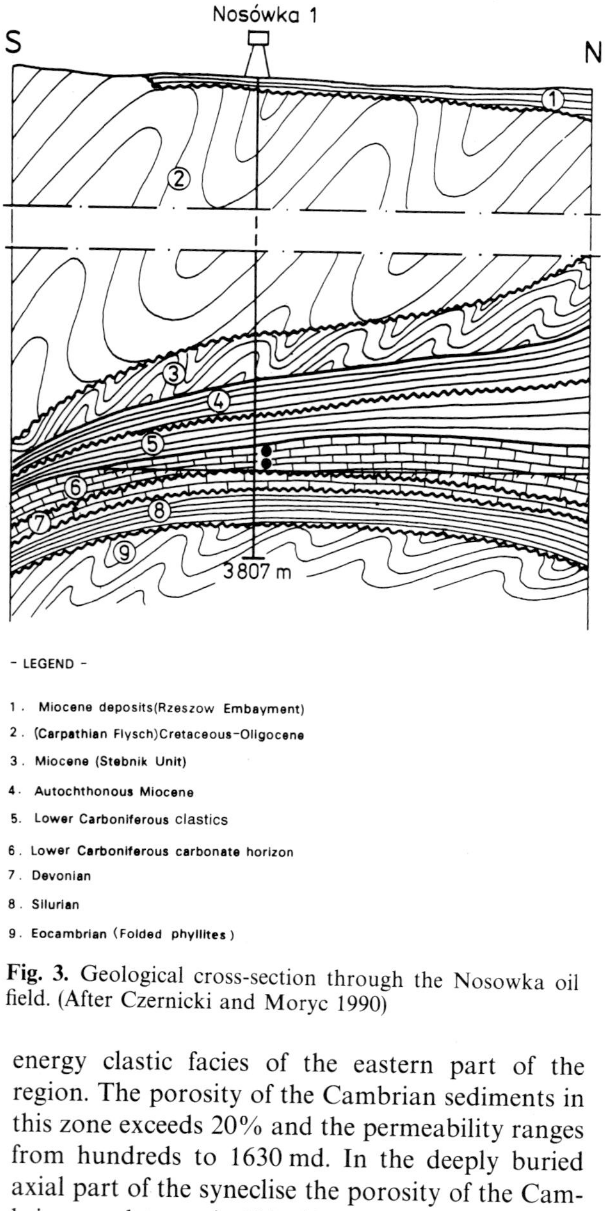

Fig. 3. Geological cross-section through the Nosowka oil field. (After Czernicki and Moryc 1990)

energy clastic facies of the eastern part of the region. The porosity of the Cambrian sediments in this zone exceeds 20% and the permeability ranges from hundreds to 1630 md. In the deeply buried axial part of the syneclise the porosity of the Cambrian sandstones is 1%–6% and the permeability 0.1–8.6 md (e.g. in well Nowa Koscielnica 1).

The clayey-carbonate Ordovician sediments form good cap rocks. However, the Tremadocian glauconitic sandstone lenses occurring at the base of the sequence may, together with Cambrian sediments, form a single reservoir rock complex. Silurian sediments, mostly claystones, are generally considered as the main cap rock series of the Cambro-Ordovician reservoirs. Taking into account higher geothermal gradients for the Cambrian, Ordovician and Silurian (2.5–3.0 °C/100 m), it can be assumed that the initial phase of oil generation in the deeper zones began in the Devonian-Carboniferous.

The Devonian deposits occur mainly in the western Pomerania and Lublin areas (Figs. 6, 7). The prospective horizons are the Lower Devonian Old Red Sandstone and the limestones and sandstones of the Middle and Upper Devonian. This is confirmed by the occurrence of many oil and gas shows, as well as by the discovery of three gas accumulations in the Devonian sediments of the Lublin area. The Lower Devonian sedimentary sequence consists of a clayey series covered by a siltstone-sandstone series. The sandstones are fine- to medium-grained, with quartz overgrowth and clay minerals acting as cement. The porosity varies between 0 to 18% and permeability from 0 to 6.1 md.

The majority of the Upper Devonian carbonate formations have very poor reservoir properties. Possible improvement in parameters may occur in the areas of well developed organogenic facies or in zones of extended fracturing where the porosity can reach up to 10% and the permeability varies from 12 to several hundred md. In the Lublin area, the occurrence of barrier and atoll reef facies is likely. Present investigations are concentrating on carbonate paleofacies analysis and tectonic evolution of the basin. The Upper Devonian sandstones in the Lublin area have better reservoir properties, but unfortunately are characterized by a great facies variability.

The Carboniferous gas and oil-bearing horizons are best known from the Lublin Fore Sudetic Monocline and western Pomerania (Fig. 8). A considerable variation in lithologies is responsible for the regional disparity in reservoir rocks. Good sealing is provided either by evaporites or by claystone-mudstone caprocks of the Zechstein. The reservoir horizons are as follows:

Lublin Trough. Prospective horizons are Upper Visean, Namurian and Westphalian sandstone beds sealed by clay-mudstone and locally carbonate layers (porosity1–22%, permeability 0.1–400 md). These formations have undergone relatively little diagenesis, but their reservoir properties are somewhat variable due to great facies variations. In other regions, these formations have usually undergone diagenesis to a greater degree, thus the presence of fractures plays an important part in influencing the reservoir properties.

The Fore Sudetic Monocline. Reservoirs include quartz sandstones, whose reservoir properties are enhanced by fissuring and weak cementation,

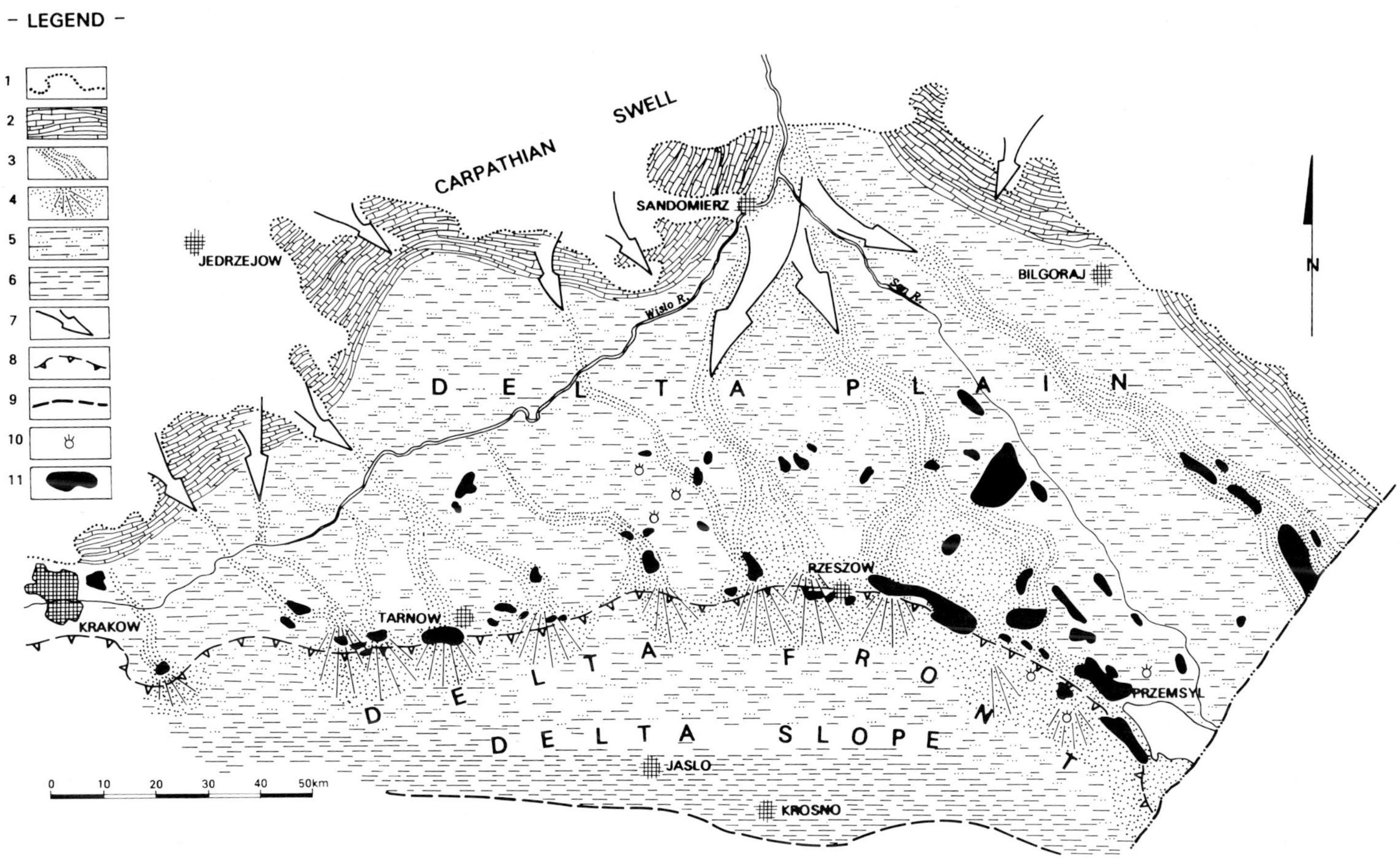

– LEGEND –

1. Northern extent of the marine Miocene deposits
2. Lithothamnion, serpulid and detrital limestone facies
3. Sandy deposits of channel facies
4. Facies with domination of sandstones
5. Clayey-sandy facies
6. Clayey facies
7. Directions of water run-off
8. Boundary of the orographic rim of Carpathians
9. Presumable southern extent of the Sarmatian deposits
10. Boreholes with gas shows
11. Major gas fields in Sarmatian and partially Badenian deposits

Fig. 4. Paleofacies map of the Mid-Miocene deltaic deposits in the eastern part of Carpathian Foredeep. (After Karnkowski 1989)

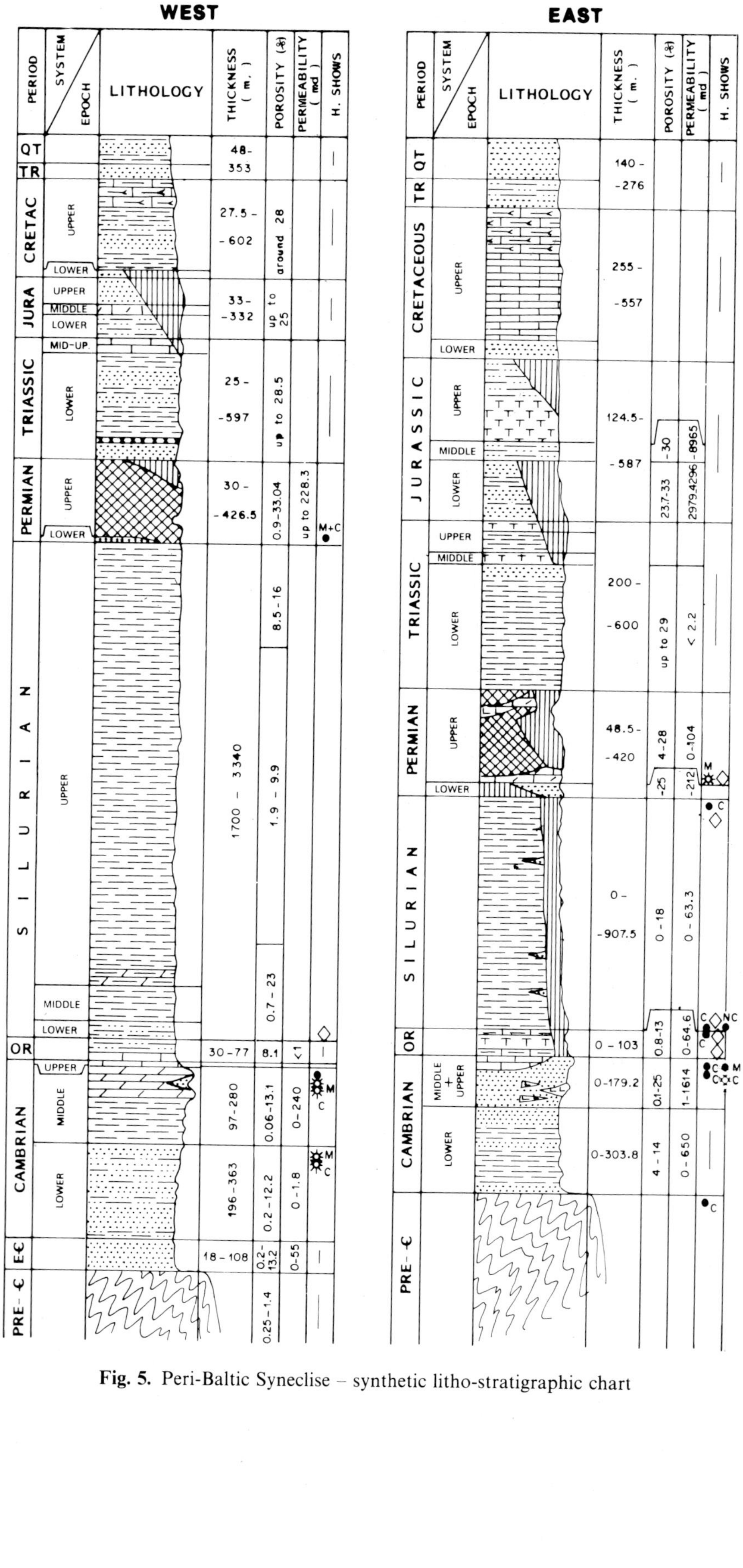

Fig. 5. Peri-Baltic Syneclise – synthetic litho-stratigraphic chart

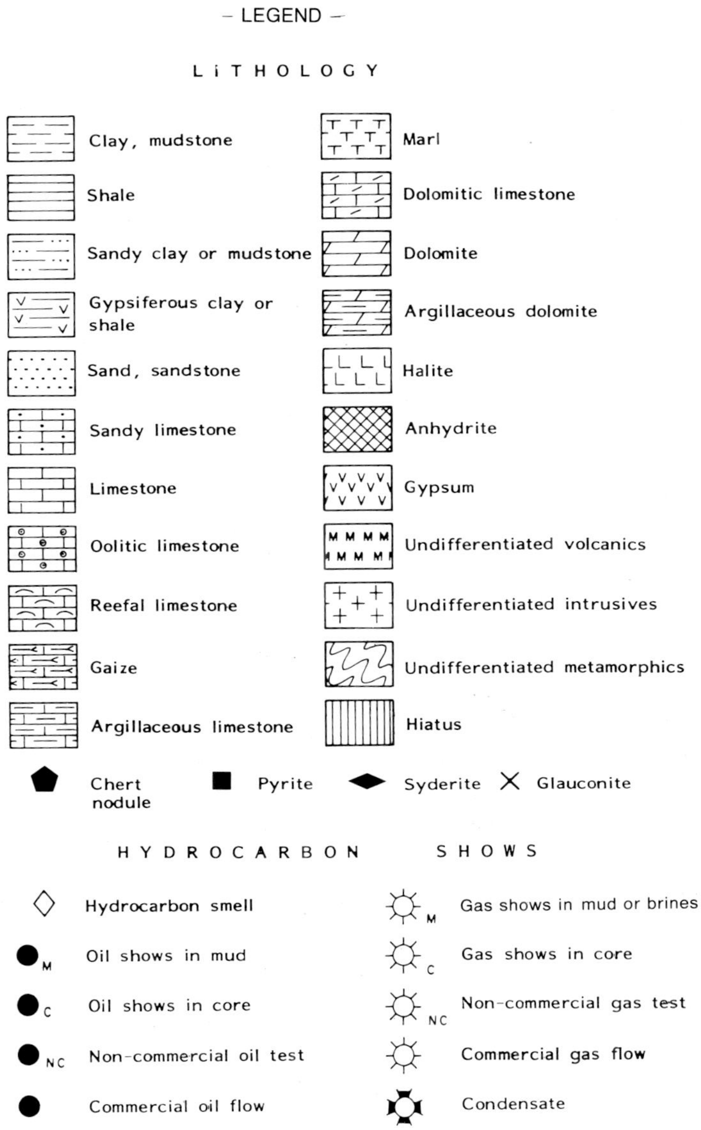

(porosity up to 10.5%, permeability 0.1–1.7 md, locally up to 20 md). The Carboniferous in this area is also considered as the source rock for the gas in the Rotliegendes sandstone reservoirs.

The Marginal Zone of the Pomeranian High, Upper Tournaisian and Visean Sediments. Reservoirs include:

- Arkosic and greywacke sandstones, the good parameters of which are related to cement dissolution and fissuring (porosity up to 22%, permeability from less than 1 to several hundred md).
- Oolitic and detrital limestones with reservoir properties enhanced by dolomitization and fissuring (porosity up to 10%, permeability up to 2 md).
- quartz sandstones with porosity up to 23%, permeability from 1.6 to over 200 md.

Middle and Upper Westphalian sediments are represented by quartz sandstones with porosity between 5–15%, but often exceeding 20%, and permeability up to several hundred md.

It is widely known that below the base of the west and northwest parts of the West European Permian Basin (WEPB), there are thick underlying Upper Carboniferous deposits containing rich organic matter and coal facies. However, in the Polish part of the WEPB, the Upper Carboniferous deposits contain only small amounts of scattered organic

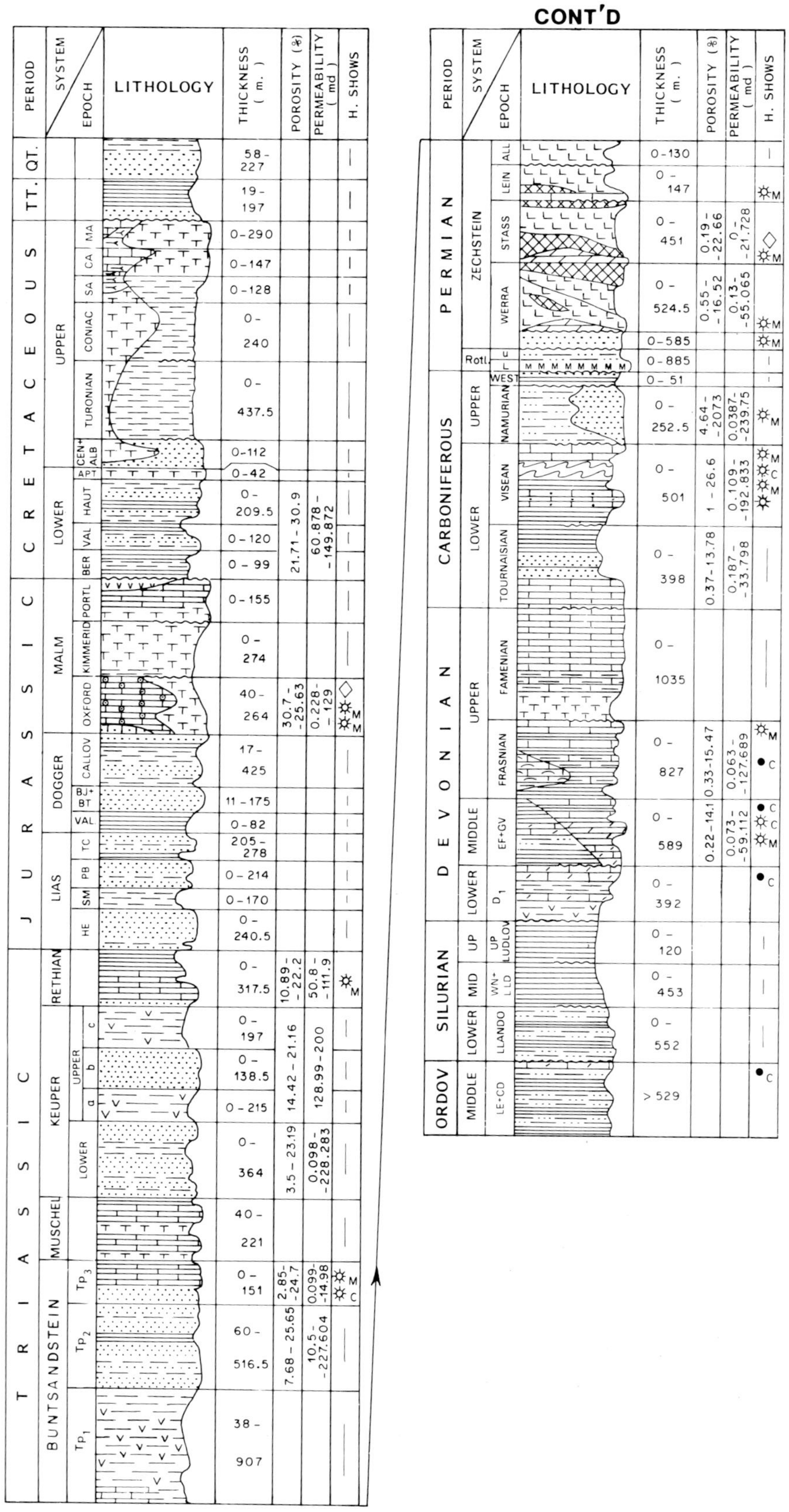

Fig. 6. Pomeranian Trough – synthetic litho-stratigraphic chart

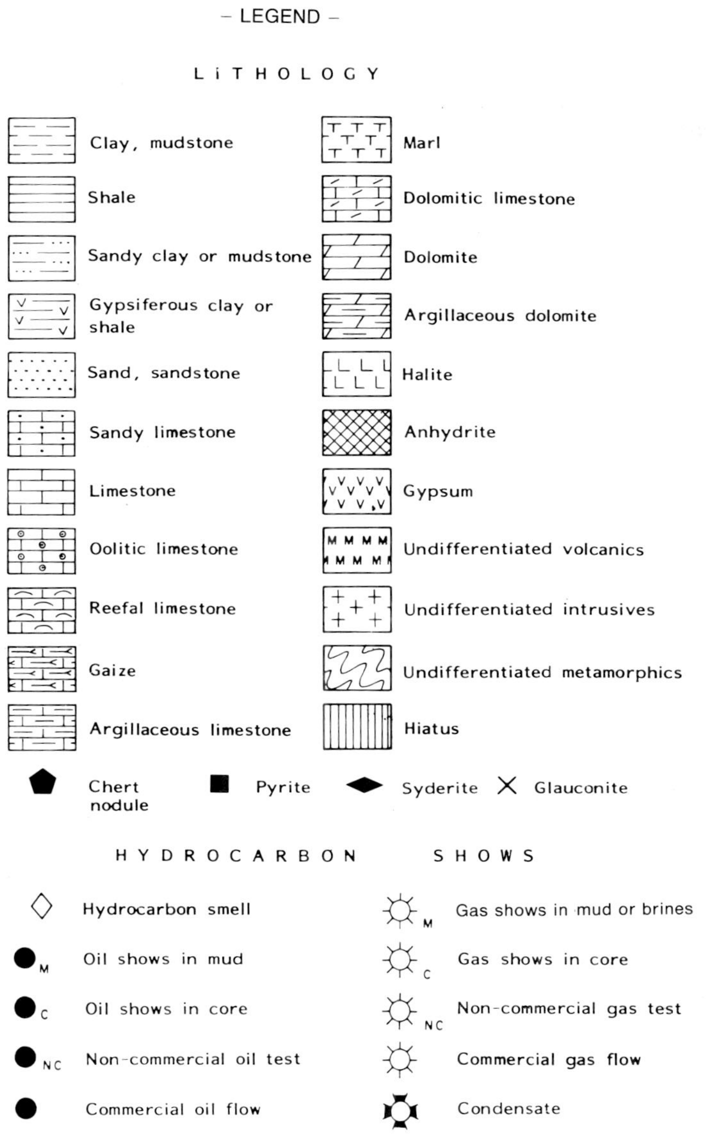

matter, and occasional isolated very thin coal layers, which are recognized as gas-prone source rocks. As a rule, the Upper Carboniferous is thin or often completely eroded. In the greater part of the WEPB, Lower Carboniferous deposits are folded. This folding is related to the Variscan orogeny.

Carboniferous formations (excluding the Lublin area) are very difficult to explore using seismic methods. In the Polish Lowlands, the Carboniferous formations occur mainly below the salt and anhydrite layers of Zechstein, which results in poor seismic records with intermittent reflections. Additionally, the wave field is disturbed by tectonic and erosional events. Comparatively better defined are the seismic markers in Carboniferous formations of the Lublin area. However, the main problem is definition of the internal structure within the sediments between markers, particularly in view of the identification of traps which may exist there. So far, eight small gas fields and one oil field have been found in the Carboniferous formations.

In the central part of the Lublin Trough, the total thickness of Devonian and Carboniferous sediments is 3.5–4.0 km with the geothermal gradient increasing to about 3.5–4.0°C/100 m. The oil generation process from Lower Famennian source rocks could have begun in the Early Carboniferous. It is highly probable that just before the Asturian phase, these source rocks went through their gas generation phase.

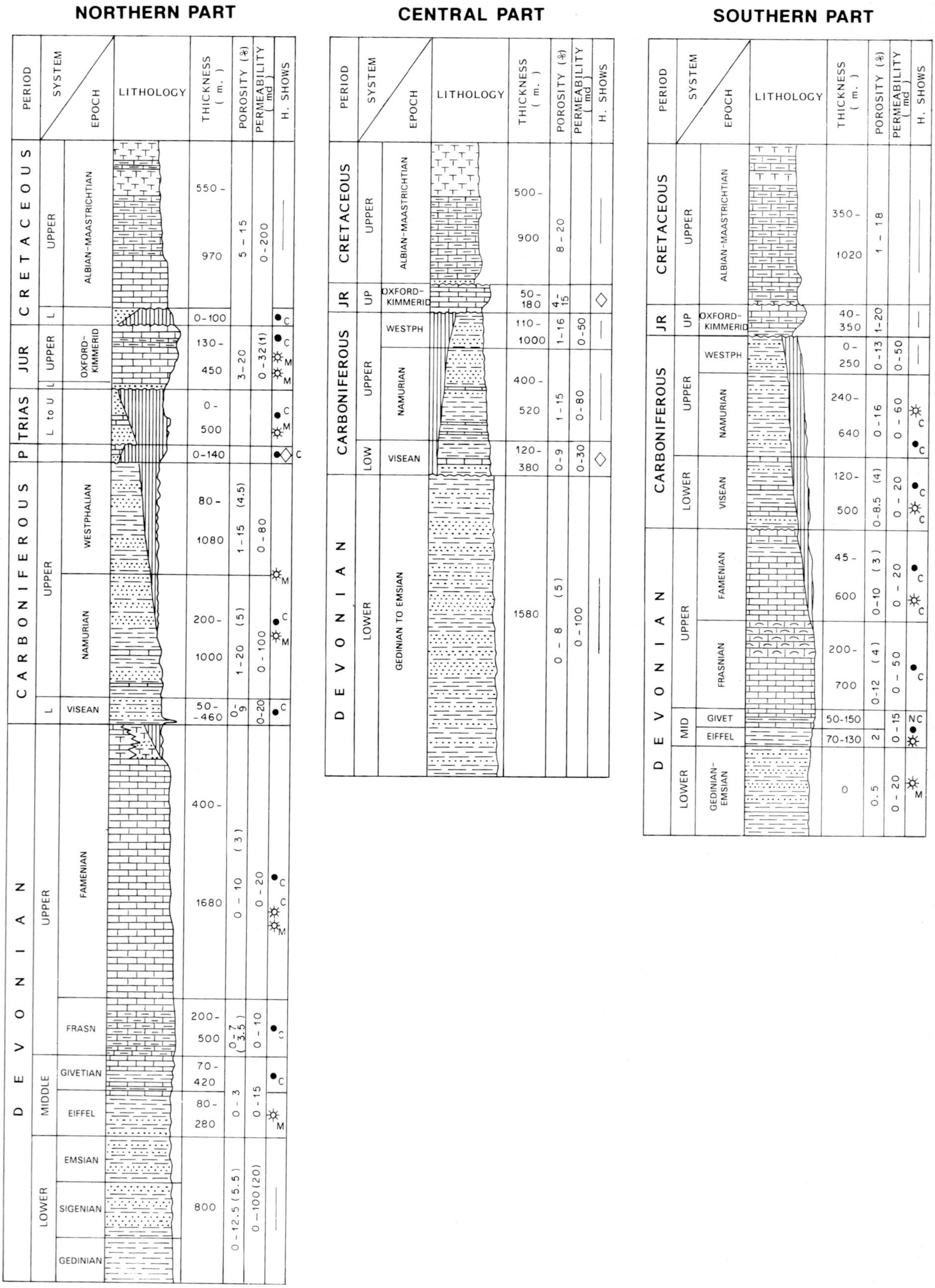

Fig. 7. Lublin Trough – synthetic litho-stratigraphic chart

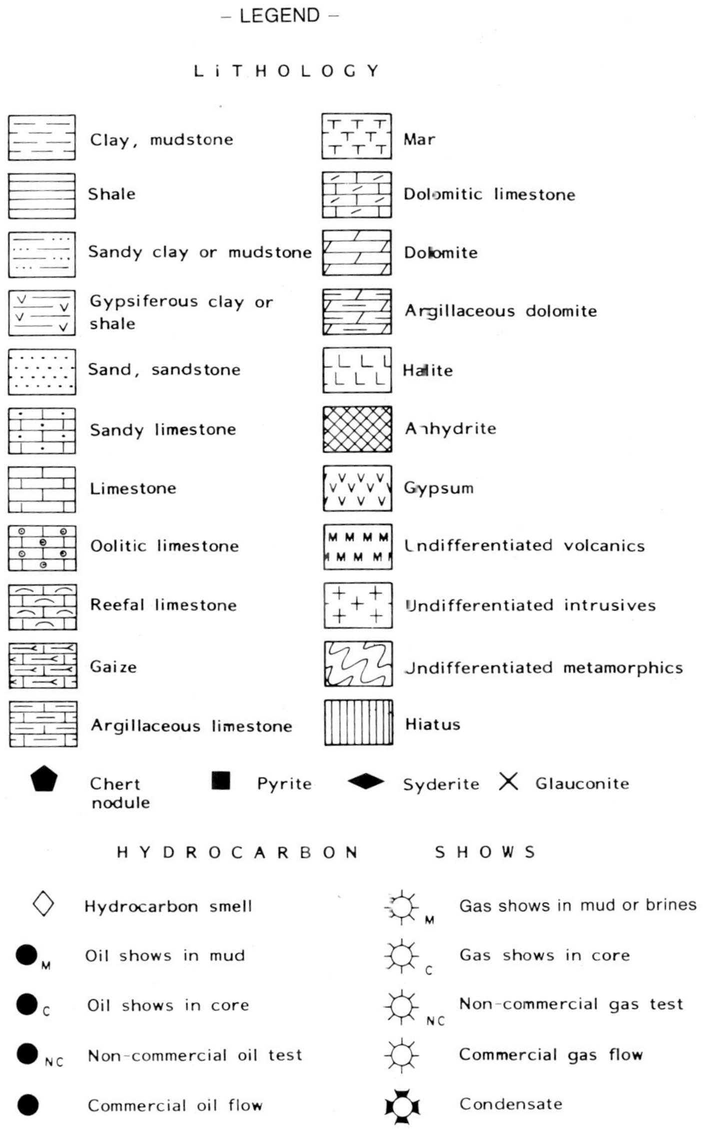

After the Late Carboniferous uplifting, a long period of denudation took place in Permian and Triassic. Subsequent burial related to the deposition of Jurassic and Cretaceous sediments was probably not enough to cause the commencement of a new phase of hydrocarbon generation. In the Pomeranian area the Upper Devonian marine sediments may have been in the oil window during the Carboniferous period (e.g. in well Debrzno IG-1).

Carboniferous source rocks in the central part of the WEPB could have begun oil generation in the Triassic or Lower Jurassic through to the end of the Jurassic. Dry methane gases were generated during the Cretaceous, as suggested by data from wells Budziszewice IG-1, Byczyna 1, Zabartowo 2, Moracz IG-1 and Wrzesnia IG-1. In the areas where Carboniferous sediments lie in comparatively shallow depth (e.g. in the Pomeranian, Warsaw and Lublin troughs) the oil generation process began in Carboniferous but did not reach the main gas generation phase.

The main gas- and oil-bearing formations of the platform areas in Poland are Permian. Gas accumulations have been discovered in Saxonian and Zechstein Limestone Member and several oil and gas accumulations found in the Zechstein Main Dolomite.

In the Fore Sudetic Monocline, Lower Permian (Rotliegendes) deposits, up to several hundred meters thick, are essentially sandstones, conglomerates with mudstone and claystone intercalations of a desert, fluvial, lacustrine and eolian facies.

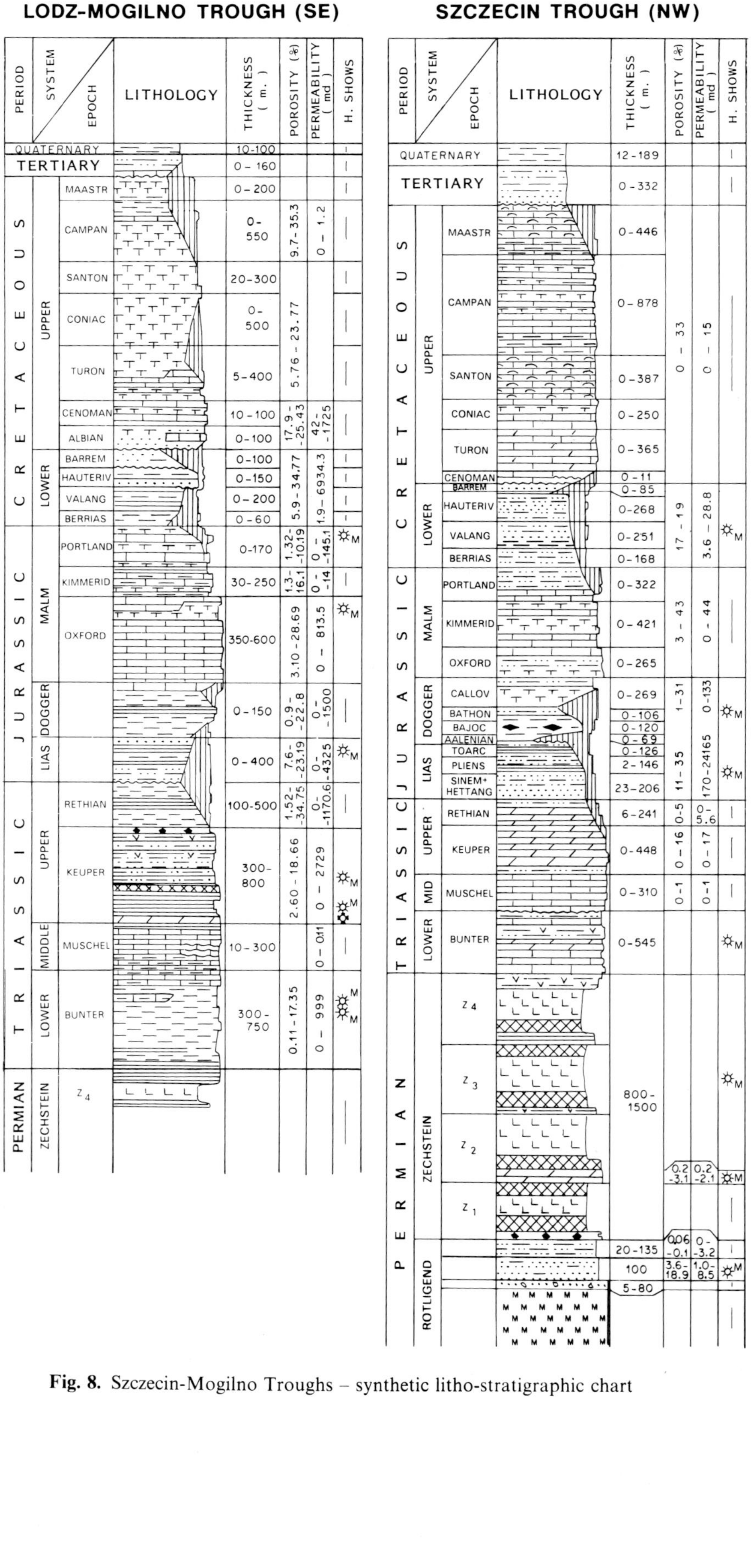

Fig. 8. Szczecin-Mogilno Troughs – synthetic litho-stratigraphic chart

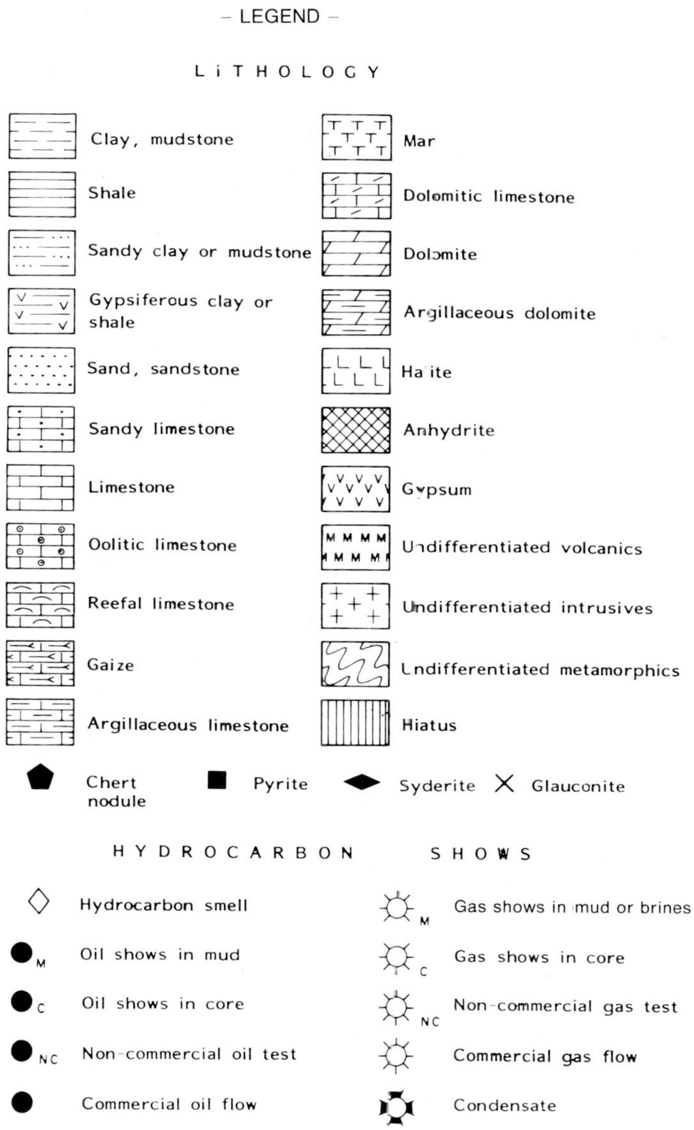

Thick effusive rocks appear locally. In the central part of the basin, the Rotliegendes deposits have unfavourable reservoir lithofacies development (mudstone and claystone).

The best reservoir parameters exist in the western and central part of the Fore Sudetic Monocline and in the NE part of the Wolsztyn Uplift slope. The porosity varies from 10 to 21.8%, and the permeability from 2 to over 100 md.

On the Eastern European Precambrian Platform, the Rotliegendes conglomerates and sandstones have thicknesses not exceeding tens of meters. Geological examination of this formation has brought to light three major gas-bearing areas:

The Southern Fore Sudetic Monocline. This contains some large fields – Zuchlow (22 billion m^3), Bogdaj-Uciechow (20 billion m^3), Zalecze (20 billion m^3) and Wierzchowice (12 billion m^3). The average depth of pay zones in 1400 m. The gas has a significant nitrogen content (28–40%), but the presence of (0.2–0.4%) helium still gives a valid reason for economic exploitation.

The Northern Fore Sudetic Monocline. Extends south and southwest from Poznan. More than 30 small fields have been discovered in this area. Their distribution is primarily determined by the Wolsztyn Uplift. The average depth of pay zones is

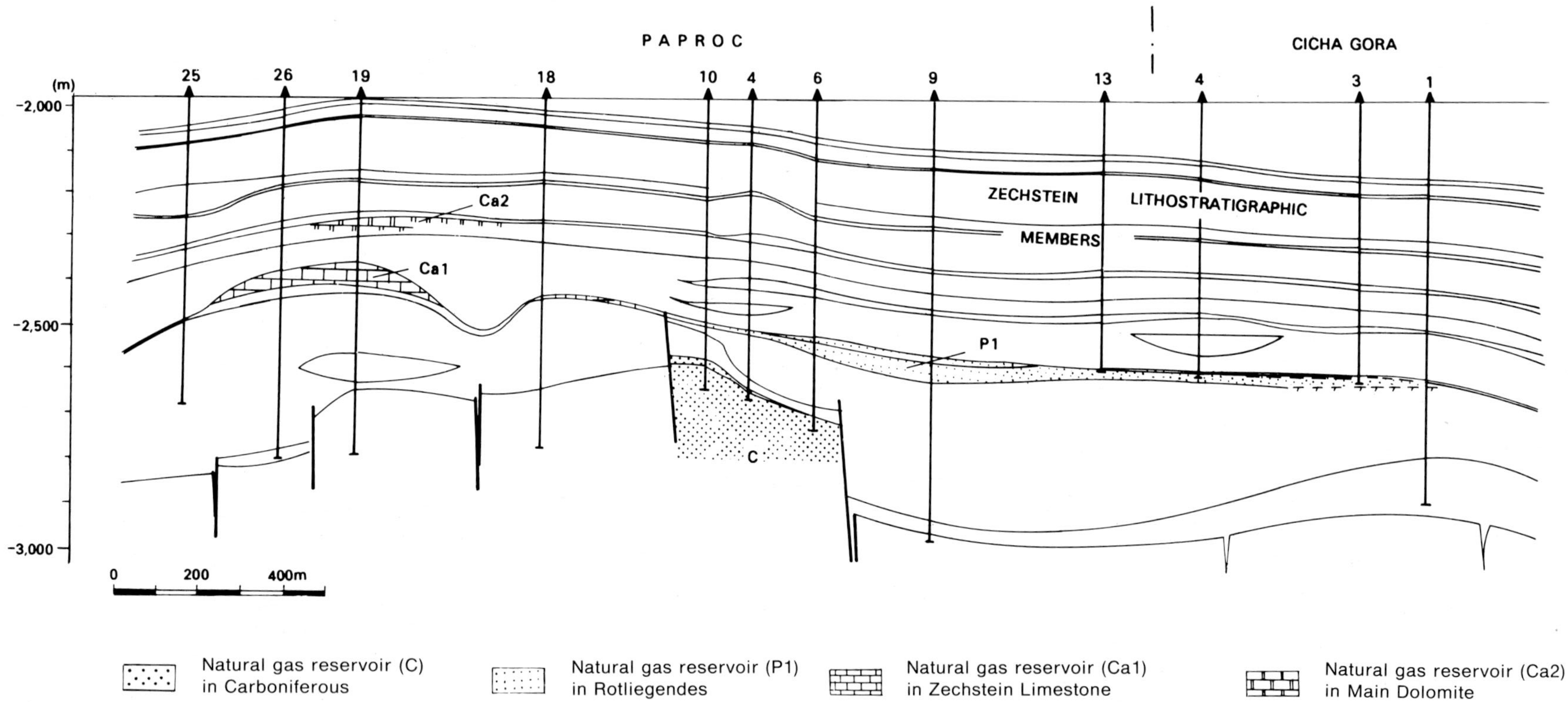

Fig. 9. Geological cross-section through the Paproc field. (Kulczyk and Zolnierczuk 1988)

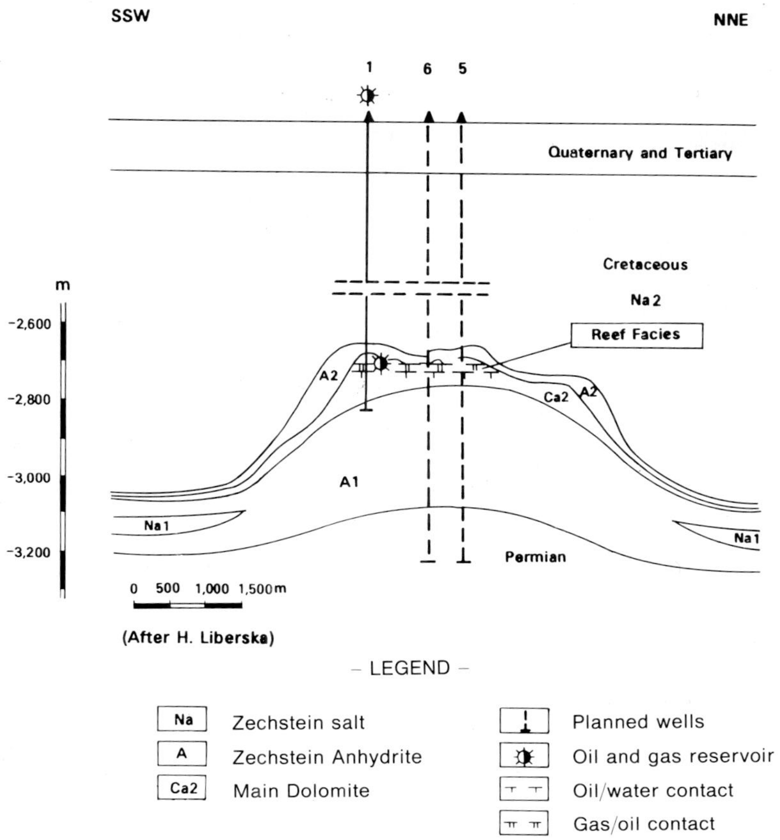

Fig. 10. Geological cross-section through the Gorzyca field. (After Kulczyk and Zolnierczuk 1988)

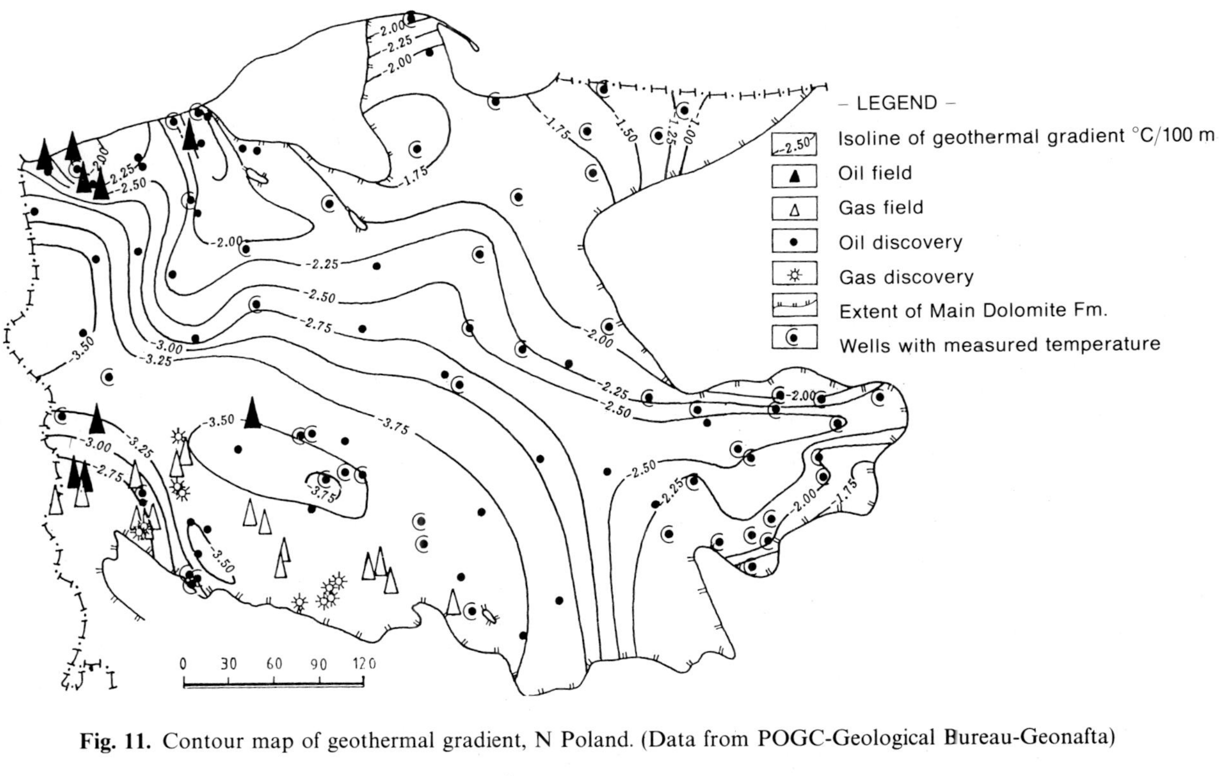

Fig. 11. Contour map of geothermal gradient, N Poland. (Data from POGC-Geological Bureau-Geonafta)

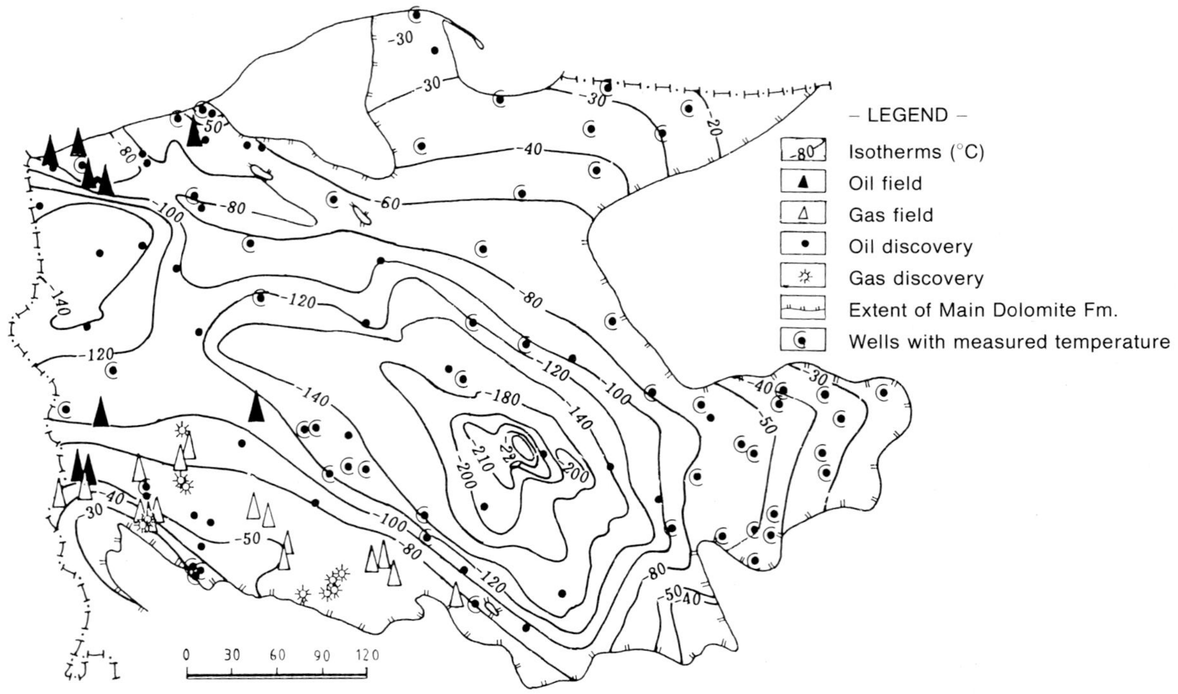

Fig. 12. Contour map of isotherms, N Poland. (Data from POGC-Geological Bureau-Geonafta)

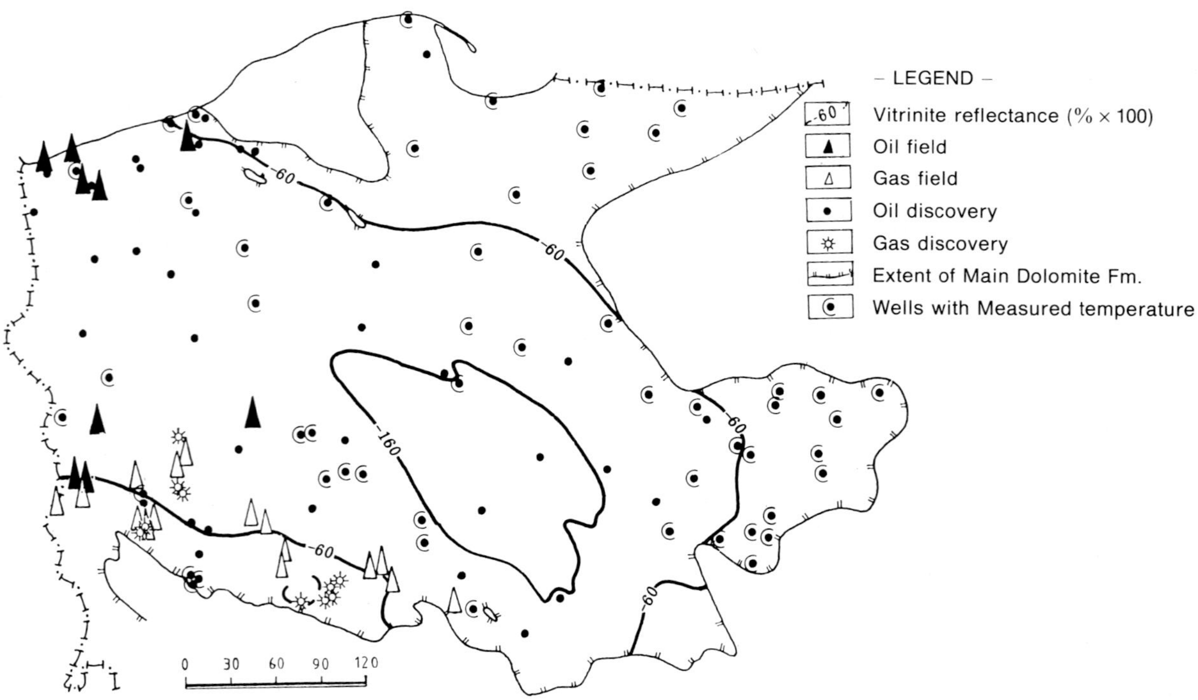

Fig. 13. Contour map of vitrinite reflectance, N Poland. (Data from POGC-Geological Bureau-Geonafta)

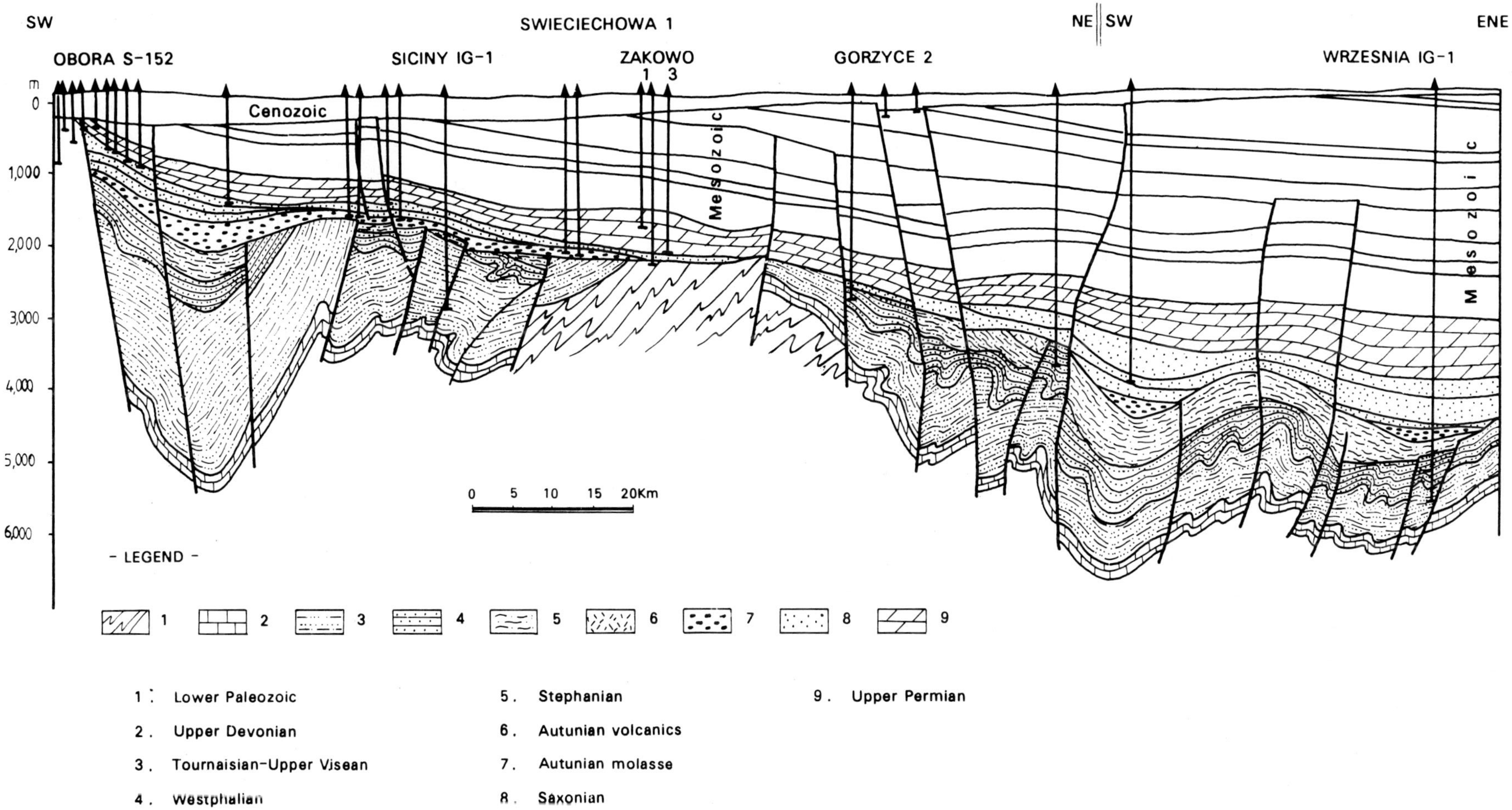

Fig. 14. Foresudetic Block and Foresudetic Monocline – geological cross-section. (After Wierzchowska-Kicułowa 1984)

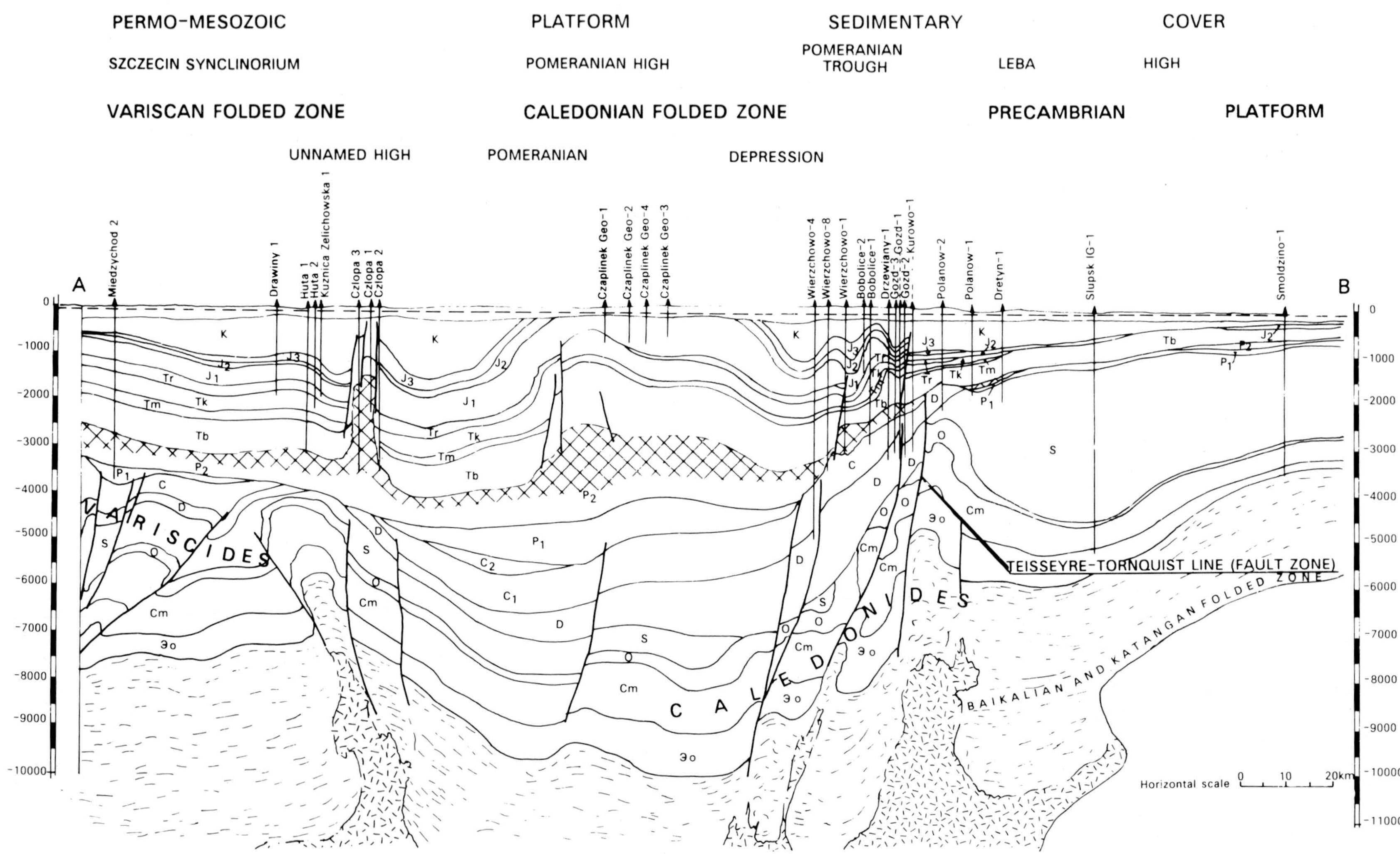

Fig. 15. Polish Lowlands – geological cross-section (Wegliniec IG-1-Smoldzino 1). (After Karnowski 1980)

Legend to Fig. 15

QT	Quaternary + Tertiary	C_2	Upper Carboniferous
K	Cretaceous	C_1	Lower Carboniferous
J_3	Upper Jurassic	D	Devonian
J_2	Middle Jurassic	S	Silurian
J_1	Lower Jurassic	O	Ordovician
Tr	Triassic-Rhaetian	Cm	Cambrian
Tk	Triassic-Keuper	Эo	Eocambrian
Tm	Triassic-Muschelkalk		Cristaline rocks
Tb	Triassic-Bunter		Extrusive rocks (diabase, ryolite, porphyrite basalt etc..)
P_2	Permian-Zechstein	↑	Well location
P_1	Permian-Rotliegendes		

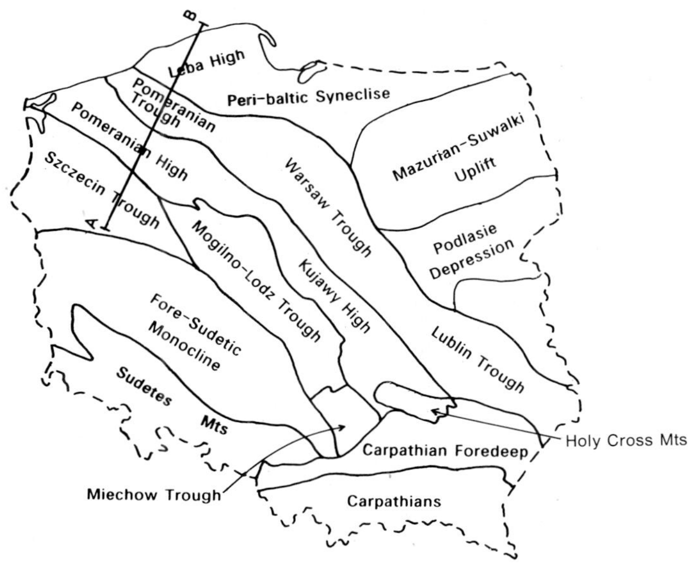

2600–3400 m. The composition of gas is close to that of the Groningen gas field, with an 83% methane content. However, westwards, the nitrogen levels increase once again, reaching 75% at times.

Western Pomerania. Poorly explored; only a few accumulations have been found. The average depth of reservoirs is 3200 m.

All gas accumulations are related to sandy facies. The distribution and size of fields is determined by the character and volume of the trap, reservoir properties, migration and accumulation processes which define the ratio of the trap infilling.

Gas accumulations are found in various traps from brachyanticlinal structures with small amplitudes, to lithological wedging of the Saxonian sandstones on the slopes of the Wolsztyn Uplift. The brachyanticlinal structures are connected with intervalley crests, sometimes additionally complicated by faults (Fig. 9). The Rotliegendes gas fields have close to hydrostatic initial reservoir pressure and passive edge waters.

The sedimentation of the Upper Permian (Zechstein) was developed in four cycles each consisting of a sequence of shale, carbonate, anhydrite and halite. Carbonates of the first cycle, forming the Zechstein (Basal) Limestone, with a thickness up to 40 m, are known to be gas-bearing. In the marginal zone of the basin, the Zechstein Limestone is in reef facies and contains debris of algae, bryozoan, foraminifers, clams, gastropods, ostracods and other marine organisms. Better reservoir properties (porosity exceeding 25%) are found in the bryozoan-

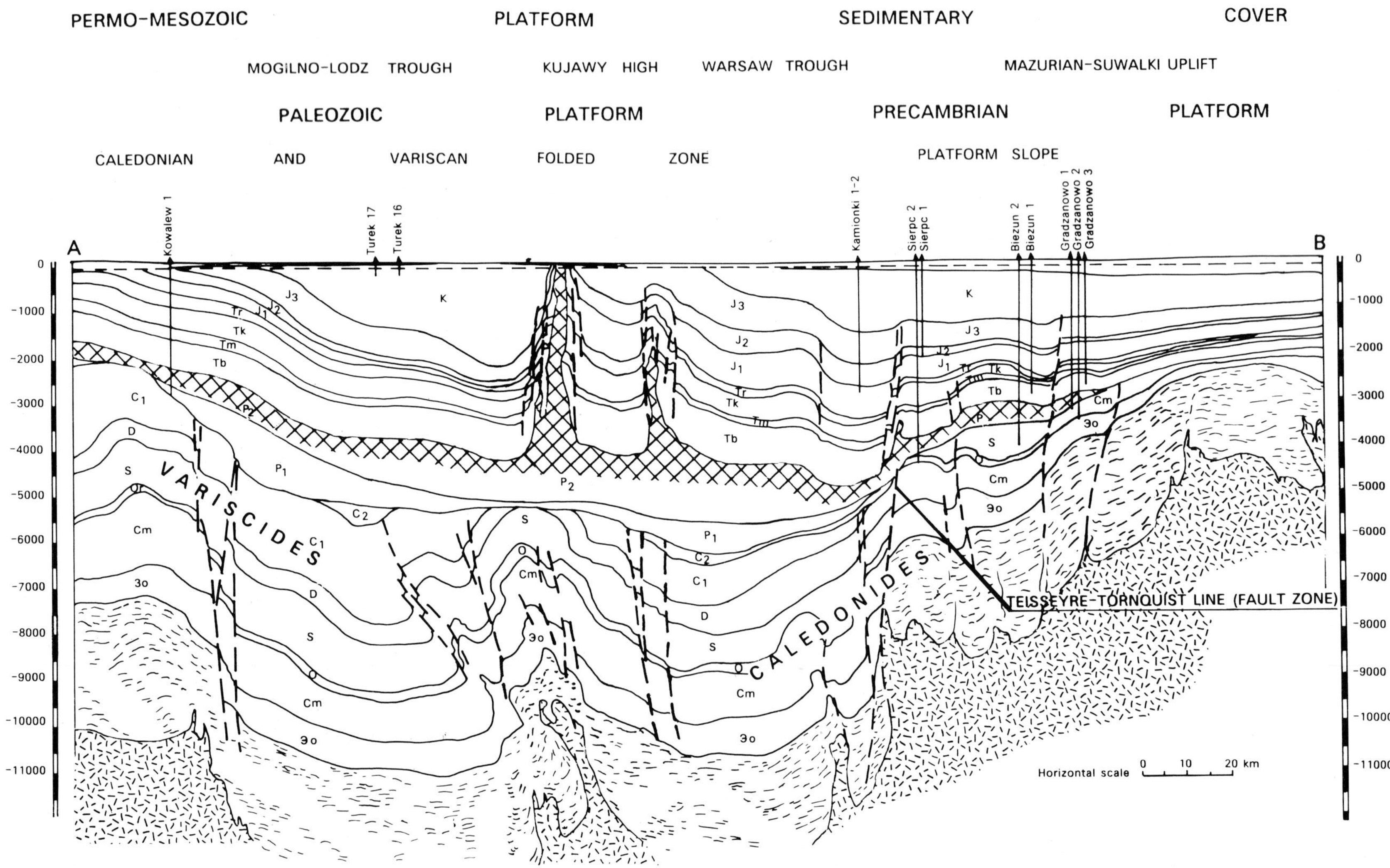

Fig. 16. Polish Lowlands – geological cross-section (Orpiszew 1-Gradzanowo 3). (After Karnkowski 1980)

Legend to Fig. 16

QT	Quaternary + Tertiary	C_2	Upper Carboniferous
K	Cretaceous	C_1	Lower Carboniferous
J_3	Upper Jurassic	D	Devonian
J_2	Middle Jurassic	S	Silurian
J_1	Lower Jurassic	O	Ordovician
Tr	Triassic-Rhaetian	Cm	Cambrian
Tk	Triassic-Keuper	Эo	Eocambrian
Tm	Triassic-Muschelkalk		Cristaline rocks
Tb	Triassic-Bunter		Extrusive rocks (diabase,ryolite, porphyrite basalt etc..)
P_2	Permian-Zechstein	↑	Well location
P_1	Permian-Rotliegendes		

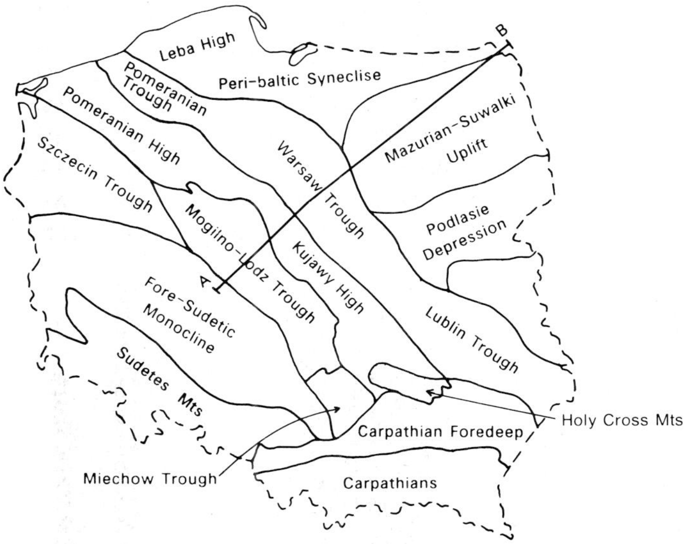

oncolite facies. In the inner part of the basin, the Zechstein Limestone is in a pelagic facies represented by dark grey marly-limey sediments.

The depositional stage of the Main Dolomite, which is the carbonate member of the second cyclothem was controlled by the vast, flat, Werra Platform, on which shallow water platform lagoonal conditions prevailed.

The slope of the platform is marked by the development of barrier reef facies. The development and extent of the Werra Platform, and consequently the distribution of the Main Dolomite depositional zones, has been influenced by the distribution of fault zones bordering the basin.

The thickness of the Main Dolomite carbonates in this zone varies from 40 to 190 m. The barrier reefs are composed mainly of algal and oncolite carbonates. These sediments thin out towards the basin and change to marly carbonates, marls and bituminous shale facies. The presence of primary porosity reservoirs is associated mainly with the barrier reef zone, where porosity is intergranular and vuggy. The best secondary porosity properties have been found in dolomite intervals, where dissolution occurred under subaerial conditions. Porosity is lower where there has been anhydritization and kaolinization. Horizons with the best properties have a porosity of 25% or more, and permeability of several tens md. Increased fracturing in zones of tectonic disturbance, or diapirism of the Oldest Salt, sometimes considerably improves the reservoir properties, however, these are local and very variable phenomena. Over the last 10 years, seismic reflection in conjunction with facies and thickness

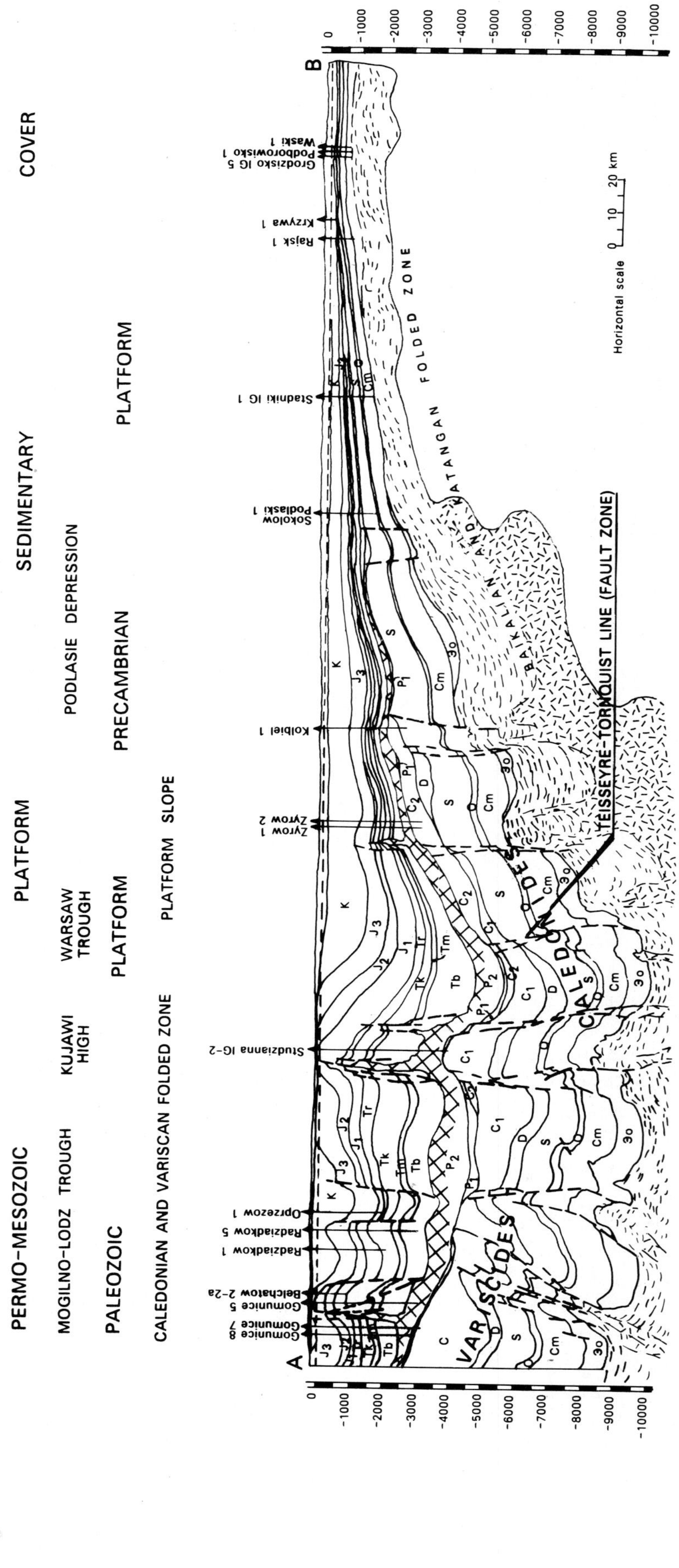

Fig. 17. Polish Lowlands – geological cross-section (Gomunice8-Waski 1). (After Karnkowski 1980)

Legend to Fig. 17

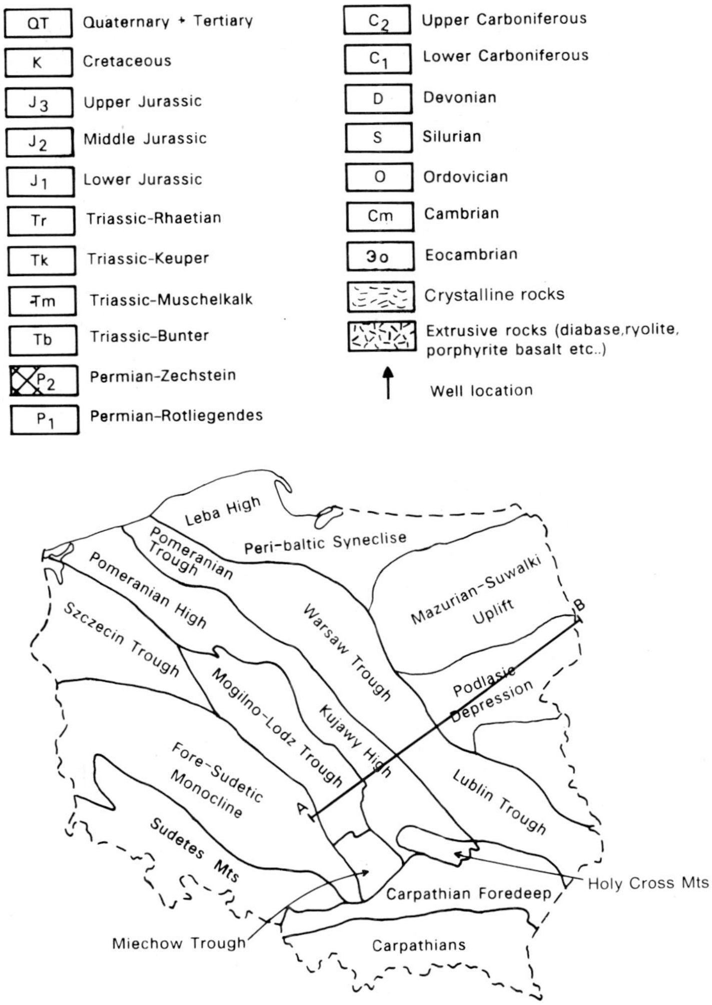

analysis has provided an effective exploration tool for recognizing buried reefs.

The reef facies also developed outside the Werra Platform in the area of deep, stagnant water basins. Two reef types have been defined: individual pinacle reefs and atolls with lagoonal facies in their centre. These reefs have large dimensions, and salt sequences provide an excellent hydrocarbon seal. They meet all requirements of a good prospective trap. At present, about 15 oil fields have been discovered in various types of reef reservoirs.

The largest oil field discovered so far in the Polish Lowlands – Kamien Pomorski with reserves of 2 million t – is contained in a fractured reservoir, in the lagoonal facies zone. Two oil-gas fields found recently, Chartow and Gorzyca, are in a barrier facies reservoir (Fig. 10). The crude oil contains methane, sulphur and associated gases that include hydrogen sulphide and significant amounts of nitrogen. The natural gas is methanic-nitric, or nitric-methanic, with a high content of heavy hydrocarbons, and commonly a high hydrogen sulphide content (1–17%). On the whole, all of the hydrocarbon accumulations are characterized by an unusually high reservoir pressure, 30–40% higher than normal.

In the Upper Permian, the source rock is probably the Zechstein Limestone with a Total Organic Carbon (TOC) between 0.1–0.3%. In grey- and black-coloured horizons, TOC increases from 0.6 to 7.0%. The source rocks are of sapropel-humic type and attained the phase of oil generation after deposition of over 2.5 km of Zechstein, Triassic and Jurassic sediments or, alternatively, at the end of the Upper Cretaceous when their burial exceeded 3 km, in areas with a geothermal gradient of 2 to 3.5 °C/100 m and an average R_0 of 6% (Figs. 11–13).

The youngest prospective formations in the Polish Lowlands are of a Mesozoic age. Hydrocarbon shows have been encountered in over 100 boreholes, but at present no commercial accumulations have been discovered in the Mesozoic. Over the last few years, an integrated study of the Mesozoic sequences has been carried out. The geological setting of the Mesozoic formations in the Polish Lowlands has been compared with that of the North Sea and NW German Basin. Many differences and some similarities have been revealed.

In the Polish Lowlands, the Laramide orogenic paroxism (Upper Cretaceous-Paleocene) caused the vertical uplifting of the Central Polish Swell and

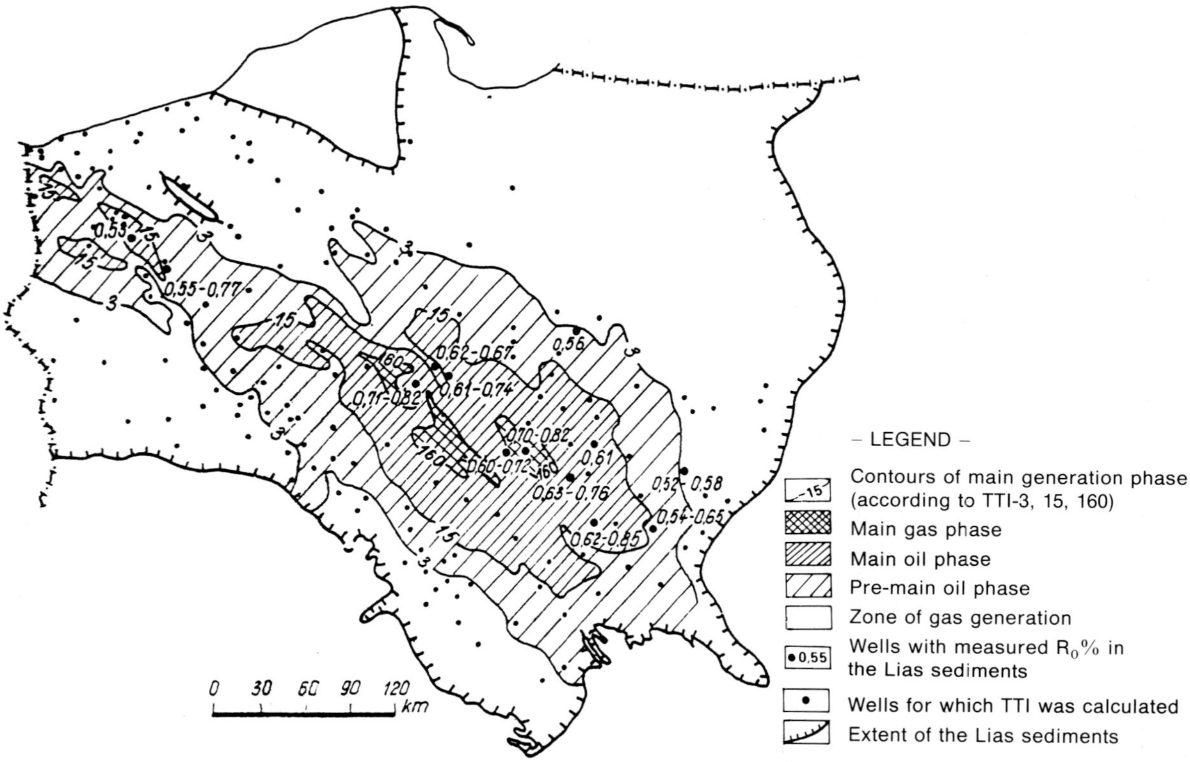

Fig. 18. Map showing the catagenesis of the Lower Lias sediments. (After Wilczek and Merta 1992)

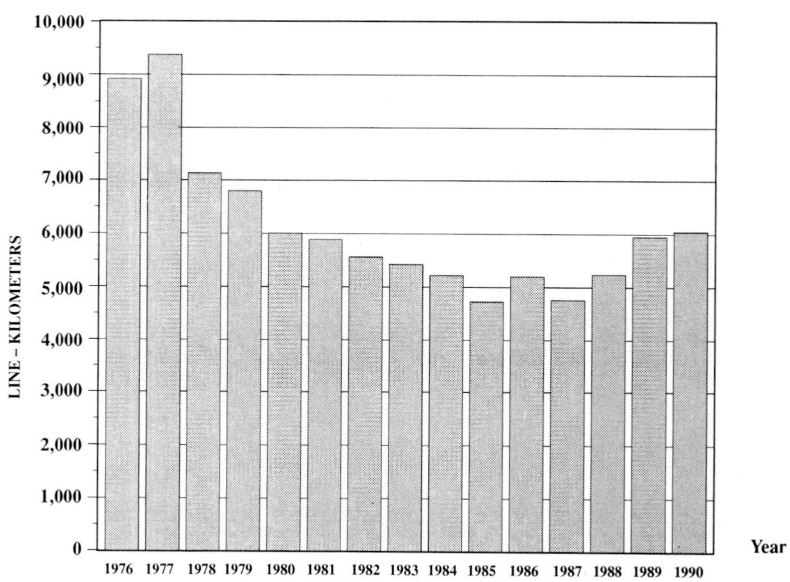

Fig. 19. Annual seismic acquisition (1976–1990)

Fore Sudetic Monocline. In the same period, the Szczecin-Mogilno-Lodz troughs, to the southwest and the Pomeranian-Warsaw troughs, to the northeast formed on both sides of the Central Polish Swell. They are the most probable hydrocarbon bearing Mesozoic basins of the Polish Lowlands province with possible reservoirs in the Triassic and Jurassic units (Figs. 14–17).

The lithology of the Triassic sequence is extremely diverse. The Middle Buntsandstein and Keuper series are the main target for exploration. They are composed of thick complexes of sandstone, claystone, mudstone, with carbonate intercalations – marly and oolitic limestone, marly dolomitic mudstone, gypsum and anhydrite (Figs. 5–8). These rocks have excellent reservoir properties, with porosity of the sandstones reaching 34%, and permeability up to 2390 md. For the Middle Buntsandstein, the seal is the Rhöt clay-carbonate member that extends throughout the basin. For the Keuper, the seals are clayey sediments with anhydrite belonging to Upper Gypsum Beds. However, these are deposits of a relict, dismembered basin and do not form a continuous cover.

The thickness of Triassic sediments is four to five times greater in the central part of troughs than on the adjoining areas. Unfortunately, Triassic sediments were mostly deposited in an inland basin under oxidizing conditions and therefore no real source rocks formed. Only in the Middle Triassic and Lower Keuper, marine dark limestones and marly claystones containing 0.5–2.1% TOC, occur. Gas shows were reported from wells which penetrated this sequence.

The lithologies of the Lower Jurassic are rather monotonous and include sandstones, mudstones and claystones. On the whole, reservoir rocks with porosity exceeding 30% and permeability of several thousand md outnumber potential seal rocks. Middle Jurassic sediments are mostly clastic sediments with favourable reservoir properties. In the Upper Jurassic there is a greater occurrence of limestones, occasionally in an oolitic-oncolite facies. The best reservoirs are the Oxfordian formations with Kimmeridgian marls acting as cap rocks.

Source rocks in the Lower Jurassic sediments include some claystone series which, especially in deeper parts of the basin, contain between 0.6 to 7.6% TOC (Fig. 18). The humic kerogen is dominant. Significant amounts of vegetable detritus and, locally, coal intercalations have been found.

There exist three main source rock horizons in the Middle Jurassic: (1) in the Upper Aalenian (about 150 m thick), (2) the Middle and Upper Bajocian (about 200 m thick), (3) the Bathonian (about 200 m thick). They contain between 0.7 to 9.7% TOC. The analysis of kerogen, and distribution of n-alkanes shows a predominance of type III kerogen. In the Upper Jurassic sequence marine carbonate deposits prevail. Marly limestone and marly claystone of Upper Kimmeridgian and Lower Portlandian age are considered to be the source rock. The thickness is about 120 m, and TOC content varies from 0.6 to 11%.

Potential source rocks of the Lower Keuper and the lower clayey series of the Liassic attained their oil generation phase at the turn of the Jurassic and Cretaceous when they were buried to over 3000 m.

The lower source rock horizon of the Dogger could have begun their oil generation phase shortly before the Laramide inversion but only in zones buried around 3000 m deep. After inversion, due to uplifting, the upper members of the Dogger could not attain adequate hydrocarbon generative potential.

However, this possibly was attained only in a narrow zone between the Damaslawek 22 and Uniejow 1 wells. Upper Kimmeridgian and Lower Portlandian marine sediments containing rich source rocks reached the effective oil generation phase during the transition from the Cretaceous to Tertiary.

Potential Lower Cretaceous source rocks either did not mature to generate oil or reached maturity in a very limited area not earlier than the end of the Tertiary. Recently, all geological data regarding prospects of the Mesozoic formations have been re-examined.

4 Exploration History, Production, Reserves

4.1 Surface Exploration

Geophysical prospecting has been carried out by two geophysical companies located in Krakow and Torun, subsidiaries of the Polish Oil and Gas Company (POGC). Fourteen fields crews complete roughly 5700 km of seismic lines yearly (Fig. 19). Examples of the seismic recordings are seen in Figs. 20 and 21.

Average density of gravity measurements is 2.5 points km^{-2}, for the whole of Poland. This density is considered sufficient for the purpose of most analysis. Magnetic surveying has not been carried out uniformly. For instance, the northern

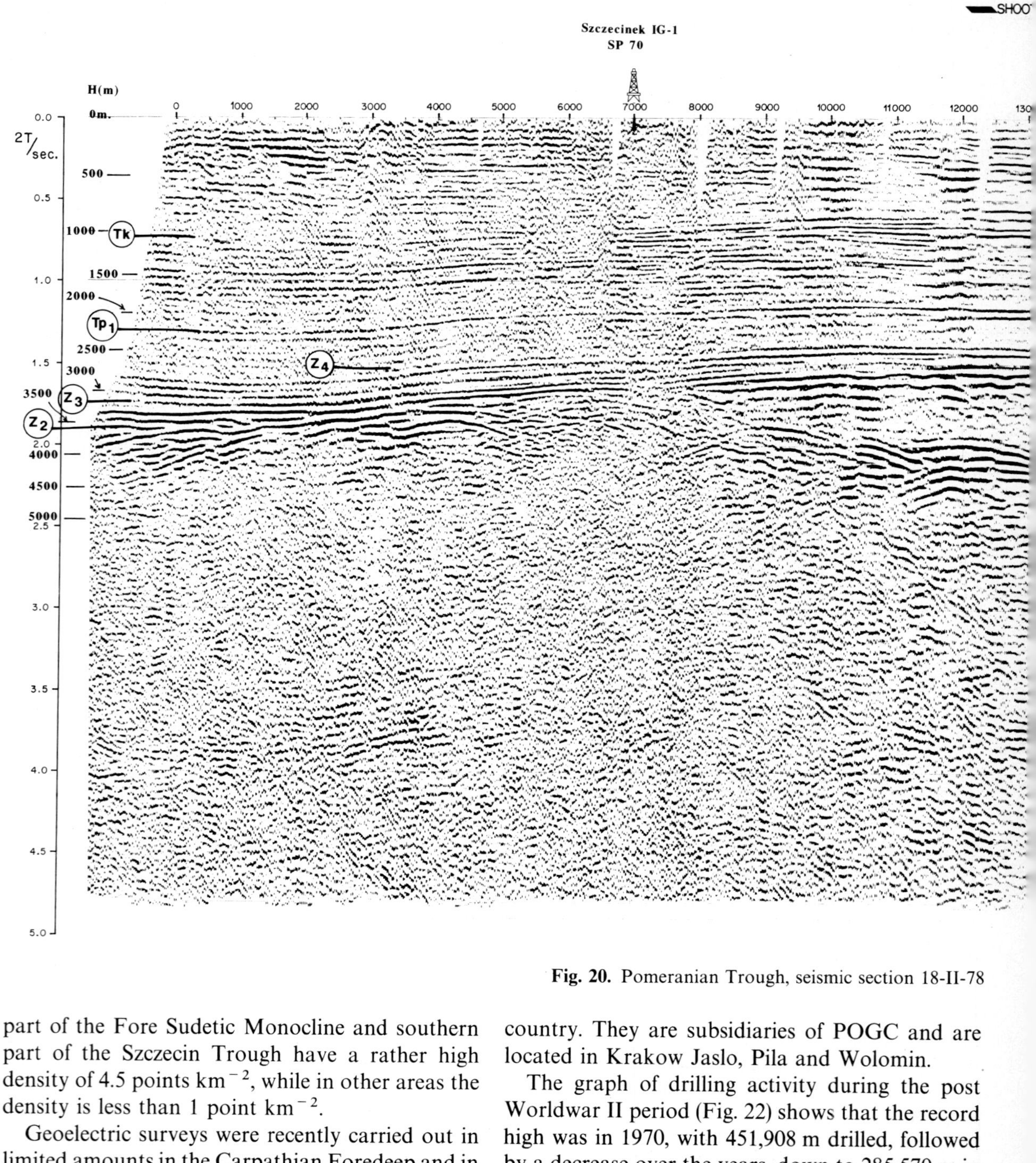

Fig. 20. Pomeranian Trough, seismic section 18-II-78

part of the Fore Sudetic Monocline and southern part of the Szczecin Trough have a rather high density of 4.5 points km^{-2}, while in other areas the density is less than 1 point km^{-2}.

Geoelectric surveys were recently carried out in limited amounts in the Carpathian Foredeep and in the Pomeranian area. A proprietary WEGA-D method is applied which measures the components of the electromagnetic field.

4.2 Drilling

During the past few years, four drilling companies operating some 80 rigs have been active in the country. They are subsidiaries of POGC and are located in Krakow Jaslo, Pila and Wolomin.

The graph of drilling activity during the post Worldwar II period (Fig. 22) shows that the record high was in 1970, with 451,908 m drilled, followed by a decrease over the years, down to 285,570 m in 1980, which may well explain the drop in production of crude oil and natural gas. Also the average depth of wells has increased with time (Fig. 23), mainly reflecting the search for deeper plays in the Polish Lowlands.

In 1990, POGC drilled 341,830 m of hole: 13,480 m of strat hole, 279,798 m in exploratory wells and 48,552 m in development wells. The total

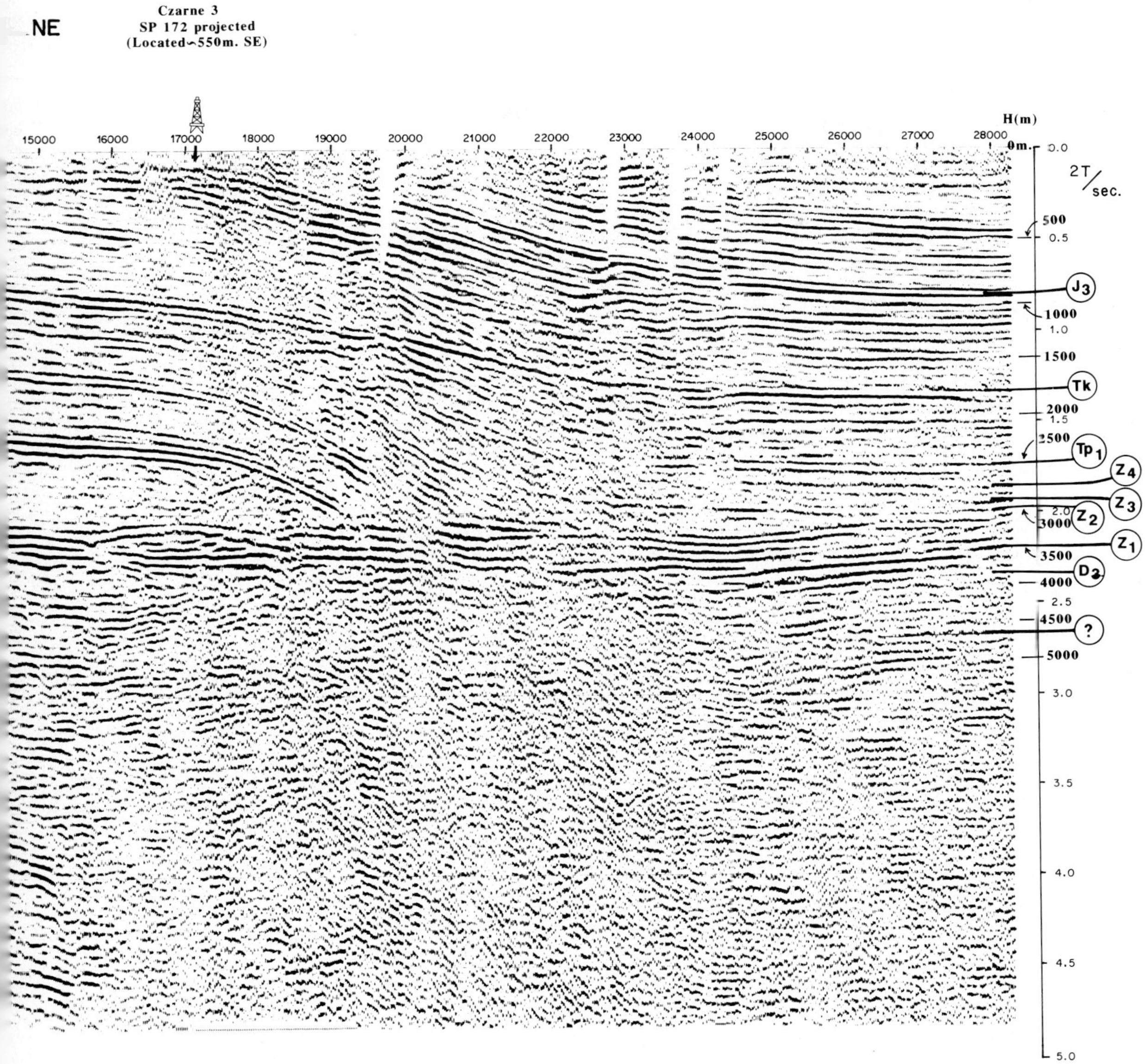

number of completed wells was 131, of which 53 were successful (42 gas, 11 oil). For 1991 it was planned to drill 362,100 m, of which 248,330 m will be exploratory drilling (44% in the Polish Lowlands, 31% in the Carpathians and 25% in the Carpathian Foredeep).

4.3 Production

As a result of the sizeable exploration effort in the early 1970s, in the mid-late years of that decade, the peak in hydrocarbon production was reached in 1978 with an output of 7.6 bln m^3 of gas and respectively in 1975 with a production of 560,000 t oil (Figs. 24, 25).

At present, domestic production of oil covers only about 2% of demand. In 1990 production was 162,640 t (Fig. 24). Output was 49.8% from the Polish Lowlands, 33.8% from the Carpathians and 16.2% from the Carpathian Foredeep and was supplied by the three POGC's production company subsidiaries: Krosno, Zielona Gora and Sanok.

The level of Polish gas production was more encouraging although only 3.4 bln m^3 was produced during 1990 (Fig. 25). Imports amounted to 7.8 bln m^3. The largest part of the gas output comes

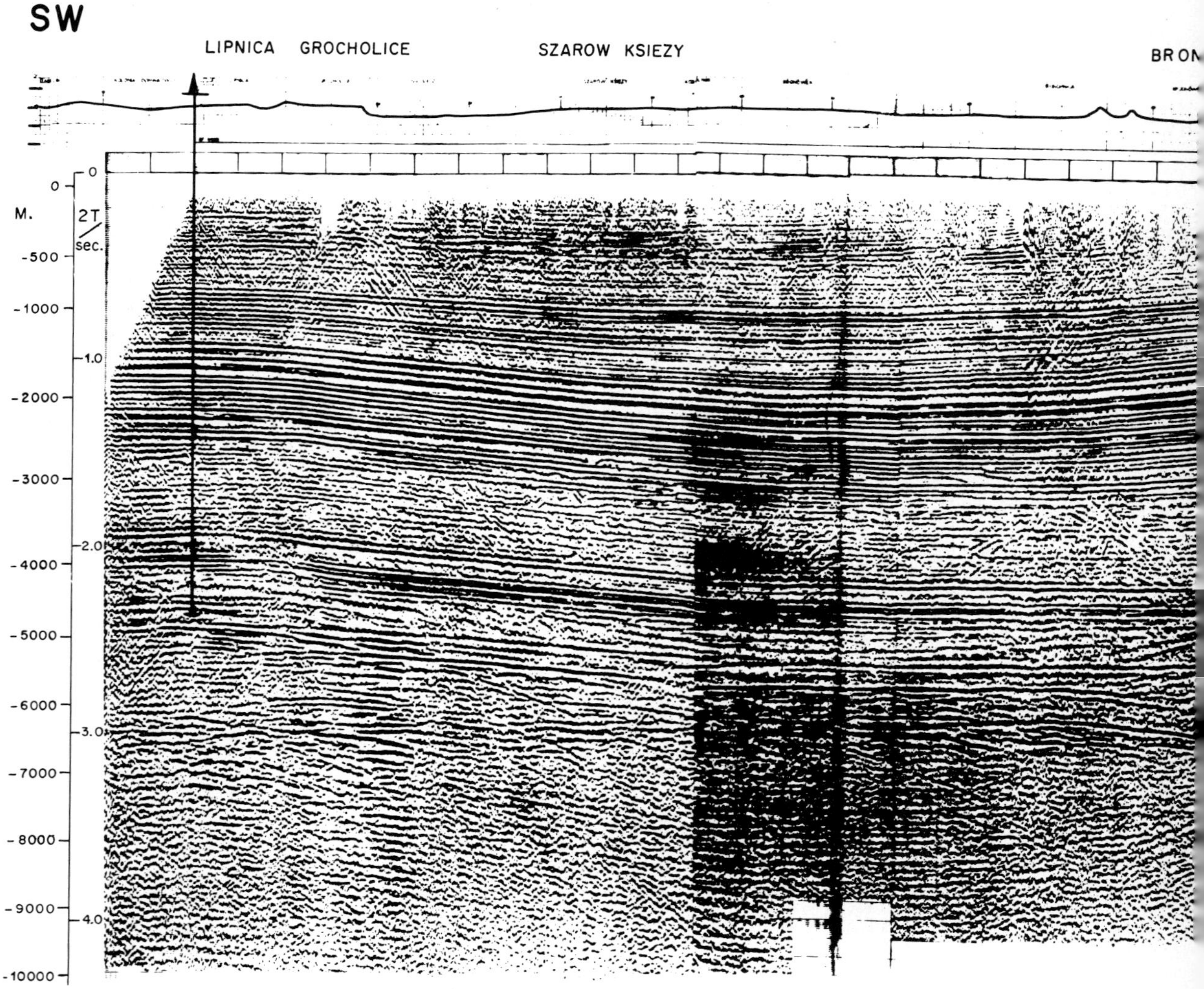

Fig. 21. Szeczecin-Lodz Troughs, seismic section 12b-12-12a-IV-84 K (Mesozoic). Signatures of the seismic horizons:
J_3 – Upper Jurassic
T_k – Triassic-Keuper
Tp_1 – Triassic-Lower Buntsandstein
Z_4 – Zechstein-Aller
Z_3 – Zechstein-Leine
Z_2 – Zechstein-Stassfurt-Main Dolomite
Z_1 – Zechstein-basal part of Werra
P_1 – Rotliegendes
D_3 – Upper Devonian (possibly)

from the Polish Lowlands (75.9%), and from the Carpathian Foredeep (22.9%).

The decline in petroleum production is the result of several factors. Besides the depletion of existing fields, the lack of exploration and development investment has been felt acutely.

4.4 Reserves

According to a recent disclosure by POGC, the recoverable reserves as of 01.01.91, or A + B + C1 + C2 categories in the Polish classification, were 35.8 million bbl and 5800 BCF. The remaining recoverable oil reserves were mostly located in the Carpathian Flysch (14.5 MM bbl) and Polish Lowlands (13.4 MM bbl). Polish Lowlands had 3220 BCF while Carpathian Foredeep had only 2520 BCF of remaining recoverable gas reserves.

Cumulative production at same date was 134 million bbl and 5040 BCF, therefore the ultimate recovery would be some 170 MM bbl and 11,000 BCF.

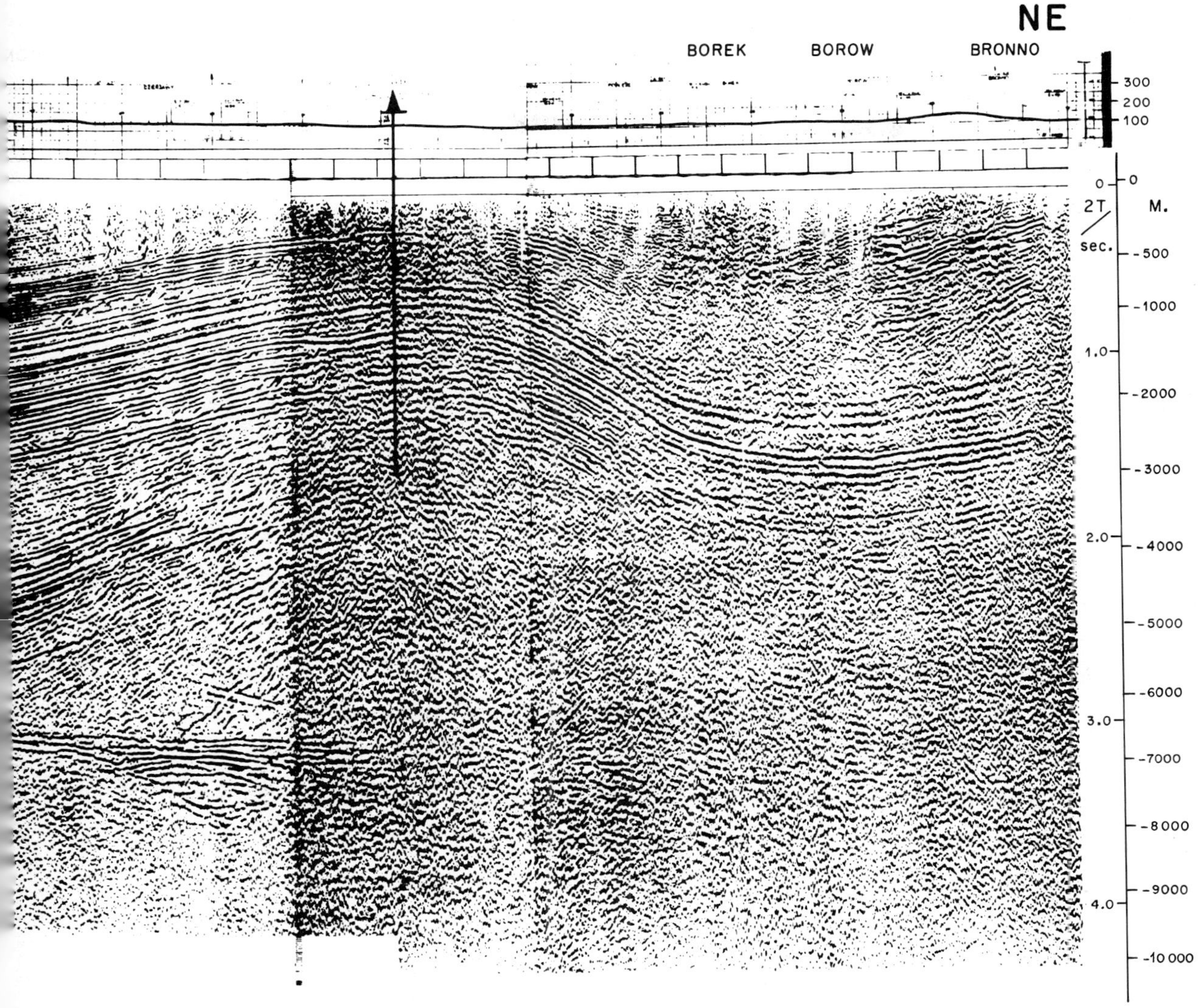

5 Hydrocarbon Potential

The results of exploration effort show that most of the territory of Poland is considered as prospective (Fig. 26). Only the Precambrian Mazury-Suwalki Uplift, the Paleozoic Sudetic Massif and the Holy Cross Mts. were classified as non-prospective. The hydrocarbon potential in Poland is estimated periodically (Figs. 27, 28). According to the geological staff of the Polish Oil and Gas Company, assuming that 1000 m^3 of gas equals 1 t of oil, the following in-place reserves are estimated as of end of 1990:

Natural gas	: 630.8 billion m^3
Associated gas	: 8.25 billion m^3
Crude oil	: 82.5 million t
	721.55 million toe in categories D1 and D2

This calculation is related to Polish classification of reserves and includes categories D1 and D2, where D1 = possible reserves in zones where fields were already discovered, D2 = possible reserves in other zones containing probable oil and gas fields. They

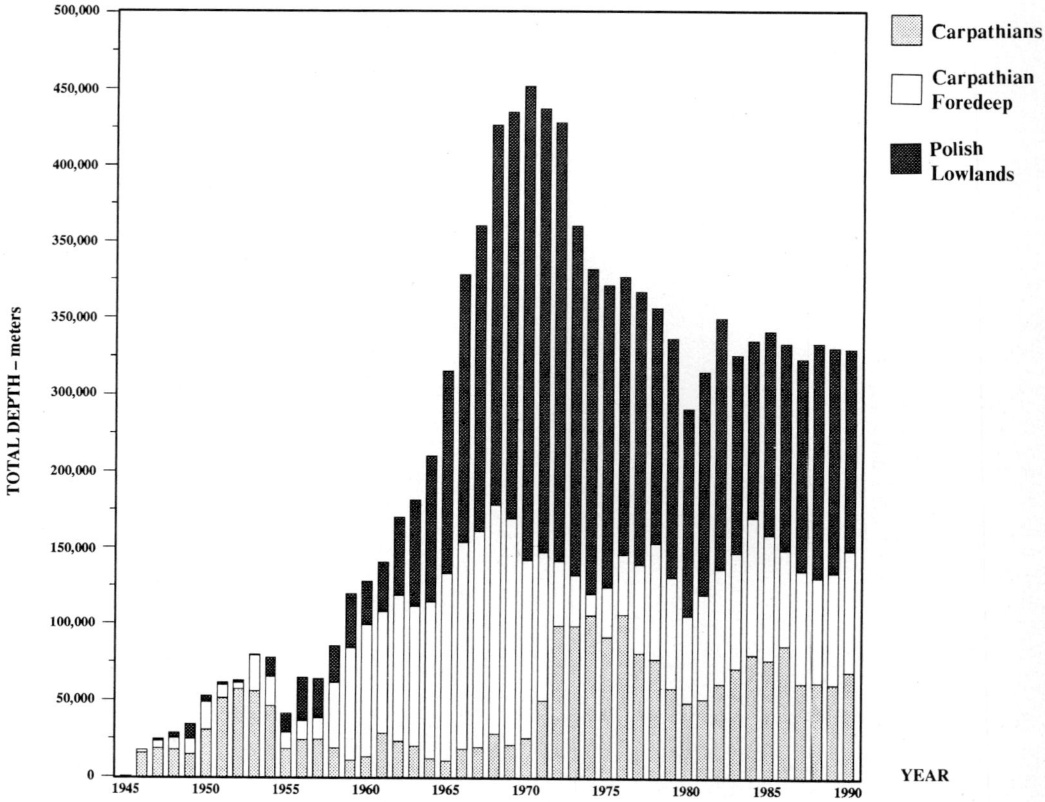

Fig. 22. Exploratory drilling activity in Poland 1945–1990

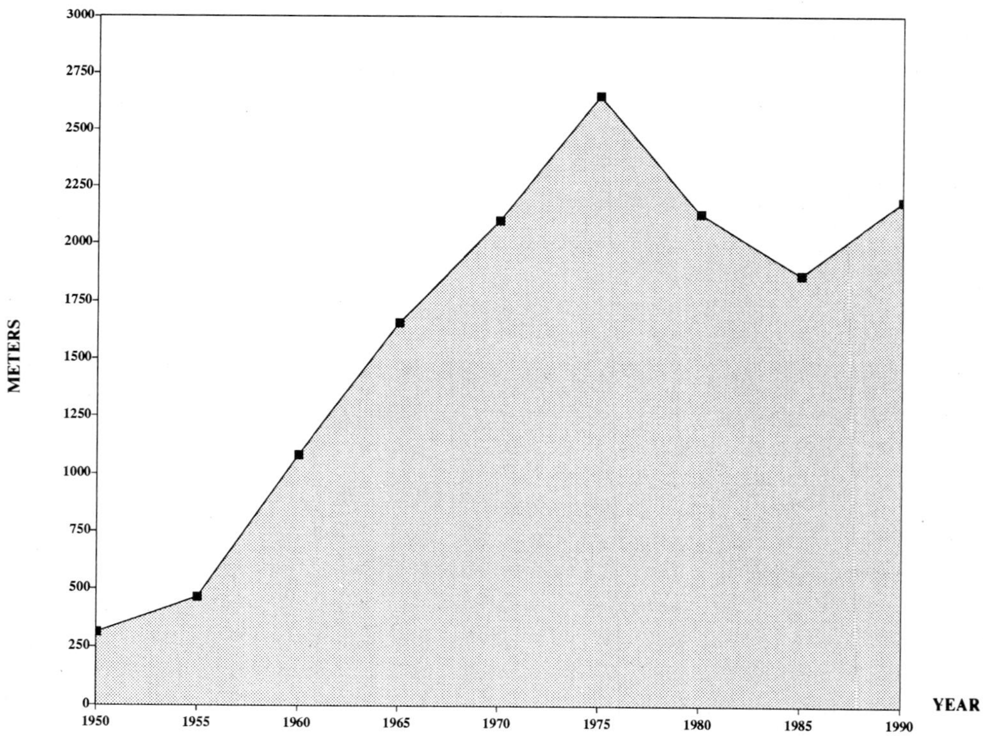

Fig. 23. Average depth of wells (1950–1990)

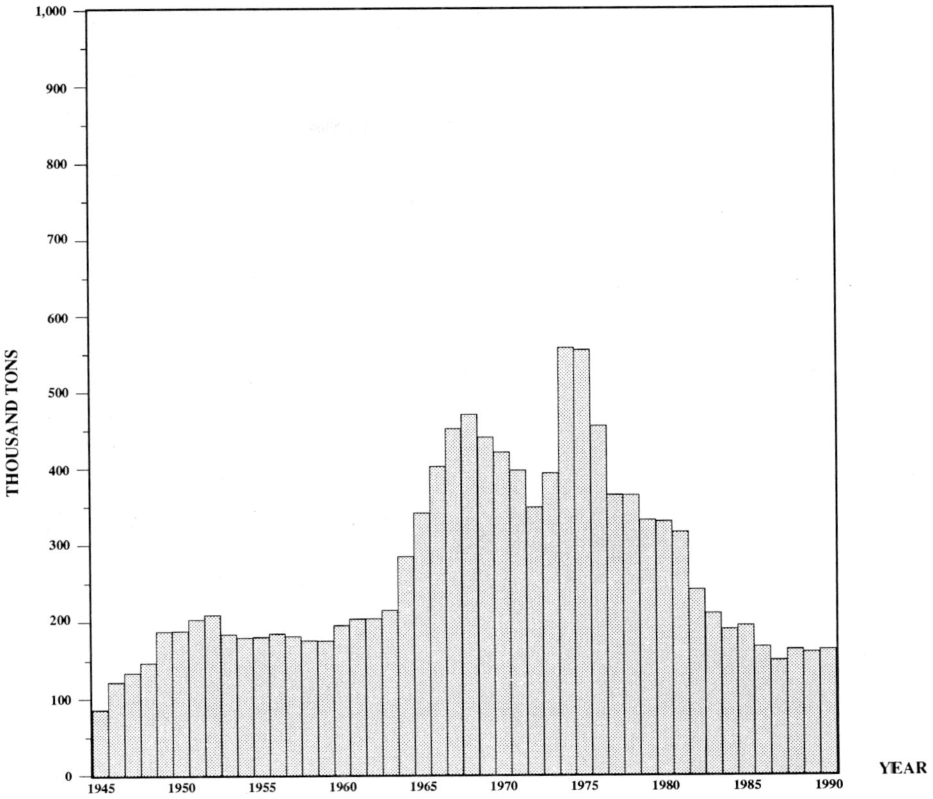

Fig. 24. Oil production in Poland (1945–1990)

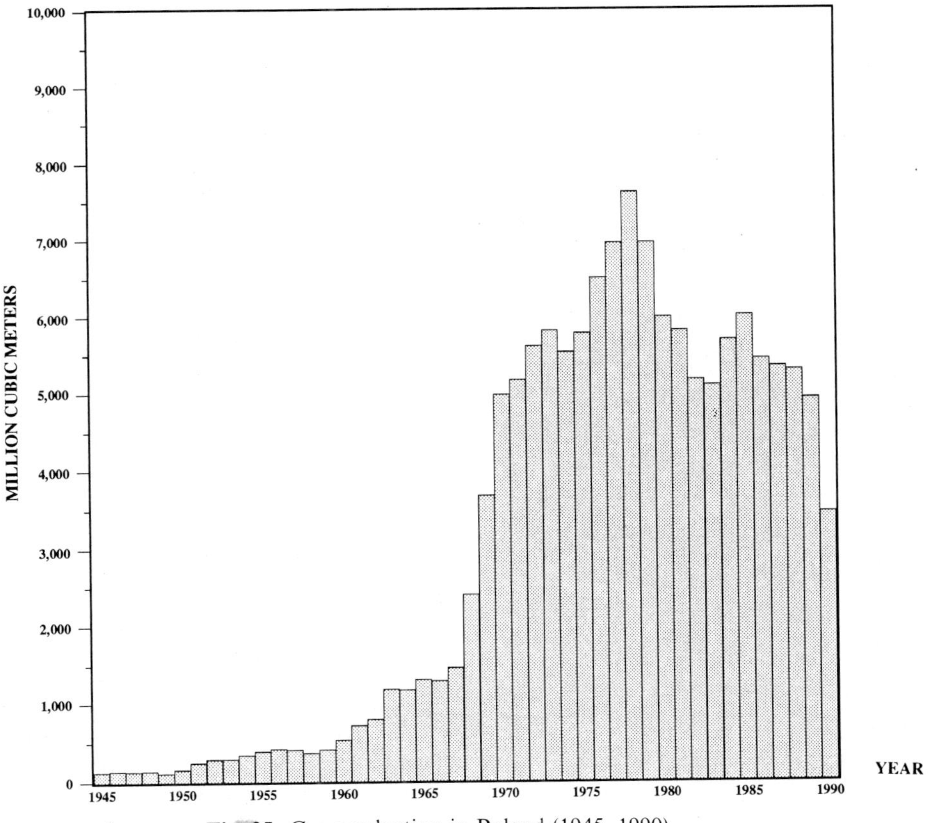

Fig. 25. Gas production in Poland (1945–1990)

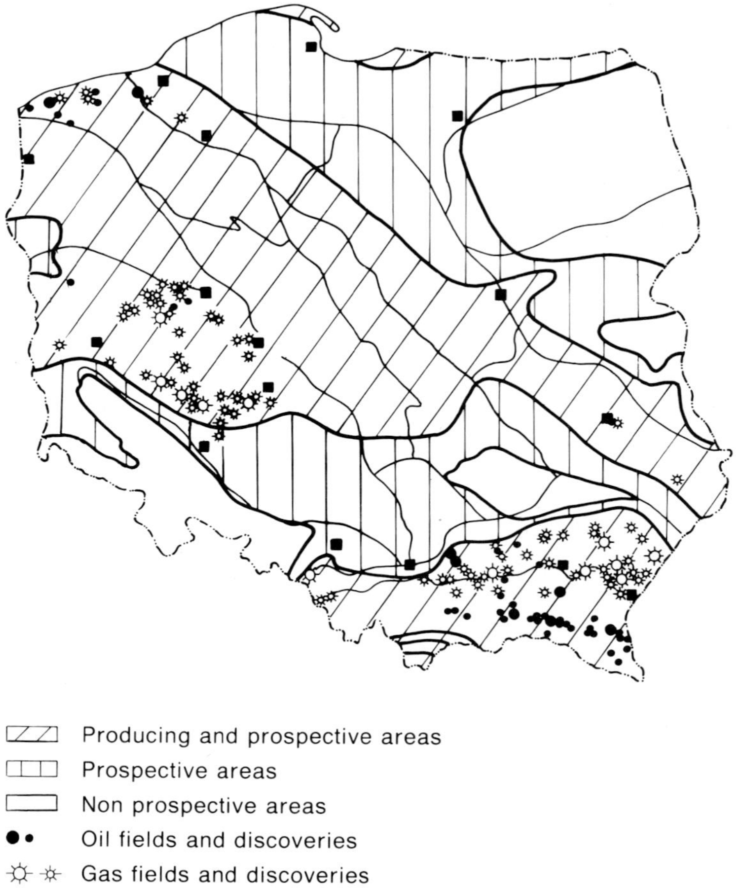

Fig. 26. Hydrocarbon prospective areas. (Data from POGC/Geonafta Bureau)

roughly correspond to the undiscovered potential recovery as defined by the World Petroleum Congress.

Another, more optimistic estimation prepared in 1987 by a group under Prof. J. Sokolowski's supervision, calculated total potential reserves as:

893 – 1378 billion m^3 of gas
227 – 352 million t of oil
1120 – 1730 million toe

5.1 Carpathian Province

The first oil discovery in the Carpathians dates back to 1854. Some 2,420,000 m have been drilled since in the *Carpathian Flysch* resulting in a density of 127.4 m km^{-2}. Seismic surveys were limited by the difficult operational conditions in mountainous terrain, inadequate equipment and more urgent demands for seismic work in other regions. Only 12,000 km of seismic lines have been recorded and 566 wells were drilled in the province. Evidence shows that a deeply buried platform extends below the Carpathian Flysch.

The Flysch is thought to be a continental rift zone which was active mainly in post-Jurassic. Many petroleum geologists consider rift zones as highly prospective for oil and gas. Most of the fields (6 oil and 12 gas) are producing from 200–2000 m depth interval (Table 1). As of 01.01.91, the D1 + D2 oil equivalent reserves were estimated at 39.2 million toe.

The figures for the *Carpathian Foredeep*, also in categories D1 and D2, are 139.7 million toe. Some 42 fields, 6 oil and 36 gas, have been discovered so far (Table 2). About 2,430,000 m of hole have been drilled, which give a quite high density of 142.9 m km^{-2} in this province, which has reached a mature stage of exploration.

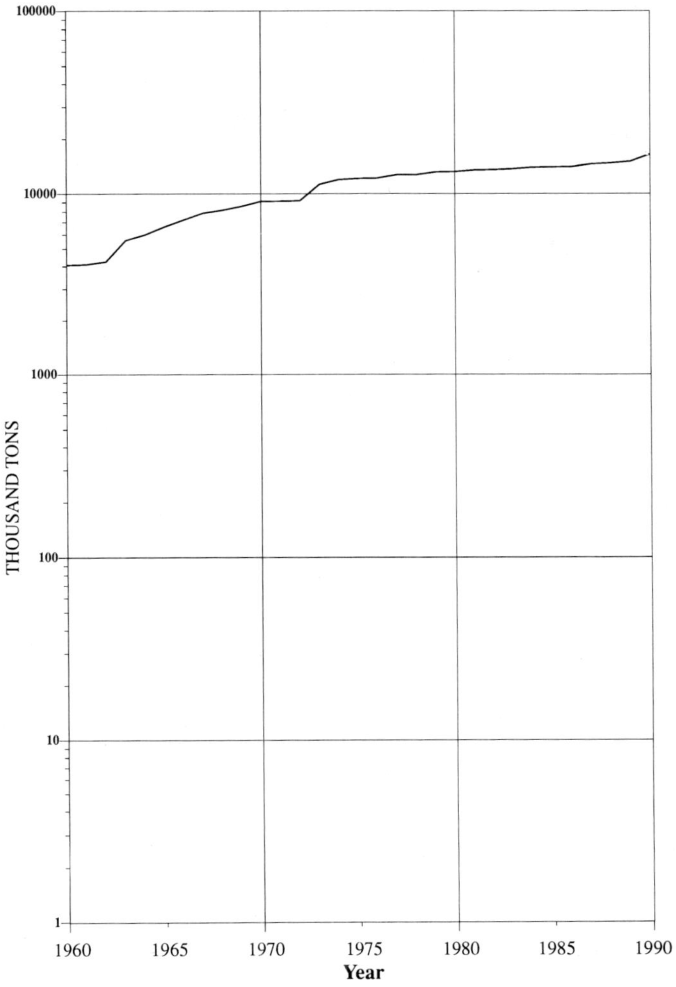

Fig. 27. Oil ultimate recoverable reserves (1960–1990)

5.2 Polish Lowlands

The geophysical surveying and exploratory drilling in the Polish Lowlands started in 1955. At the end of 1990 a total of 5,350,000 m of hole had been drilled in this province, resulting in a drilled density of 24.2 m km^{-2}. It should be mentioned, however, that the same index calculated for the depths below 3000 m does not exceed a value of 3 m km^{-2}. These density figures indicate that the Polish Lowlands could be the area in which exploration may lead to new discoveries.

Initial discoveries of oil and gas deposits in the Polish Lowlands occurred after 1960. The Rybaki oil field was discovered in 1961 and the Bogdaj-Uciechow gas field was discovered in 1964. At present, production comes from 21 oil fields and 45 gas fields (Table 3). Total hydrocarbon reserves in categories D1 and D2 are 542.65 million toe.

6 Conclusion

From comments on exploration history and hydrocarbon potential, it is evident that there is urgent need to increase the efficiency of exploration and production. This target is strictly dependent on:

- Further modernization of geophysical field equipment and of the processing centre.
- Increasing the amount of seismic surveys, especially 3-D recording.
- Maintaining the level of exploration drilling at some 300,000 m yearly, and development drilling up to 100,000 m.

Table 1. Characteristics of oil and gas fields in the Carpathian Flysch

No.	Field name	Discov. year	Depth pay m	Stratigr.	Reservoir	Field type	Prod. status	Reserves
1.	Bóbŕka-Rogi	1858	850–1050	Flysch	Sandstone	Oil	Producing	A = 110 th.
2.	Dąbrówka Tuch.	1966	280–600	Cretac.	Sandstone	Gas	Prod. 1992	R = 134.4 mln
3.	Gorlice-Glinik	1976	549–571	Cretac.	Sandstone	Gas	Producing	R = 22 mln
4.	Grabownica Wieś	1962	1125	Flysch	Sandstone	Gas	Not prod.	R = 322 mln
5.	Jaszczew-Męcinka	1972		Flysch	Sandstone	Gas	Producing	R = 2663 mln
6.	Jurowce-Srogów	1979	1227–1530	Flysch	Sandstone	Gas	Producing	R = 72 mln
7.	Łąkta	1971	1900	Jurassic	Sandstone	Gas	Producing	R = 6100 mln
				Cretac.	limest + dol.			
8.	Mrukowa	1954	350–600	Flysch	Sandstone	Oil	Producing	R = 84 th.
9.	Nosówka	1989	3390–3540	Carbonif.	Limestone	Oil	Producing	n.a.
10.	Osobnica	1953	500–1000	Flysch	Sandstone	Oil	Producing	R = 590 th.
11.	Roztoki-Sobniów	1971	713	Flysch	Sandstone	Gas	Producing	R = 4610 mln
12.	Rudawka Rymanow.	1950	250–550	Flysch	Sandstone	Gas	Producing	R = 4.8 mln
13.	Sanok-Zabłotce	1950	820	Flysch	Sandstone	Gas	Producing	R = 914 mln
14.	Słopnice	1973	1470–4251	Fl. + Cretac.	Sandstone	Cond.gas	Producing	R = 2850 mln
15.	Strachocina	1927	392–788	Flysch	Sandstone	Gas	Not prod.	G = 382.4 mln
16.	Szalowa	1946	425	Flysch	Sandstone	Gas	Producing	R = 215 mln
17.	Wańkowa	1963	210–400	Flysch	Sandstone	Oil	Producing	n.a.
18.	Węglówka	1956	250–1000	Cretac.	Sandstone	Oil	Producing	R = 2360 th.

Note: Figures are m^3 for gas, tons for oil and toe for oil and gas fields.

R = recoverable: A = available (remaining) G = geological (in place) n.a. = data not available.

Table 2. Characteristics of oil and gas fields in the Carpathian Foredeep

No.	Field name	Discov. year	Depth pay m	Stratigr.	Reservoir	Field type	Prod. status	Reserves
1.	Bochnia-Gdów	1946	704	Miocene	Sandstone	Gas	Producing	n.a.
2.	Borek	1968	515–600	Miocene	Sandstone	Gas	Producing	n.a.
3.	Brzeźnica	1966	285–580	Miocene	Sandst. shal.	Gas	Producing	R = 300 mln
4.	Brzezowiec	1976	341–838	Cretac.	Sandstone	Gas	Producing	R = 68 mln
5.	Brzezówka	1964	1940–1990	Mioc. + Juras.	Sandst. limest	Oil	Producing	R = 185 th.
6.	Cetynia	1958	780–1037	Miocene	Anhydr. sandst.	Gas	Producing	R = 136.2 mln
7.	Czarna Sędziszow.	1967	625–1000	Miocene	Sandstone	Gas	Producing	R = 688 mln
8.	Dąbrówka	1966	546–601	Miocene	Sandstone	Gas	Not prod.	R = 593 mln
9.	Dąbrowa Tarnowska	1962	661–693	Jurassic	Limestone	Oil + gas	Not prod.	R = 43.4 th.
10.	Grabina-Nieznanowice	1971	282–745	Miocene	Sandstone	Gas	Producing	A = 397 mln
11.	Grobla-Pławowice	1962	360	Juras. Cretac.	Lim. sandst.	Oil	Producing	R = 2685 th.
12.	Husów-Albigowa-Krasne	1963	43–2248	Miocene	Sandstone	Gas	Producing	A = 3528 mln
13.	Jarosław	1959	835–1470	Miocene	Sandstone	Gas	Producing	R = 7626 mln
14.	Jodłówka	1980	2870	Miocene	Sandstone	Gas	Prod. 1993	n.a.
15.	Jasprząbka Stara	1986	1207–1366	Cretac.	Sandstone	Oil	Producing	G = 1231.3 th.
16.	Jaśniny	1980	781–863	Miocene	Sands. mudst.	Gas	Producing	R = 527 mln
17.	Kańczuga	1959	1102–1498	Miocene	Sandst. shale	Gas	Producing	G = 765 mln
18.	Kielanówka	1978	2307	Cretac.	Sandstone	Gas	Producing	A = 1331 mln
19.	Korzeniów	1968	158–1230	Miocene	Sandstone	Oil + gas	Producing	A = 155 th.
20.	Lubaczów	1957	634–993	Juras. Mioc.	Limest. sandst. anhydrite	Gas	Producing	A = 1533 mln
21.	Maćkowice	1962	785–2098	Miocene	Sandstone	Gas	Producing	R = 298 mln
22.	Marklowice	1950	500–600	Carbonif.	Sandst. shale	Gas	Not prod.	R = 887 mln
23.	Mirocin	1962	650–1450	Miocene	Sandstone	Gas	Producing	R = 5154 mln
24.	Niwiska	1956	1076	Triassic	Limestone	Gas	Not prod.	R = 30 mln
25.	Partynia	1958	800–840	Jurassic	Limestone	Oil	Not prod.	R = 115 th.
26.	Pruchnik-Pantalowice	1966	430–1000	Miocene	Sandstone	Gas	Not prod.	A = 2796 mln
27.	Przemyśl	1958	400–2570	Miocene	Sandstone	Gas	Producing	R = 74 210 mln
28.	Przeworsk	1972	273–420	Miocene	Sandst. shale	Gas	Producing	A = 259 mln
29.	Pilzno	1980	170–755	Miocene	Sandstone	Gas	Producing	n.a.
30.	Raciborsko	1971	528–685	Miocene	Sandstone	Gas	Producing	R = 460 mln
31.	Rzeszów	1978		Miocene	Sandstone	Gas	Producing	A = 595 mln
32.	Sarzyna	1966	420–663	Miocene	Sandstone	Gas	Producing	R = 220 mln
33.	Smęgorzów	1962	462	Jurassic	Limestone	Gas	Producing	n.a.
34.	Swarzów	1958	620–680	Cretac.	Sandstone	Gas	Not prod.	R = 322 mln
35.	Święte-Zadąbrowie	1971	1649	Miocene	Shale	Gas	Producing	R = 150 mln
36.	Tarnów	1966	1677–1725 462–739	Jurassic Miocene	Dolom. limest. sandstone	Gas	Producing	R = 5680 mln
37.	Uszkowce	1959	850–1220	Miocene	Sandstone mudstone	Gas	Producing	R = 1130 mln
38.	Wojnicz-Zakrzów	1964	320 1036	Miocene	Sandstone	Gas	Producing	R = 132 mln
39.	Zalesie	1982	2070	Miocene	Sandstone	Gas	Producing	n.a.
40.	Zagorzyce-Sędziszów	1964	678–1567	Miocene	Sandstone	Gas	Not prod.	R = 413 mln
41.	Żołynia	1962	195–540	Miocene	Sandstone	Gas	Producing	R = 4396 mln
42.	Żukowice	1978	1072–1303	Miocene	Sandstone	Gas	Producing	A = 325 mln

Note: Figures are m^3 for gas, tons for oil and toe for oil and gas fields. R = recoverable: A = available (remaining) G = geological (in place) n.a. = data not available.

Table 3. Characteristics of oil and gas fields in the Polish Lowlands

No.	Field name	Discov. year	Depth pay m	Age of reservoir	Lithology	Field type	Production status	Reserves
Foresudetic Monocline								
1.	Antonin	1969	1288	Zechst. limest.	Lim. dol	Gas	Producing	A = 845 mln
2.	Bogdaj-Uciechów	1964	1297–1313	Rotliegendes zechstein	Sandstone carbonates	Gas	Producing	R = 19800 mln
3.	Borzęcin	1969	1295	Zechst. lim. rotliegendes	Limestone	Gas	Producing	A = 1482 mln
4.	Brzostowo	1976	1387	Perm. + carbon.	Sandstone	Gas	Producing	A = 426 mln
5.	Buk	1981	2900	Rotliegend.	Sandstone	Gas	Producing	A = 122 mln
6.	Bukowiec	1976	2690	Rotliegend.	Sandstone	Gas	Producing	A = 504 mln
7.	Ceradz-Kalwy	1977	3207	Rotliegend.	Sandstone	Gas	Producing	A = 626 mln
8.	Chraplewo	1976	3020–3039	Rotliegend.	Sandstone	Gas	Not prod.	A = 319 mln
9.	Czeszów (Trzebnica)	1967	1300	Zechst. lim.	Limest. dolom.	Gas	Producing	A = 748 mln
10.	Czerwieńsk	1971	1881–1896	Main dolom.	Dolomite	Oil	Producing	n.a.
11.	Chartów	1990		Main dolom.	Carbonates	Oil	Producing	n.a.
12.	Duszniki	1980	3341	Rotliegend.	Sandstone	Gas	Producing	R = 90 mln
13.	Góra	1977	1339	Rotliegend.	Sandstone	Gas	Producing	A = 1816 mln
14.	Gaj	1980	3041	Main dolom.	Dolomite	Oil	Producing	n.a.
15.	Górzyca	1989	2711–2746	Main dolom.	Dolomite	Oil + gas	Not prod.	R = 970 th.
16.	Grodzisk-Granowo	1975	2682–2709	Rotliegend.	Sandstone conglomerate	Gas	Producing	R = 2500 mln
17.	Henrykowice	1960	1399	Zechst. limest. rotliegend.	Sandstone	Gas	Producing	A = 150 mln
18.	Jarocin	1977	2654	Rotliegend.	Sandstone	Gas	Producing	A = 138 mln
19.	Janowo	1972	1565	Zechst. limest.	Limestone	Gas	Producing	A = 273 mln
20.	Grochowice	1980	1614–1671	Rotliegend.	Sandstone	Gas	Prod. 1996	R = 2400 mln
21.	Kaleje	1974	3016	Rotliegend.	Sandstone	Gas	Producing	A = 850 mln
22.	Jastrzębsko	1990		Main dolom.	Carbonates	Oil	Producing	n.a.
23.	Jeniniec	1986		Main dolom.	Carbonates	Oil	Producing	n.a.
24.	Klęka	1975	2877	Rotliegend.	Sandstone	Gas	Producing	A = 869 mln
25.	Kopanki W	1977	2700	Rotliegend.	Sandstone	Gas	Producing	A = 258 mln
26.	Niemierzyce	1979	2757	Rotliegend.	Sandstone	Gas	Producing	A = 55 mln
27.	Maszewo	1970	1612–1651	Main dolom.	Dolomite	Oil	Producing	n.a.
28.	Nowa Sól	1966	1008–1060	Main dolom.	Dolomite	Oil	Producing	n.a.
29.	Otyń	1963	1073–1112	Zechst. limest	Dolomite	Gas	Producing	A = 106 mln
30.	Kargowa	1972	2071–2170	Main dolom.	Limestone	Gas	Prod. 1995	R = 2650 mln
31.	Podrzewie	1978	3311	Rotliegend.	Sandstone	Gas	Producing	A = 477 mln
32.	Porażyn	1977	2725	Rotliegend.	Sandstone	Gas	Producing	A = 225 mln
33.	Paproć	1980	2490–2630	Rotliegend. carbonifer.	Sandstone	Gas	Producing	R = 4450 mln
	Paproć E		2571				not prod.	R = 80 mln
	Paproć W		2558				not prod.	n.a.

34.	Radlin	1982	3066–3131	Rotliegend.	Sandstone	Gas	Not prod.	R = 6700 mln
35.	Radziądz	1969	1445	Zechst. limest.	Dolomite	Gas	Producing	A = 634 mln
36.	Rybaki-Połęcko	1961	1600–1850	Main dolom.	Dolomite	Oil	Producing	R = 140 thous.
37.	Piekary	1982	2940	Rotliegend.	Sandstone	Gas	Producing	n.a.
38.	Sątopy	1978	2750	Rotliegend.	Sandstone	Gas	Producing	A = 70 mln
39.	Stęszew	1979	2845	Rotliegend.	Sandstone	Gas	Producing	A = 198 mln
40.	Strykowo	1980	2820	Rotliegend.	Sandstone	Gas	Producing	R = 290 mln
41.	Strzępiń	1982	2750	Rotliegend.	Dolomite	Gas	Producing	n.a.
42.	Sulęcin	1973	2911–2961	Main dolom.	Dolomite	Oil	Producing	R = 77 th.
43.	Tarchały	1966	1330	Zechst. limest.	Sandstone	Gas	Producing	A = 2126 mln
			1507	rotliegend.				
44.	Ujazd	1976	2595	Rotliegend.	Sandst. congl.	Gas	Producing	R = 3660 mln
45.	Wierzchowice	1971	1324	Zechst. limest.	Carbonates	Gas	Producing	A = 5392 mln
				rotliegend.	sandstone			
46.	Wiewierz E	1971	1325	Rotliegend.	Sandstone	Gas	Not prod.	A = 144 mln
47.	Wilków	1980	1447	Rotliegend.	Sandstone	Gas	Producing	R = 4253 mln
48.	Załęcze	1971	1231–1354	Rotliegend.	Sandstone	Gas	Producing	R = 20400 mln
49.	Zbąszyń	1974	2181–2212	Main dolom.	Dolom. limest.	Gas	Prod. 1995	R = 2400 mln
50.	Żakowo	1968	1697–1740	Main dolom.	Limest. dolom.	Gas	Not prod.	R = 2150 mln
51.	Żuchlów	1978	1208–1342	Rotliegend.	Sandstone	Gas	Producing	R = 22000 mln
Pomeranian High								
1.	Daszewo	1980	2775	Main dolom.	Dolomite	Oil	Producing	R = 225 th.t
2.	Białogard	1982	3149–3184	Perm. + Carbon.	Sandstone	Gas	Producing	A = 540 mln
3.	Błotno	1980	3181	Main dolom.	Limestone	Oil	Producing	n.a.
4.	Gorzysław N + S	1976	2692–2832	Carbonifer.	Sandstone	Gas	Producing	R = 1810 mln
5.	Jarkowo	1986		Main dolom.	Dolomite	Oil	Producing	n.a.
6.	Kamień Pomorski	1972	2331	Main dolom.	Dolomite	Oil	Producing	R = 1900 th.
7.	Międzyzdroje	1971	2843	Main dolom.	Dolomite	Oil	Producing	n.a.
8.	Petrykozy	1977	2735	Main dolom.	Dolomite	Oil	Producing	n.a.
9.	Rekowo	1975	2665	Main dolom.	Dolomite	Oil	Producing	n.a.
10.	Trzebusz	1978	2845	Rotlieg + Carb.	Sandstone	Gas	Producing	A = 97 mln
11.	Wapnica	1979	2805	Main dolom.	Dolomite	Oil	Producing	n.a.
12.	Wysoka Kamieńska	1978	3036	Main dolom.	Limestone	Oil	Producing	R = 300 th.
Peri-Baltic Syneclise								
1.	Dębki-Żarnowiec	1972	2695–2765	Cambrian	Sandstone	Oil	Producing	R = 100 th.
Lublin Trough								
1.	Ciecierzyn	1984	3558	Devonian	Dolom. limest.	Gas	Not prod.	R = 1360 mln
2.	Świdnik	1970	892	Carbonifer.	Sandstone	Oil	Producing	R = 28 th.

Note: Figures are m^3 for gas, tons for oil and toe for oil and gas fields.

R = recoverable: A = available (remaining) G = geological (in place) n.a. = data not available.

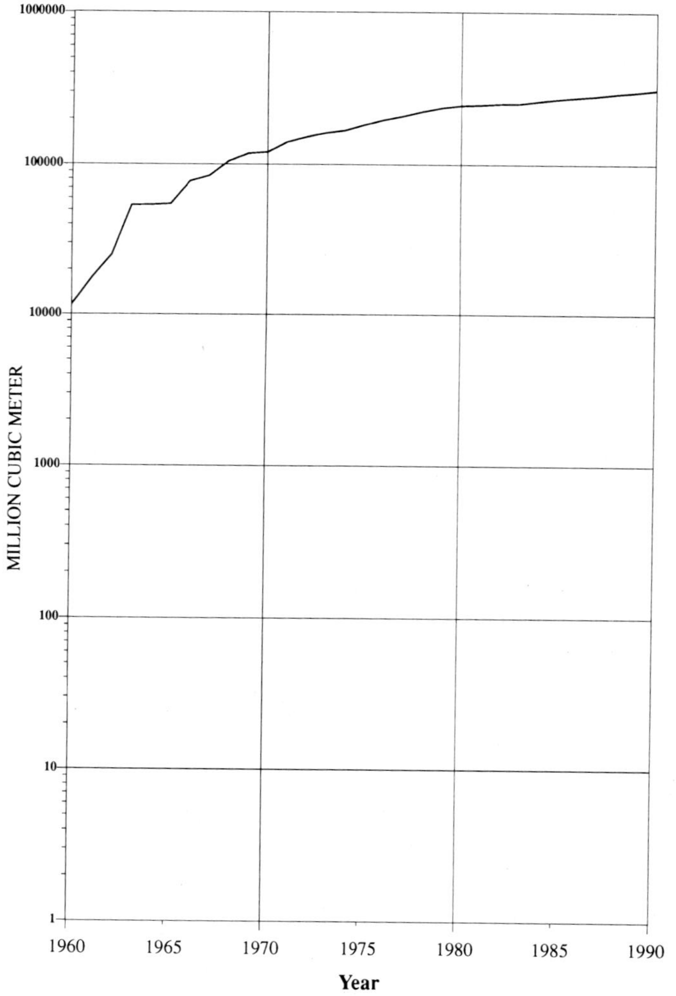

Fig. 28. Gas ultimate recoverable reserves (1960–1990)

- Increasing the volume of sophisticated laboratory analysis.
- Modern study of sedimentary basin development and basin modelling.
- Application of new exploration technologies.
- Preparation of underground storage sites.
- Implementing the exploitation technology for coal-bed methane resources.

Acknowledgements. The author would like to acknowledge the assistance of P. Karnkowski, W. Weil and S. Radecki from the Polish Oil and Gas Company for their support. Major Polish contributions are listed in the following "Suggested Reading". Unfortunately, they are all in Polish and not always accessible to foreign readers. I would also like to thank Bogdan Popescu from Petroconsultants S.A., whose comments and suggestions greatly improved the initial version of this paper.

References

Borys Z, Cisek R, Czernicki J (1989) Nowe perspektywy poszukiwań słóż weglowodorów w piaskowcacn dolnej kredy jednostki skolskiej w Karpatach. Nafta 10–12: 151–157.

Karnkowski P (1980) Wgłębne przekroje geologiczne przez Niż Polski. Wyd. Geologiczne Warszawa.

Karnkowski P (1989) Utwory deltowe przeógórza Karpat. Przegl. Geol. 1: 28–32.

Kulczyk T, Zołnierczuk T (1988) O najważnicjszych problemach górnictwa nafty i gazu w zachodniej częsci Niżu Polskiego. Nafta 11–12: 276–280.

Wierzchowska Kicułowa K (1984) Budowa geologiozna utworów podperinskich monokliny przedsudeckiej. Geol. Sudetica 1: 121–142.

Wilczek T, Merta H (1992) Wstepne wyniki badań pirolitycznych metodą Rock-Eval. Nafta Gaz 5–6: 100–115.

Suggested Reading

Antonowicz L, Knieszner L (1975) O rafach dolomitu głównego w Polsce. Nafta 3–4

Czernicki J, Moryc W (1990) Złoże ropy naftowej Nosówka koło Rzeszowa. Nafta 45, 49–54

Depowski S, Tyski S, Stolarczyk F (1979) Ropo-i gazonośność paleozoiku polskiej cześci syneklizy perybałtyckiej. Przegl. Geol 11: 593

Górecki W (1991) Petroleum opinions in the years 1978–1991. Academy of Mining and Metallurgy, Kraków (manuscript)

Jawor E (1983) Utwory miocenu między Krakowem a Dębicą. Przegl. Geol 11

Karnkowski P (1962) Uwagi o roponośności i gazonośności polskich Karpat fliszowych i ich przedgórza. Przegl Geol 7: 333–338

Karnkowski P (1978) Paleodelta w miocenie Przedgórza Karpat. Przegl Geol 11

Karnkowski P, Cisek R, Palij A, Borys Z (1988) Budowa geologiczna polskich Karpat wschodnich i perspektywy ropogazonośności. Nafta 44, 1–2

Karnkowski P (1990) Stan i perspektywy rozwoju geologii naftowej w Polsce. Technika Poszukiwań Geologicznych, Geosynoptyka i Geotermia 3–4: 1–5

Kuśmierek J, Ney R (1988) Problemy tektoniki podłoża a rozwój struktur pokrywy wschodniej części Karpat polskich. Przegl Geol 6

Nowotarski Cz (1985) Geologiczna efektywność badań geofizycznych Zakładu Geofizyki Kraków i jej znaczenie dla poszukiwań naftowych. Nafta 41, 10

Program prac poszukiwawczych Polskiego Górnictwa aftowego i Gazownictwa na rok 1991 i na lata 1991–1995 (1990) Biuro Geologiczne GEONAFTA, Warszawa (unpublished)

Rudzik M, Zagórski J (1986) Udział Biura Geologicznego GEONAFTA w rozpoznaniu sejsmicznym Polski. Nafta 42, 11

Sokołowski J, Deczkowski Z (1977) Zmiany układów strukturalnych głównych formacji perspektywicznych obszaru przedsudeckiego jako zasadniczy czynnik decydujący o migracji i akumulacji węglowodorów. Nafta 6: 181–193

Weil W (1989) Zarys geologii naftowej Polski. Biuro Geologiczne GEONAFTA, Warszawa (manuscript)

Wilczek T (1986) Ocena możliwości powstawania węglowodorów w mezozoicznych skałach macierzystych Niżu Polskiego. Przegl Geol 9: 496–502

7 Exploration History and Hydrocarbon Prospects in Romania

NELU IONESCU[1]

CONTENTS

1 Introduction

Romania enjoys the privilege of being the first country registered in world statistics with a commercial production of 275 metric tonnes of crude oil in 1857 (Table 1), but historical documents attest the existence of rudimentary oil means of production, many centuries before, in the Carpathian foothills. The first natural gas discovery is related to a well drilled for potash salt in 1909, at Sarmasel, in the Transylvanian Depression. From the first discoveries up to now, the Romanian companies either working alone, or together with foreign companies (especially, during the early stages of exploration) discovered more than 400 oil and gas fields in all sedimentary provinces.

It is worth mentioning that the first foreign oil company was registered in Romania in 1864, but the best known were: Steaua Romana, with German capital, registered in 1896, Romano-Americana, subsidiary of Standard Oil, registered in 1904 and Astra Romana, subsidiary of Royal Dutch Shell, registered in 1910.

The oil production of Romania has increased continuously, with two periods of decline after the World Wars, and peaking in 1976, with a year by production of 14.7 MM tons (294,000 bbl/day). The present oil production in Romania ranges between 6.0 and 6.5 MM tons per year (120–130 M b/d), of which 8.5% comes from offshore the Black Sea.

[1] 109 Str. Isbiceni, Bucharest, Romania.

2 Exploration History

The exploration for hydrocarbons in Romanian territory can be considered to have reached a mature stage. Almost all geological units, (some 140,000 km^2 of which 22,000 km^2/offshore) have been explored systematically, using several geological and geophysical methods: geological surveys, gravimetric, magnetometric, electrometric and seismic measurements. Drilling has proceeded at various depths from a few hundred metres to 7025, the national drilling record.

Geological maps have been issued at various scales: 1:1,000,000, 1:500,000, 1:200,000, 1:100,000; they were based on detailed field geological mapping at scales of: 1:50,000, 1:20,000, 1:10,000 and even 1:5,000 in some regions. Geophysical synthesis maps for gravimetric, magnetometric and electrometric studies are available at 1:100,000 scale. The isochrone or isobath maps drawn up on the basis of the seismic results, on several scales, are correlated with drilling results. Some good reflectors have been correlated across large areas, which has made possible complete seismic map synthesis at scales of 1:100,000, in many geological units.

Presenting the figures in Table 2, without comment, could lead to the conclusion that the prospects of discovering new oil and gas fields in Romania is extremely low; this is wrong. For this reason we consider it necessary to make the following important observations:

Table 1. Romania's "Firsts"

- In 1857 Romania produced 275 tons of crude, being the first country officially registered in world statistics with a commercial crude output
- In 1859, Bucharest used kerosene for public lighting purposes. Kerosene was refined locally, at Ploiesti
- In 1909, the first gas field was discovered at Sarmasel in the Transylvania Basin
- The first European gas piping system was built in Transylvania, in 1913
- Up until the second world war, Romania ranked as second oil-producing country in Europe after Russia
- The production record highs:

Oil : 1976 – 14.7 million tons (294,000 BOPD)
Gas : 1985 – 35.2 billion m^3 (3,405,692 MCFD)

Table 2. Key exploration figures (situation as of 01.01.91)

Seismic data
- Onshore: 142,612 km of seismic lines; average density: 1.21 km km^{-2}
- Offshore: 42,441 km of seismic lines; average density: 1.93 km km^{-2}

of which, using modern CDP technologies:

- Onshore: 50,798 km 6- and 12-fold; average density = 0.43 km km^{-2}
  12,309 km 24-fold; average density = 0.10 km km^{-2}
- Offshore: 13,715 km 6- and 12-fold; average density = 0.62 km km^{-2}
  27,425 km 24-fold; average density = 1.25 km km^{-2}

Drilling
- Onshore: 23,600 geological[a] wells; average density 1 well 5 km^{-2}
- Offshore: 50 geological wells; average density 1 well 440 km^{-2}
- Number of wells deeper than 4000 m
  - onshore average density 365 = 1 well 323 km^{-2}
  - offshore average density 2 = 1 well 11,000 km^{-2}

[a]According to the well classifications used in Romania, "geological wells" are understood to be wells which have contributed to the discovery and the delineation of the oil or gas fields. This would include wells classified by AAPG as new field wildcats, appraisal wells, and in part extension (or development) wells. They can be loosely called exploratory wells.

- from the total volume of seismic lines carried out onshore, only 50,798 km were recorded using CDP technique, and from these only 12,309 km were run using 24 CDP fold coverage. No higher fold coverage has been used;
- 3-D techniques were used only in surveying 175 km^2 onshore, and 760 km^2 offshore;
- the number of wells bottoming deeper than 4000 m is 365 onshore and 2 offshore.

The last point calls for a presentation of the results for each of the major geological units of Romania, to obtain a correct picture of the actual level of knowledge of hydrocarbon occurrence, and of future prospects.

After more than 100 years of exploration for and production of hydrocarbons in Romania, a few conclusions can be made regarding the results of this activity:

1. The basic conditions thought to account for the genesis, accumulation and preservation of hydrocarbons are present in almost all the Romanian geological units. Numerous oil and gas fields have been discovered in these units.
2. There is a good understanding of the geological formations situation between 0 and 3500 m, which are mainly of a Neogene age. Less well understood are the geological formations deeper than 3500/4000 m, which are basically pre-Neogene.
3. The seismic results are unsatisfactory in the regions with complex geology and hard relief. They are generally poor at depths below 3500 m, both onshore and offshore.
4. The source rocks have been identified and described mainly using old technology and less often, based on modern technology. Anyhow, the oil and gas fields are the best argument in favour of the existence of the source rocks.
5. Many rocks with good petrophysical reservoir parameters exist from the Paleozoic to the Tertiary.
6. There is a large variety of traps. The predominant type is structural.
7. The Romanian oil and gas fields can be considered as small to medium in size but occurring with a remarkable frequency.
8. Despite intense exploration activity utilizing geophysical methods and exploratory wells, many large unexplored areas of interest, both onshore and offshore, still remain.

3 Geological Framework

The main geological element of Romania is represented by the Carpathian Mountains Arc (Fig. 1). They have three main branches: East Carpathians, South Carpathians and Apuseni Mountains.

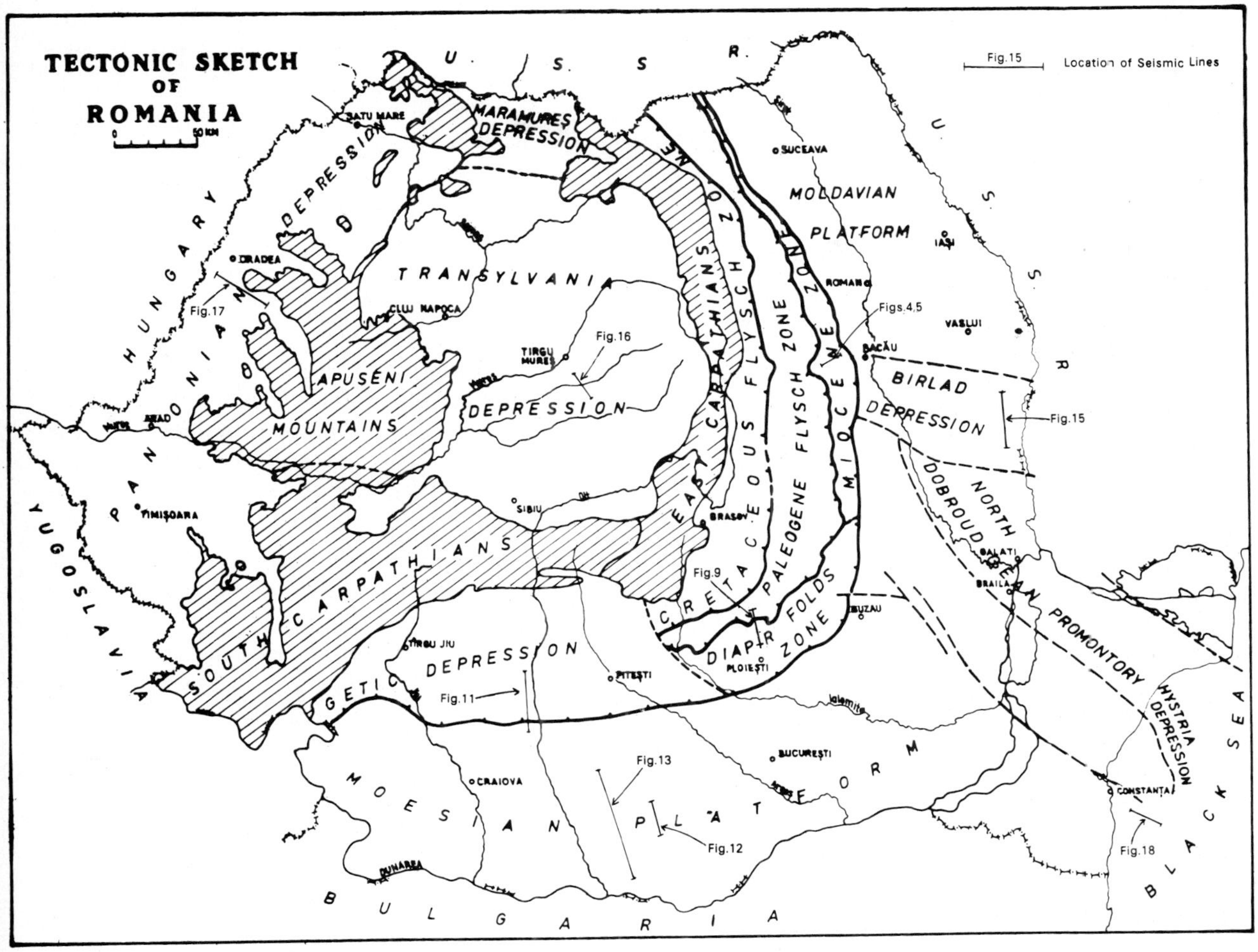

Fig. 1. Tectonic sketch map of Romania

Outside the Carpathians lie the Moldavian Platform (part of the East European Platform), North Dobrudgea and the Moesian Platform, which represent the Carpathian foreland. Overthrust relations were found between the East Carpathians and the South Carpathians on one hand, and the foreland units on the other. In the interior of the Carpathian chain lies the Transylvania Depression. The Pannonian Depression lies west of the Apuseni Mountains and the South Carpathians, in the western part of the country. In the southeast part of the country, the North Dobrudgean Promontory sinks northwestwards under the Carpathians, and actually represents vestiges of the Hercynian orogenesis. Between the North Dobrudgean Promontory and the Moldavian Platform lies the Predobrudgean (Birlad) Depression.

Recent exploration work in the Black Sea offshore within Romanian territory, shows the existence of a depression developed during the Tertiary. The basin formed over the pre-existing Hercynian paleorelief and overlies the eastward extension of the North Dobrudgean Promontory, Moesian Platform and Predobrudgean Depression. All the above-mentioned geological elements are hydrocarbon-bearing provinces, each with distinct genetic, migration and trapping characteristics.

4 Some Observations Regarding the Main Oil and Gas Provinces

Most of sedimentary provinces in Romania are petroliferous basins. They are confined to the Carpathian nappes, foredeep and foreland. Posttectonic basins, located in the Paratethys realm, offshore Black Sea in central and in the western Romania also proved to be hydrocarbon-bearing. The following description of geology per petroliferous basin is preceded by a basin data sheet showing general basin parameters and statistics as of 1.01.1991.

4.1 Precarpathian Depression

The Precarpathian Depression includes the area made up of flysch and molasse formations situated between the Crystalline-Mesozoic core of the Carpathians and the adjacent forelands. The age of these formations ranges from Cretaceous to Pliocene. The Precarpathian "depression" is the geological unit with the most numerous and significant oil and gas fields in Romania. Within the Precarpathian Depression, four distinct subunits can be separated by their stratigraphic, structural and paleogeographic features. These features have had a marked influence on the different conditions for the genesis, accumulation and preservation of hydrocarbons.

These subunits are:

1. Paleogene Flysch Zone in Moldavia;
2. Miocene Zone in Moldavia;
3. Mio-Pliocene Zone in Muntenia
4. Getic Depression

4.1.1 Paleogene Flysch Zone

1. Area: 9500 km^2
2. Seismic lines: 747 km (6 and 12 CDP) + 175 km^2 3-D
3. Number of geological wells: 1890
   deeper than 4000 m: 29
4. Number of oil and gas fields: 25
5. Main source rocks: dysodile schists and menilites (Oligocene)
6. Main reservoir rocks: Oligocene (Kliwa Sandstone)

   Average parameters:
   - porosity: 10–14%
   - permeability: 2–100 md
   - saturated thickness: 20–30 m
7. Types of traps: normal and faulted anticlines and scale folds
8. Average size of structures:
   - area: 3–6 km^{-2}
   - oil in place: 25–150 MMbbl
9. Initial average daily production per well: 40–600 bbl
10. Objective formation age: Oligocene

The Paleogene Flysch Zone in Moldavia includes the territory limited to the west by the Cretaceous Flysch Zone (Audia Unit) and by the Miocene Zone (Subcarpathian Unit) to the east. The Paleogene Flysch Zone is also called the Marginal Fold Nappe. These geological formations, generally detrital, are composed mainly of sandstones, marls, clays, dysodile schists and menilites and are over 7000 m in thickness. Tectonic relationship with the contiguous units is overthrust, with a typical amplitude of 10–12 km. The Medio-marginal Unit (Tarcau Unit), which is the most important nappe in the Carpathians, is exceptionally overthrust; it has been checked by drilling and found to exceed 35 km in horizontal displacement (Fig. 2).

The dominant features of the Paleogene Flysch Zone are the thin thrust sheets, with scale structures and the faulted anticlines (5–10 km long and 1–3 km wide). The general axial trend of these structures is north-south (Fig. 3).

Exploration of this zone has consisted mainly of field geological surveys and drilling. The geological complexity and difficult relief, frequently afforested, has meant that geophysical methods have not contributed much to solving the problems related to hydrocarbon exploration and production. The main results obtained from field geological surveys are detailed geological maps at scales 1:5000, 1:20,000 and synthesis maps at 1:50,000, 1:100,000 and 1:200,000. They are still used to locate exploratory wells.

A few seismic profiles shot along the valleys crossing the Carpathians using 6- and 12-fold CDP, show unconvincing results (Fig. 4). A 3-D experimental survey (175 km^2) shows the improved quality of the results gained with this technique (Fig. 5).

The exploration philosophy relies on the basis that the structural axial trends are more or less parallel to the Carpathian trend. Thus, using the field geological observations made in the tectonic halfwindows, drilling has been aimed at further identifying the structures in the same direction under the nappes, where there are better conditions for the preservation of hydrocarbon accumulations (Fig. 6). The results of this approach have been successful for shallow structures with normal tectonics. However, they are costly and inefficient for the complex structures located at greater depth. 1890 "geological wells" have been drilled up to the present, 29 of which are deeper than 4000 m.

The hydrocarbon accumulations are related to the Paleogene, with the main reservoirs being Kliwa Sandstone and Fusaru Sandstone. The main

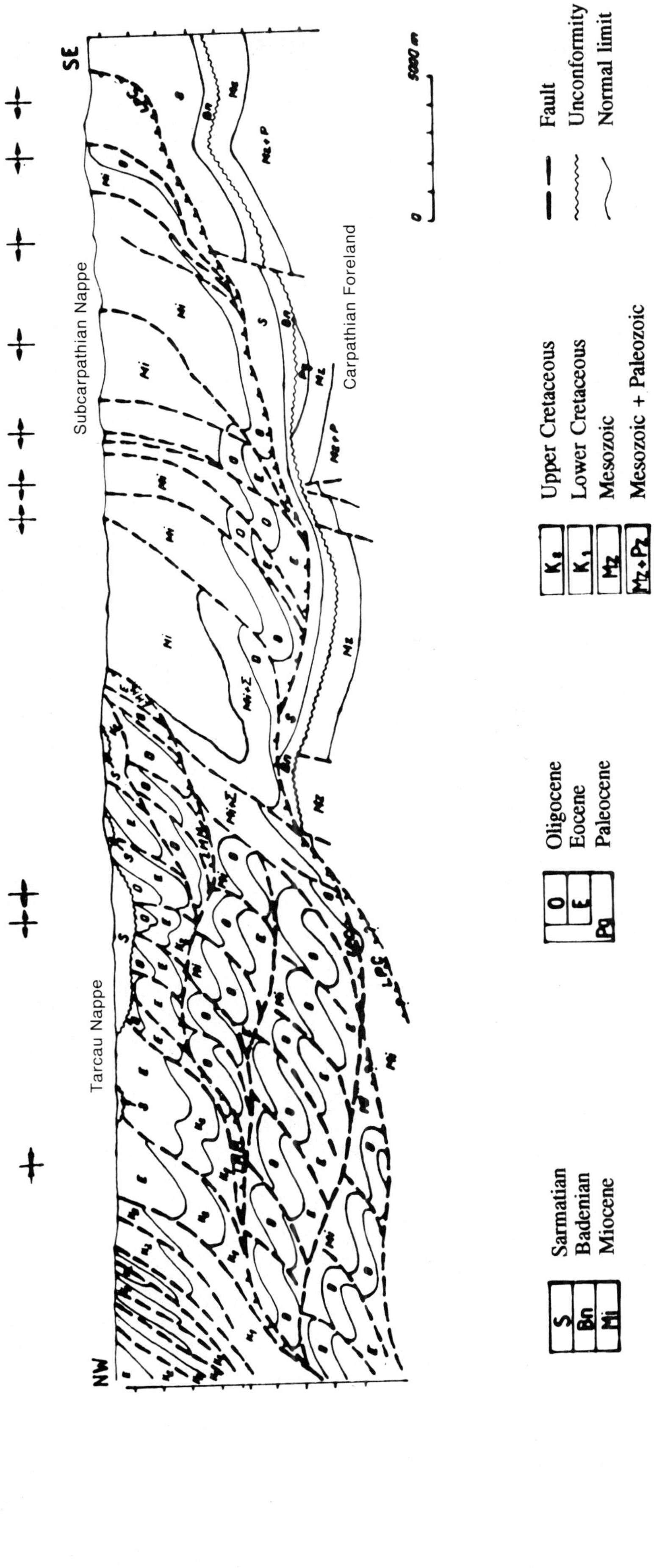

Fig. 2. Geological cross-section through the East Carpathians

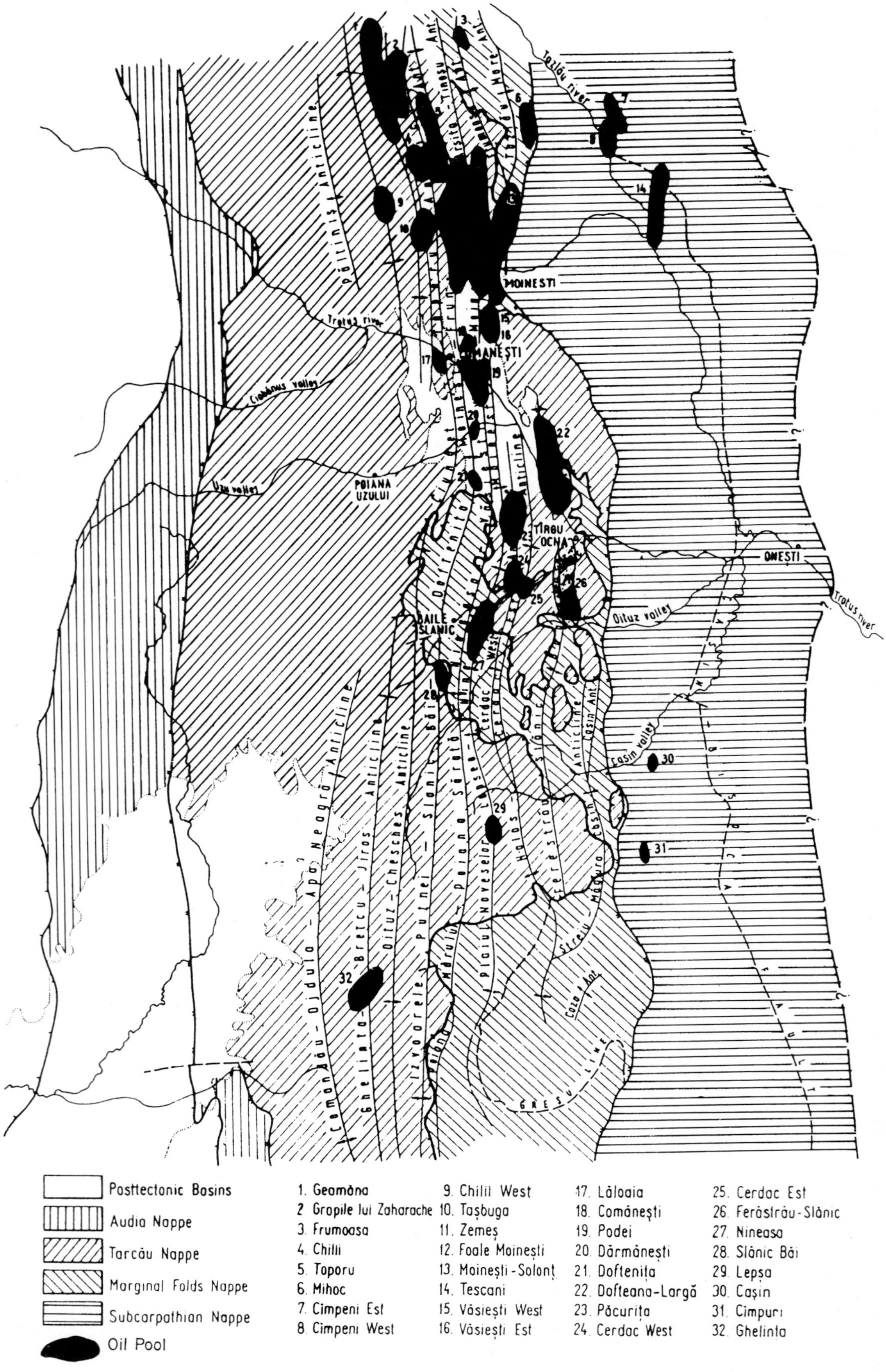

Fig. 3. Anticline trends and hydrocarbon pools in the Tazalu-Putna area

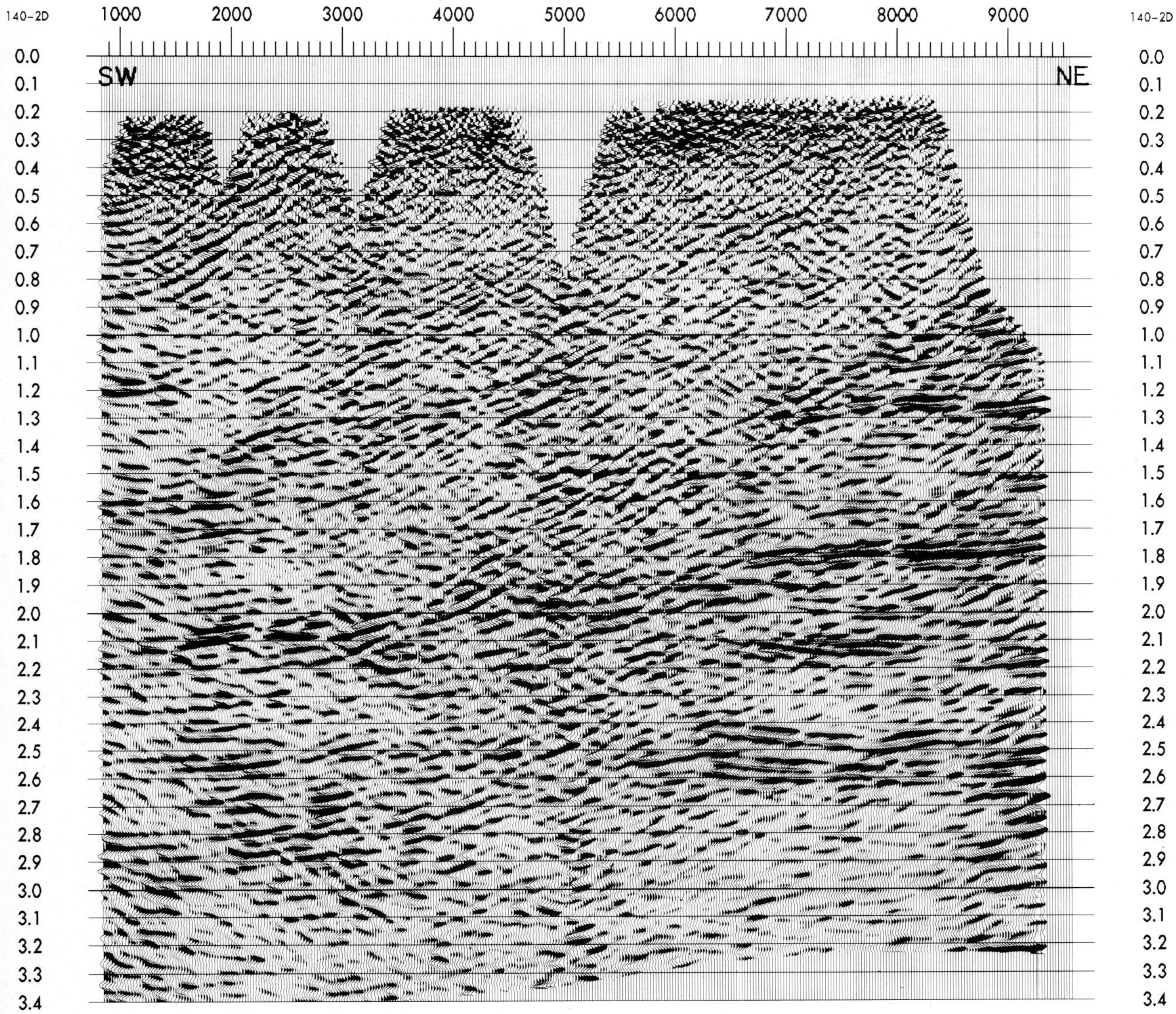

Fig. 4. 2-D seismic line across the Moldavian sector of the Paleogene Flysch Zone

source rocks are the Paleogene dysodile schists and menilites (Fig. 7). The traps are of a structural type, in faulted anticlines or scale folds. The most important accumulations have been discovered in the marginal units where they are sealed by the "Medio-marginal" nappe overthrust. The wells drilled in the Paleogene Flysch Zone in Moldavia proved that reservoirs have a good structural and lateral facies continuity. The unsuccessful results so far can be linked to the difficult location of traps in subsurfaces. The key to successful hydrocarbon prospecting in this zone is better subsurface mapping using modern seismic technologies. The complex geology, the difficult access, the rugged relief, all represent a real challenge. Overcoming the difficulties in the field should result in the discovery of new hydrocarbon accumulations.

4.1.2 Moldavian Miocene Zone

1. Area: 2500 km^2
2. Seismic lines: 550 km (CDP)
3. Number of geological wells: 612
   deeper than 4000 m: 11
4. Number of oil and gas fields: 4
5. Main source rocks: dysodile schists (Oligocene) and blackish clays (Middle Miocene)
6. Main reservoir rocks: sands and sandstones (Oligocene and Middle Miocene)

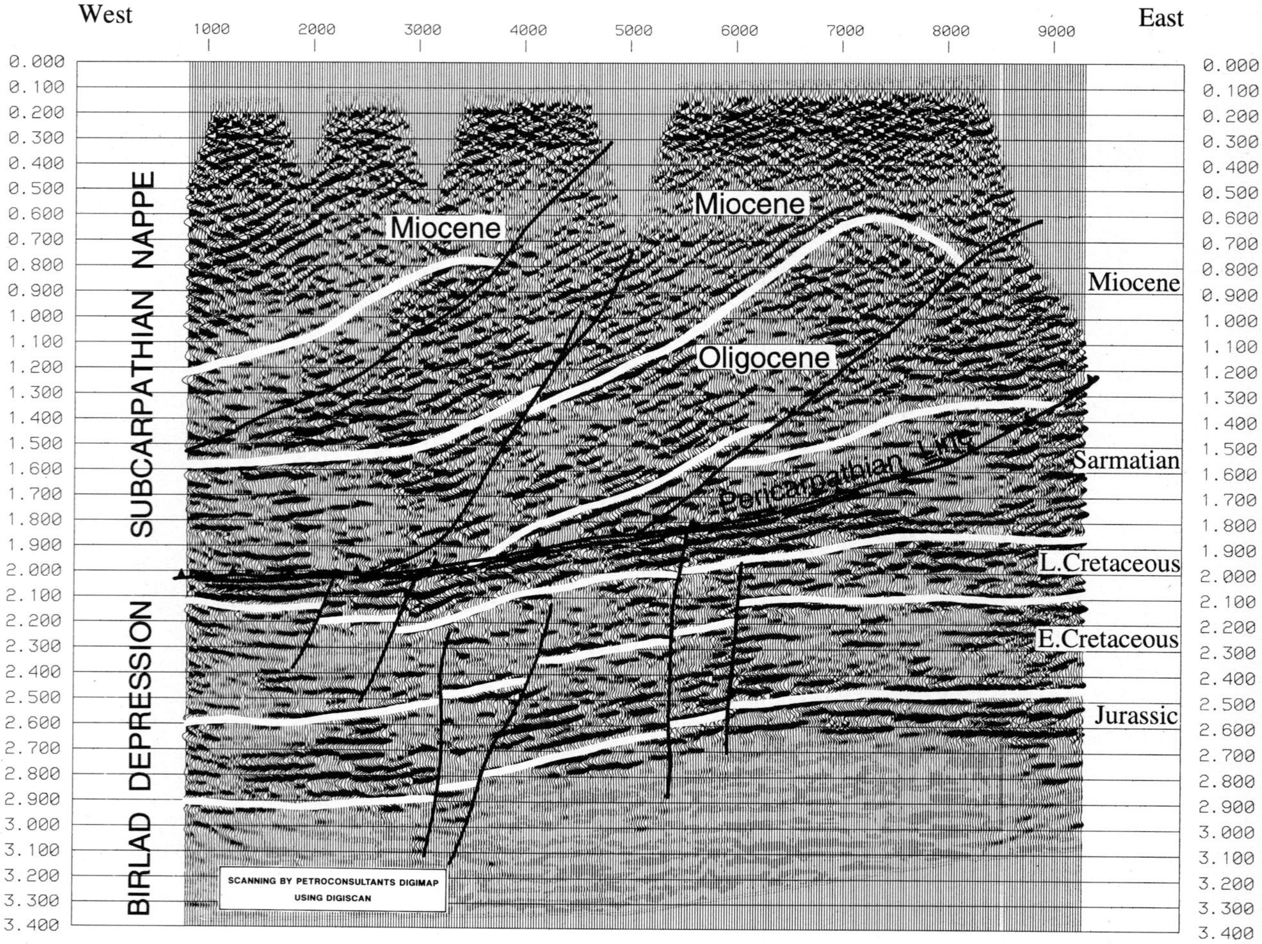

Fig. 5. 3-D seismic line across the same area as Fig. 4

Average parameters:
- porosity: 12–14%
- permeability: 5–70 md
- saturated thickness: 10–20 m

7. Type of traps: normal and faulted anticlines, scale folds
8. Average size of structures:
   - area: 1–2 km²
   - oil in place: 16–32 MMbbl
9. Initial average daily production per well: 16–40 bbl
10. Objectives of interest: Paleogene, Miocene in the Carpathian Foredeep. Cretaceous and Miocene in the Moldavian Platform

The Miocene Zone in Moldavia, also called the Subcarpathian Nappe, includes the northern sector of the Carpathian foredeep, from the northern border of Romania to the Putna Valley, between the external nappe units and the foreland. The main lithologies of this zone are detrital mollasse formations, of alternating sandstones, sands, shales, marls, belonging to Middle-Upper Miocene (Fig. 8). The most common structures are faulted, regional homoclines, and occasional gentle anticlines. The Miocene Zone is in an overthrust relationship with adjacent units along the External Line and the Pericarpathian Line. Seismic exploration has provided good results on the external flank and unsatisfactory results where the tectonics are more complex. In the Moldavian Platform, a good reflector corresponding to the Badenian anhydrite has been well correlated on the seismic sections beneath this zone. The extension of the Moldavian Platform for more than 20 km under the Miocene Zone has been documented.

Out of 612 geological wells drilled so far, only 11 are deeper than 4000 m. Exploration drilling is based mainly on field geological mapping, and has

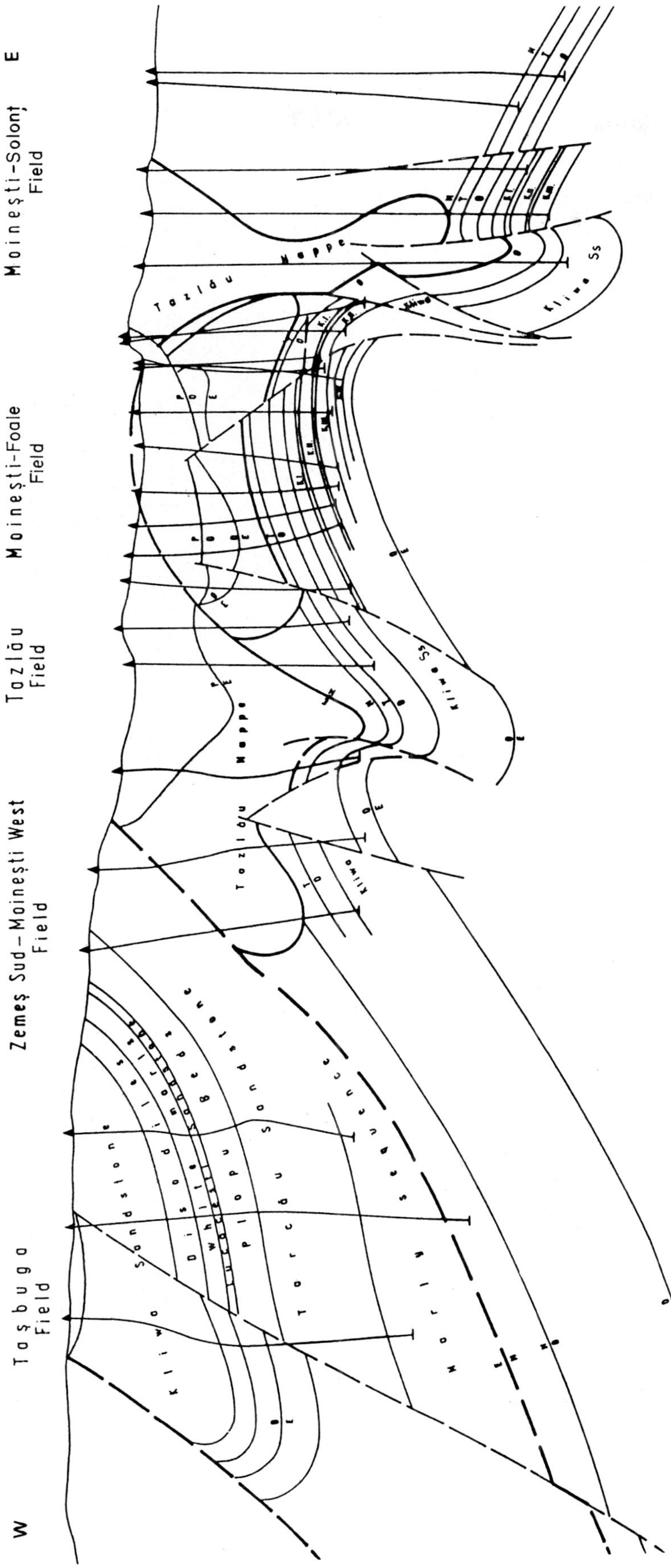

Fig. 6. Cross-section in the Moinesti area, Tarcău and Marginal Folds Nappe area

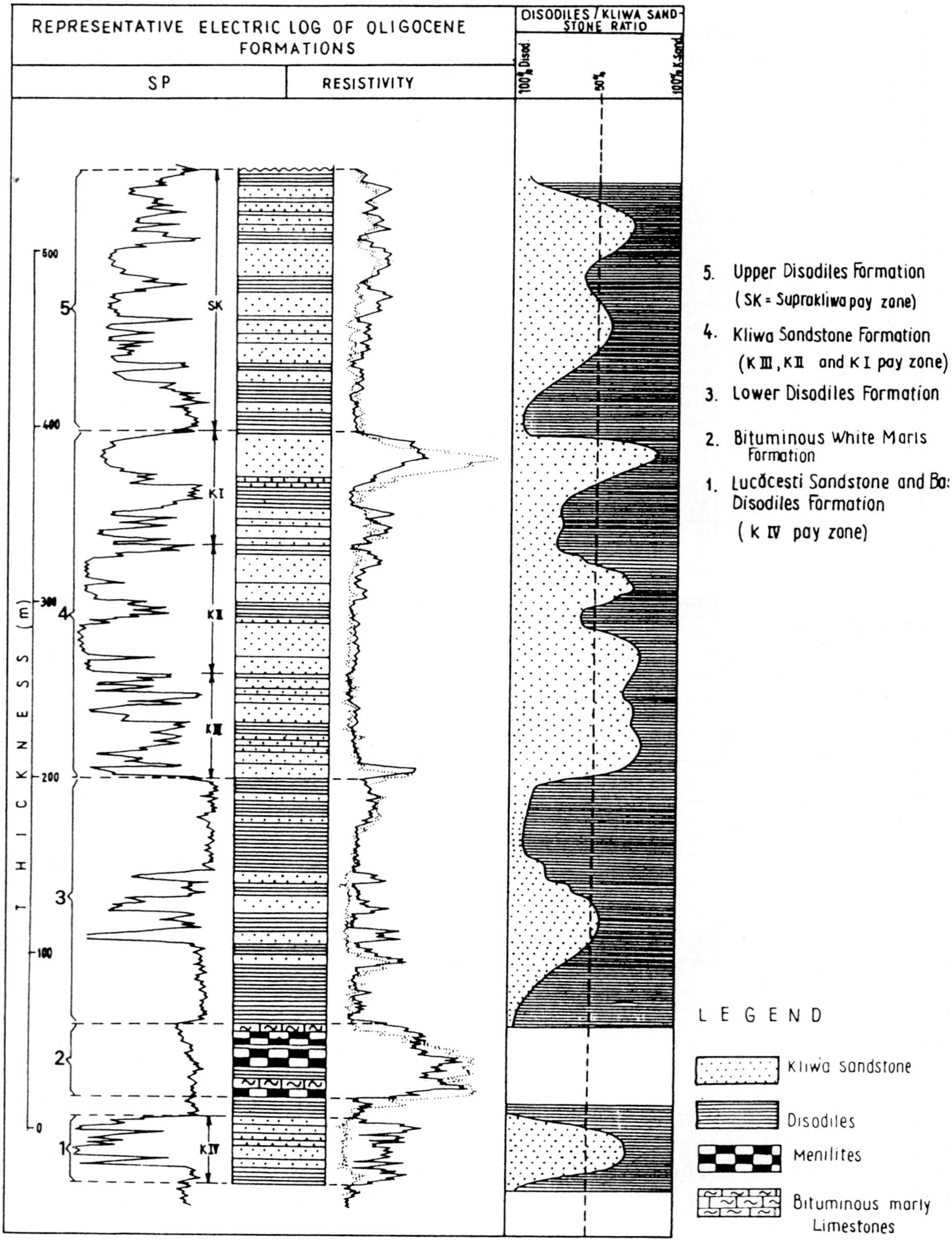

Fig. 7. Representative well log; Slanic-Oituz area, Marginal Folds Nappe

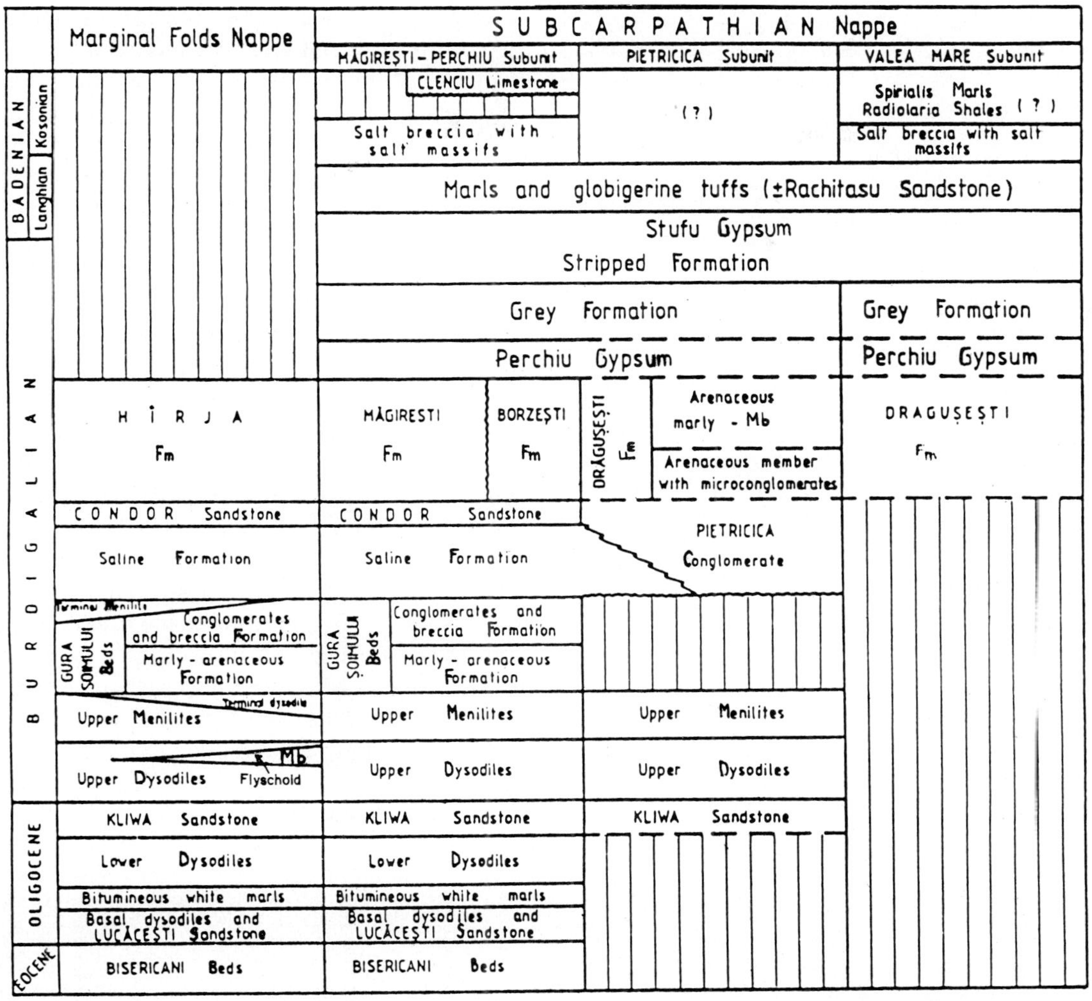

Fig. 8. Correlation chart of Tertiary formations in the Subcarpathian Nappe and the Marginal Folds Nappe

led to the discovery of 4 minor hydrocarbon accumulations in sandstones belonging to the Oligocene, Lower and Middle Miocene. The genesis of these accumulations is associated with the Oligocene Dysodile Schists but the blackish shale intercalations in the Lower Miocene must also be considered as good source rocks.

Future hydrocarbon discoveries in this zone are linked directly to the effectiveness of modern seismic technologies in locating hydrocarbon traps. It is worthwhile to consider the Miocene formations in the foredeep, and the Cretaceous or Badenian belonging to the Moldavian Platform, under the Subcarpathian Nappe.

In these formations, small wet gas accumulations have been identified at Frasin-Gura Humorului in Romania and at Lopusna in Ukraine.

4.1.3 Muntenia Mio-Pliocene Zone

1. Area: 3,000 km^2
2. Seismic lines: 1030 km (CDP)
3. Number of geological wells: 3030
   deeper than 4000 m: 14
4. Number of oil and gas fields: 46
5. Main source rocks: dysodile schists (Oligocene), blackish marls and clays (Neogene)
6. Main reservoir rocks: sands and sandstones in Oligocene, Burdigalian, Sarmatian, Meotian, Pontian, Dacian and Levantin

Average parameters:
- porosity: 14–25%
- permeability: 10–500 md
- saturated thickness: 30–60 m

7. Type of traps: normal and faulted anticlines, related to salt diapirism, pinch-outs, unconformities
8. Average size of structures:
   - area: 6–30 km^2
   - oil in place: 50–350 MMbbl
9. Initial average daily production per well: 35–400 bbl
10. Objectives of interest: Oligocene, Miocene, Pliocene

The Mio-Pliocene Zone in Muntenia, or the Zone of Diapir Folds includes the sector between the Rimnicu Sarat Valley and Dimbovita Valley. It represents the most productive, but also the oldest, and consequently most explored petroliferous zone in Romania. Salt diapirism-related traps are the important structural feature of this zone (Figs. 9, 10). They are located to the east, along five main alignments, and in general run parallel to the Carpathian chain. The salt diapirism decreases progressively from the Carpathian folds to the external flank of the foredeep which lies out on the Moesian Platform. Practically all possible exploration methods have been used. The quality of the results, especially of the seismic, is very variable. It ranges from very good to unsatisfactory, depending mainly on the geological complexity of the prospected zone, on the difficulties in acquisition (hard relief, industrial areas, forests etc.) and the depth of the horizons of interest. The total depth of geological wells ranges from a few hundred metres of 7025 m in well 7000 Băicoi, the national record. Forty-six oil and gas fields have been discovered so far in this basin.

Only 14 wells have been drilled deeper than 4000 m, from a total number of 3030 geological wells. All geological formations beginning with the Oligocene, Burdigalian, Sarmatian, Meotian, Pontian, Dacian and Levantin are oil- and gas- bearing. The reservoirs are mainly sands and sandstones with good porosity and permeability. The main source rock is considered to be the dysodile schists from the Oligocene, but also of great importance are the Middle Miocene/Pliocene blackish shales, which may be considered responsible for most of the hydrocarbons accumulated in these formations.

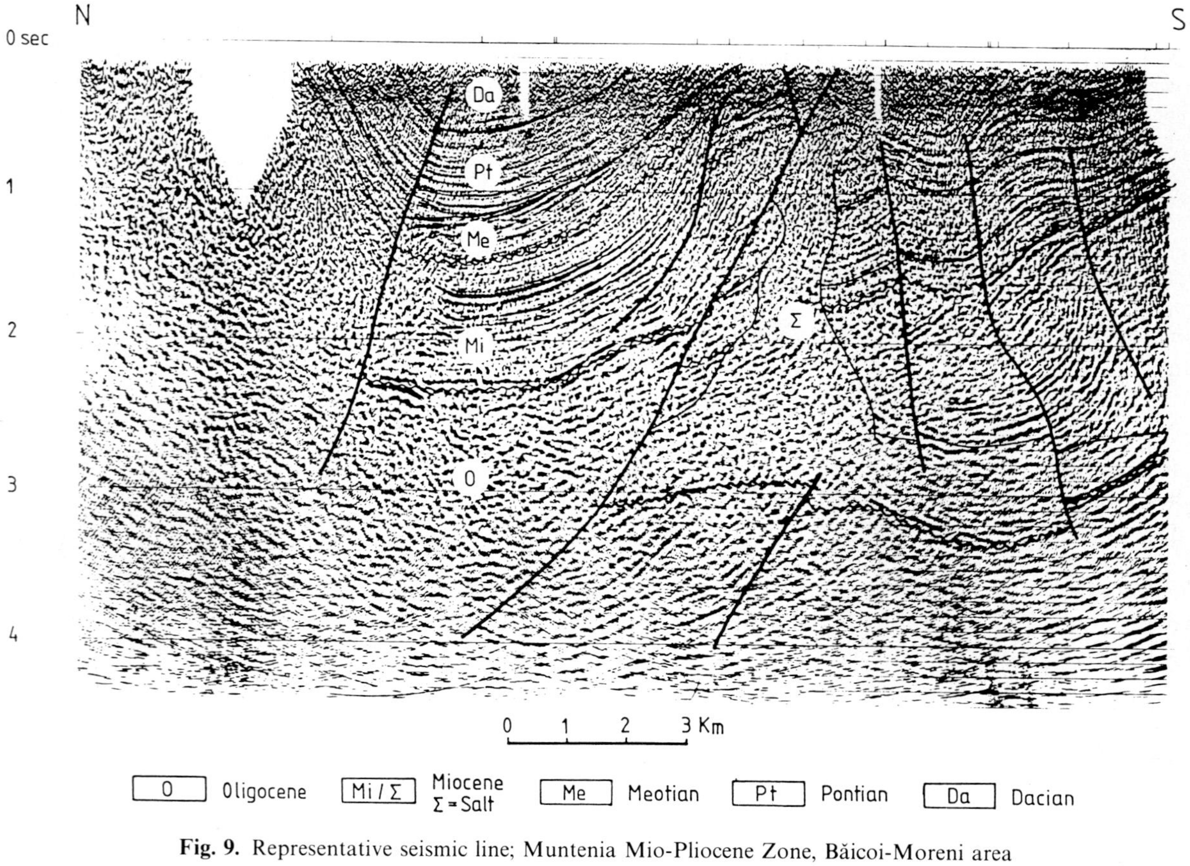

Fig. 9. Representative seismic line; Muntenia Mio-Pliocene Zone, Băicoi-Moreni area

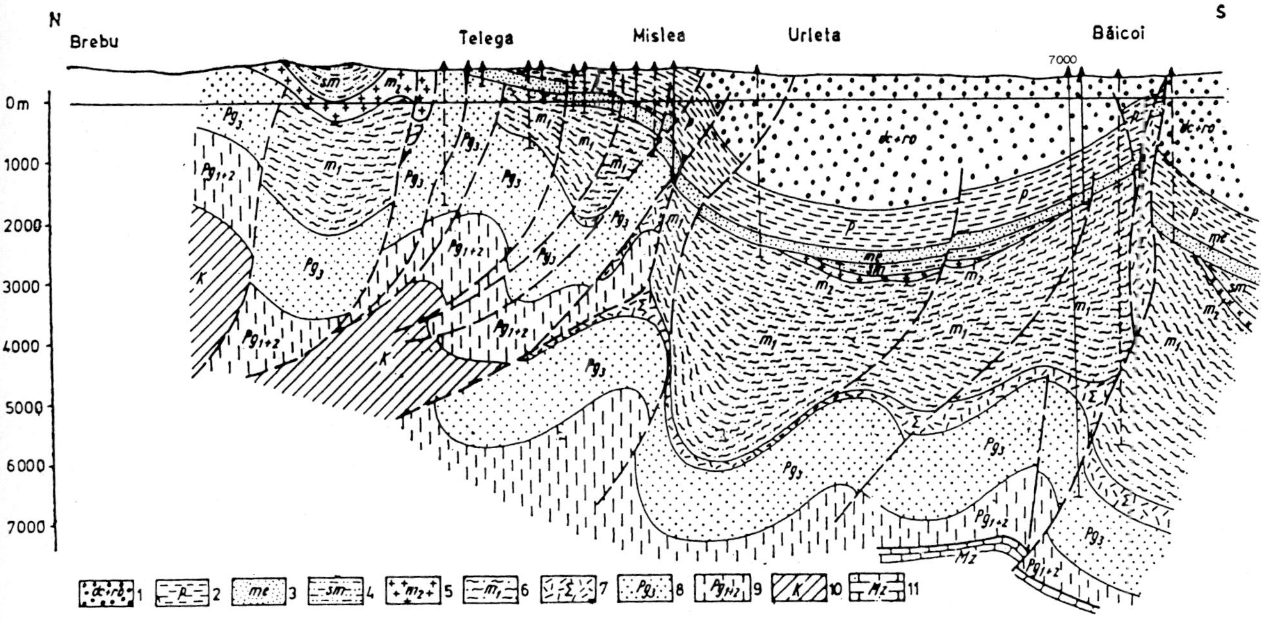

Fig. 10. Geological cross-section between Brebu and Băicoi, Diapir Folds Zone
1 Dacian and Romanian; *2* Pontian; *3* Meotian; *4* Sarmatian; *5* Badenian; *6* Lower Miocene; *7* Salt Formation; *8* Oligocene; *9* Eocene and Paleocene; *10* Cretaceous; *11* Mesozoic

All types of traps are present in the basin: from traps genetically connected with salt diapirism, to lithostratigraphic traps related to unconformities, pinch-outs of sand bodies or to the paleoreliefs. The most important structure is the Baicoi-Moreni-Gura Ocnitei alignment which has so far produced more than 120 million tons of crude or over 900 million bbl.

The mature stage of exploration would not be encouraging for exploration for new opportunities. However, each year as a result of reinterpretation of geological and production data, new producing blocks are discovered, extending the already known producing areas. Exploration of the Lower Miocene and Paleogene under the diapir folds, near the contact with the Moesian Platform, is being carried out at present. It is worth mentioning that in well 7000 Baicoi oil shows were encountered at a depth of 6191 m in Paleogene sandstone (Fig. 10). The seismic lines do not offer unequivocal interpretable results, especially for the deeper horizons. The seismic network is not homogeneous due to rough relief.

4.1.4 Getic Depression

1. Area: 8000 km^2
2. Seismic lines: 8614 km (CDP)
3. Number of geological wells: 4834
   deeper than 4000 m: 52
4. Number of oil and gas fields: 35
5. Main source rocks: blackish clays (Oligocene), blackish-grey clays and marls (Miocene)
6. Main reservoir rocks: sandstones and sands in Paleogene, Miocene and Pliocene.

   Average parameters:
   - porosity: 15–25%
   - permeability: 10–500 md
   - saturated thickness: 20–100 m

7. Type of traps: normal and faulted anticlines, pinch-outs, unconformities
8. Average size of structures:
   - area: 10–20 km^2
   - oil in place: 8–40 MMbbl
9. Initial average daily production per well: 70–800 bbl
10. Objectives of interest: Oligocene, Miocene, Pliocene

The Getic Depression represents the segment of Carpathian Foredeep between the Dimbovita Valley, the Danube, the Carpathians and the Moesian Platform. The geological formations of the Paleogene-Miocene and Mio-Pliocene cycles are developed in an almost complete stratigraphic sequence, and represent the most important hydrocarbon-bearing formations. The tectonics are less complicated compared to the Diapir Folds Zone; and salt diapirism is insignificant in the depression (Fig. 11). The oil and gas fields are located in Paleogene, Miocene and Pliocene sandstones and sands. The Oligocene shales and the blackish clays from the Miocene are considered as source rocks. The traps are both structural (normal and faulted anticlines), and lithostratigraphic (pinch-outs, unconformities and paleorelief).

The seismic survey results are generally good for the Pliocene and Miocene, which has been moderately folded and faulted and unsatisfactory for the Lower Miocene and Paleogene. As with other prospective zones in Romania, the discovery of new hydrocarbon fields is related to the direct location of traps by seismic surveys, especially for the Lower Miocene and the Paleogene.

4.2 Carpathian Foreland

Broadly speaking, the Carpathian foreland consists of two platform areas: the Moesian Platform with a Baikalian and/or Hercynian basement and the Moldavian Platform with a Proterozoic basement. The latter covers the Romanian part of the larger East European Platform. The Scythian Platform, with a Caledonian basement, occurs in central-eastern Romania in a region having a very limited areal development.

4.2.1 Moesian Platform

1. Area: 43,000 km^2
2. Seismic lines: 15,322 km
3. Number of geological wells: 7830
   deeper than 4000 m: 198
4. Number of oil and gas fields: 143
5. Main source rocks: bituminous clays and limestones (Silurian, Mid-Devonian, Dogger, Albian and Neogene)
6. Main reservoir rocks: sandstones and sands (Lower and Upper Triassic, Lower and Middle Jurassic, Neogene), limestones and dolomites (Devonian, Middle Triassic, Malm, Cretaceous)
   Average parameters:
   - porosity: 15–25%
   - permeability: 5–400 md
   - saturated thickness: 50–100 m
7. Type of traps: normal and faulted anticlines, faulted monoclines, pinch-outs, unconformities, paleoreliefs
8. Average size of structures:
   - area: 2 to 15 km^2
   - oil in place: 7 to 80 MMbbl
9. Initial average daily production per well: 70–1000 bbl
10. Objectives of interest: Paleozoic, Mesozoic, Tertiary

The Moesian Platform is developed between the Carpathian Foredeep and the Danube. It is bordered to the northeast by the North Dobrudgean Promontory. The Moesian Platform consists of Paleozoic, Mesozoic and Tertiary (Upper Miocene and Pliocene) clastics and carbonates. The local thickness of the sediments is over 10,000 m, while in the northeastern part, in the vicinity of the Focsani Depression, the thickest sedimentary sequence in Romania (over 15,000 m) is developed. Four major sedimentary cycles, separated by more or less extended unconformities, have been identified: Cambrian-Middle Carboniferous, Permian-Upper Triassic, Upper Liassic-Senonian and Badenian-Quaternary. The most important seismically mapped element is the Upper Cretaceous unconformity.

Two structural substages have been described. The first of these is the Paleozoic-Triassic with large, moderately faulted anticlines and the second, Post-Triassic, which has a northern-dipping, quasi-homoclinal structure (Figs. 11, 12). The Moesian Platform sinks under the Carpathian Foredeep, along a series of east-west faults (Fig. 13). In this general faulted homocline, numerous structural details were generated by local movements, by the variety of the paleoreliefs and by the thickness and nature of the deposits associated with the process of differential rock compaction. The first commercial oil discovery dates from 1956 and resulted from the re-interpretation of earlier seismic surveys. The seismic recordings are of good to very good quality for the Neogene and Cretaceous, good to satisfactory quality for Mesozoic sequences and of unsatisfactory quality for the Paleozoic sequences. The presence of a very important reflection coefficient (0.40–0.47) along the Neogene and Cretaceous interface considerably decreases the energy of reflections coming from deeper horizons. This creates problems related to the elimination of multiple reflections. The exploratory drilling density differs from one zone to another, with the less

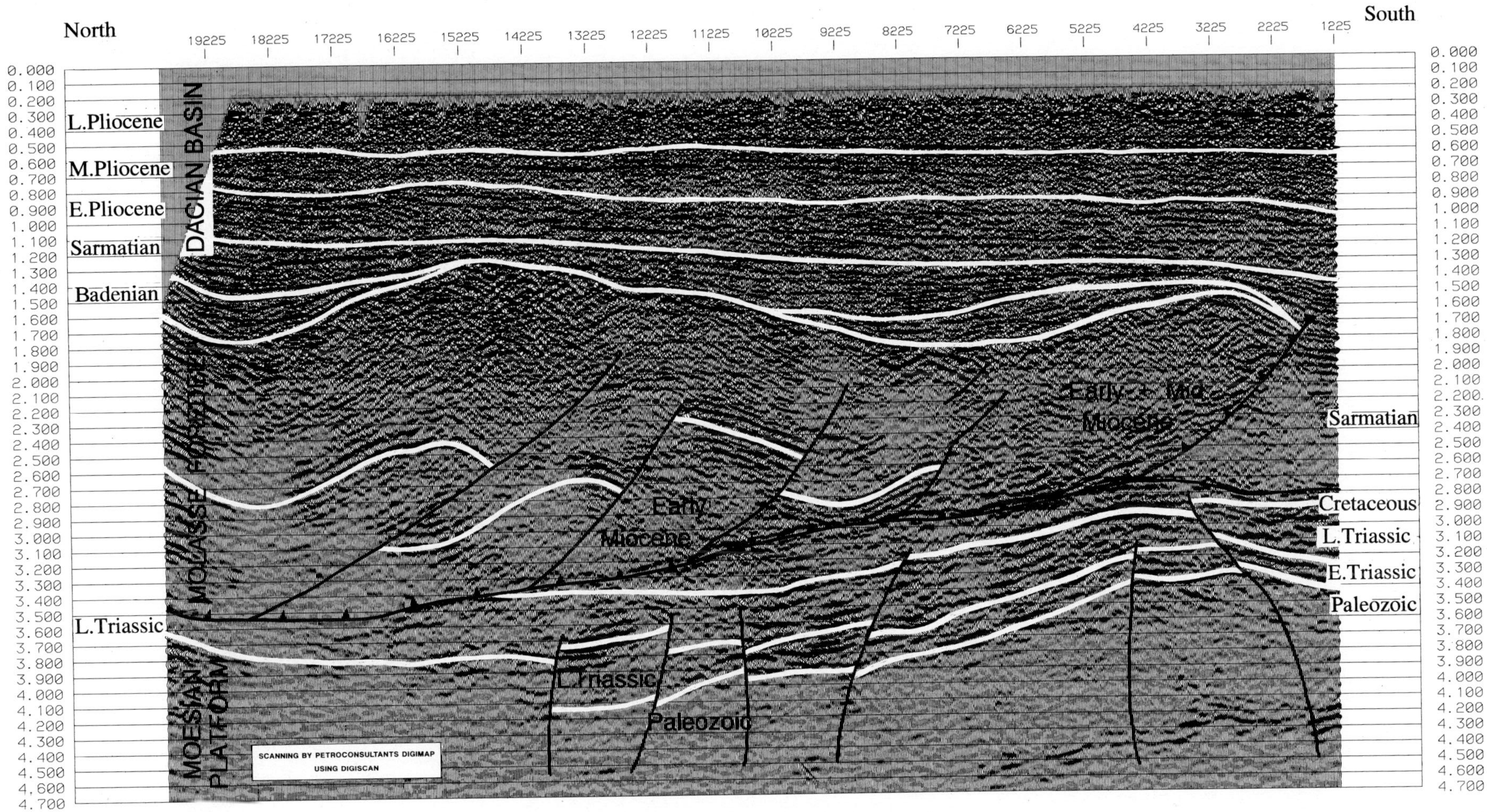

Fig. 11. Representative seismic line, Getic Depression

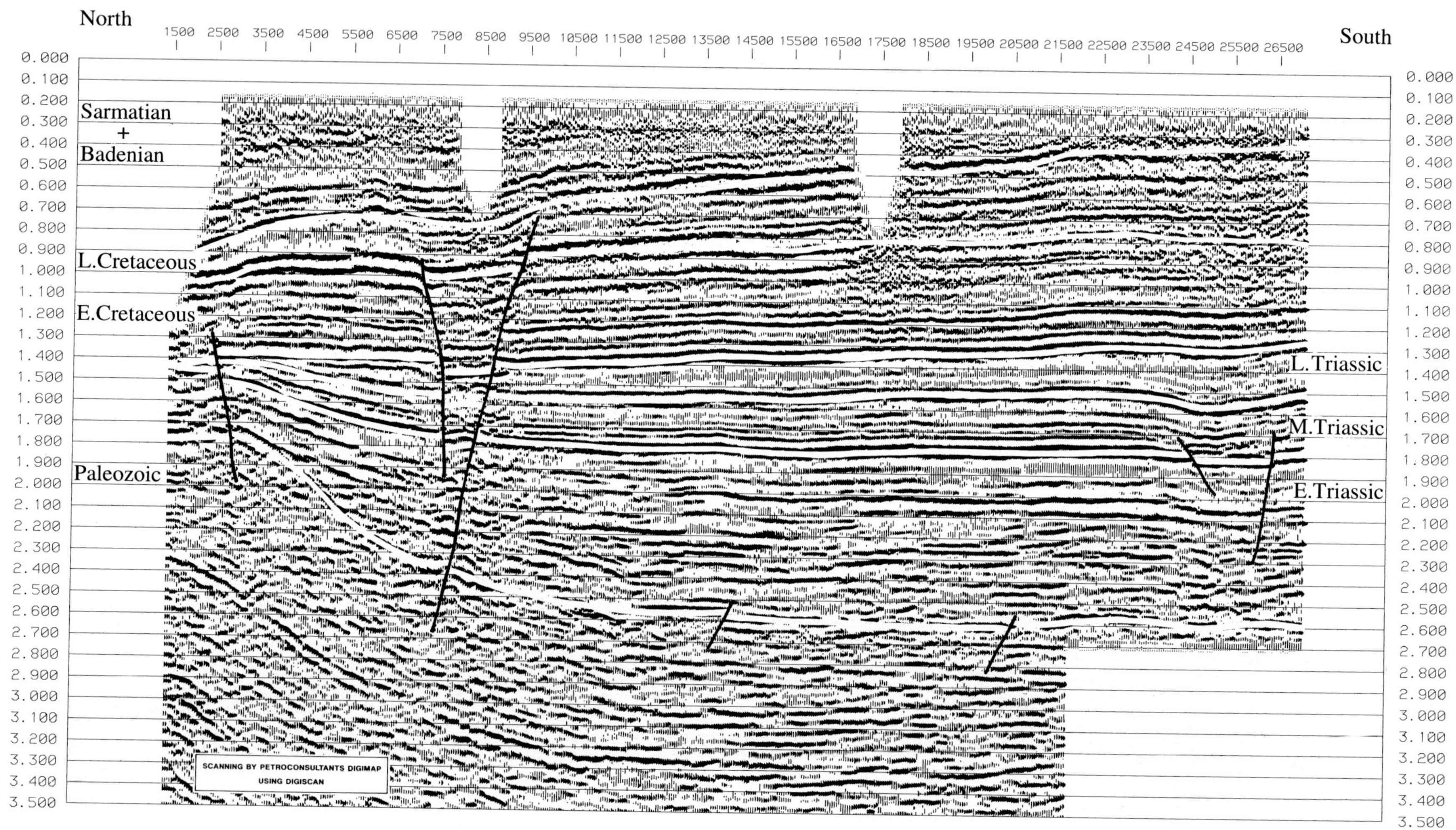

Fig. 12. Representative seismic line; central part of Moesian Platform

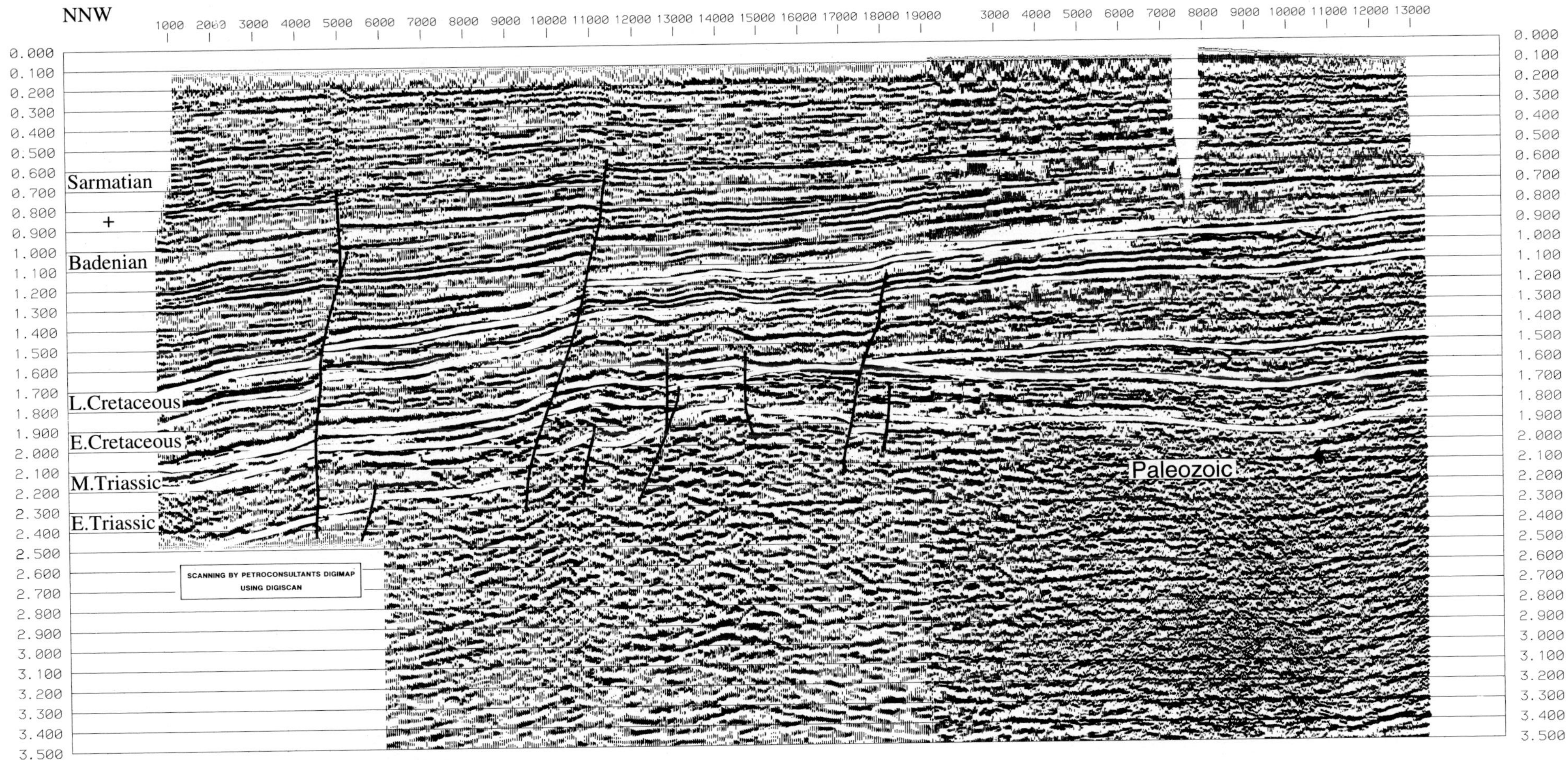

a

Fig. 13a–d (p. 233–236). Representative seismic line; Moesian Platform, same general area as Fig. 12

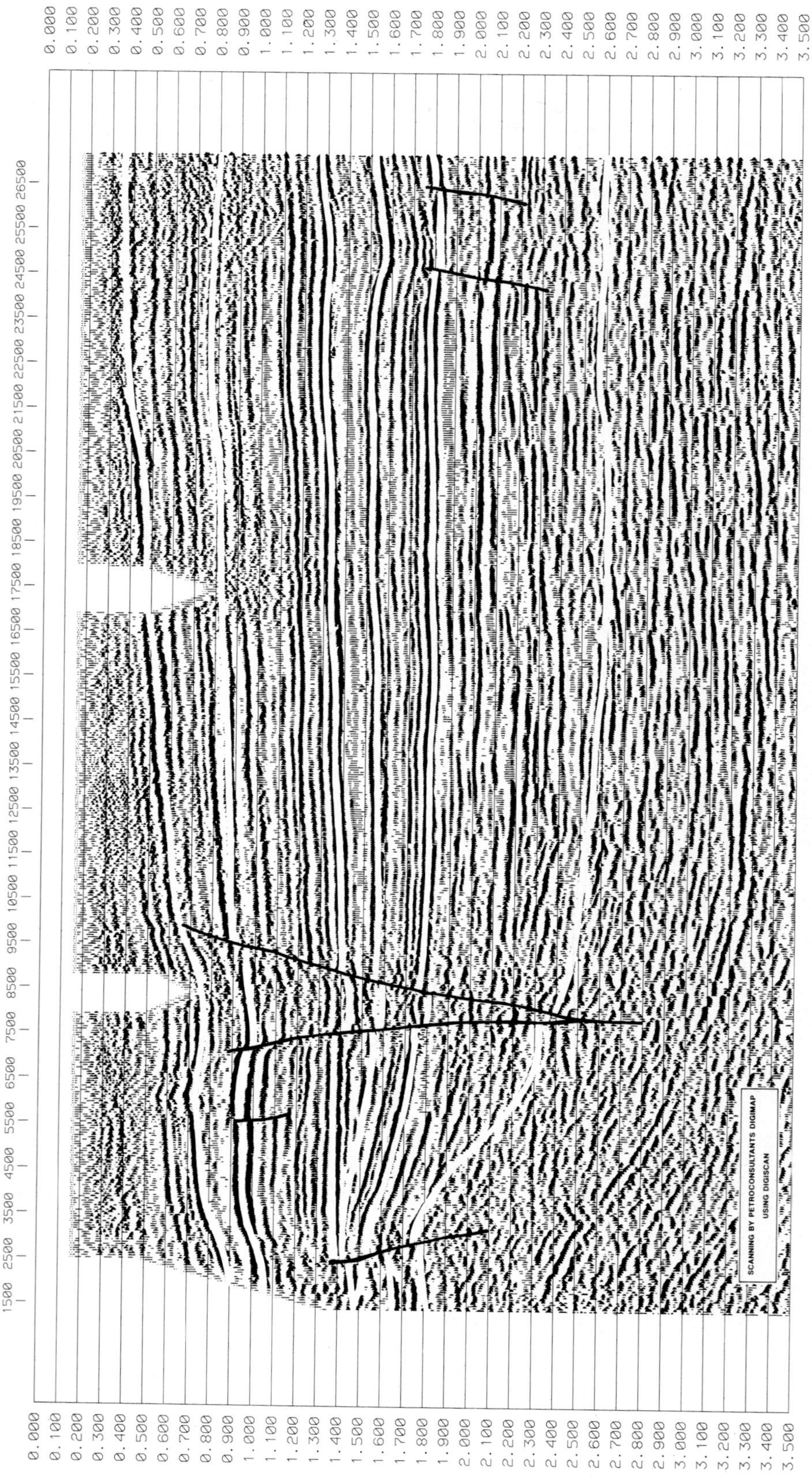
1500 2500 3500 4500 5500 6500 7500 8500 9500 10500 11500 12500 13500 14500 15500 16500 17500 18500 19500 20500 21500 22500 23500 24500 25500 26500
0.000 0.100 0.200 0.300 0.400 0.500 0.600 0.700 0.800 0.900 1.000 1.100 1.200 1.300 1.400 1.500 1.600 1.700 1.800 1.900 2.000 2.100 2.200 2.300 2.400 2.500 2.600 2.700 2.800 2.900 3.000 3.100 3.200 3.300 3.400 3.500
SCANNING BY PETROCONSULTANTS DIGIMAP
USING DIGISCAN

Fig. 13b

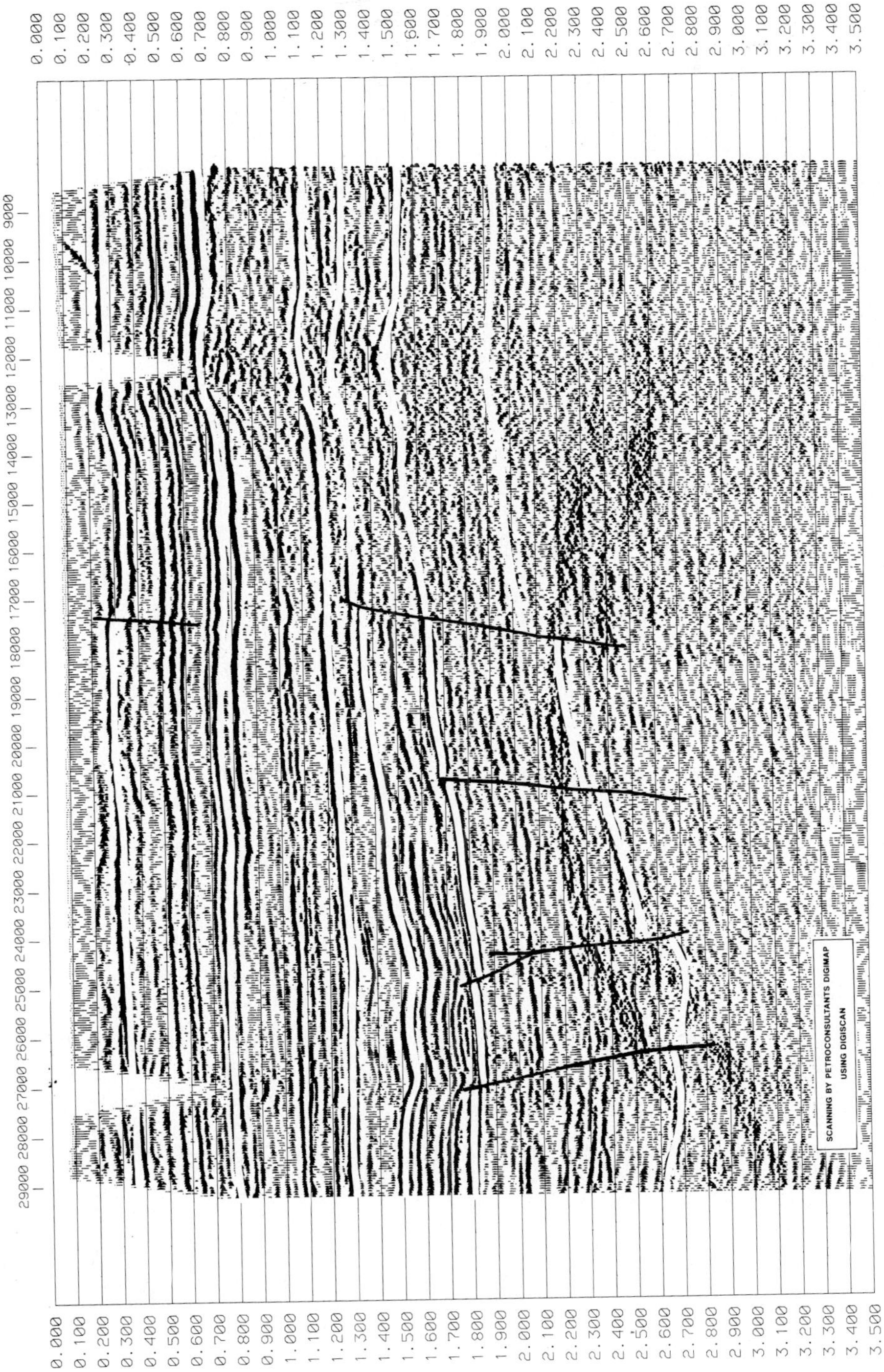
SCANNING BY PETROCONSULTANTS DIGIMAP
USING DIGISCAN
29000 28000 27000 26000 25000 24000 23000 22000 21000 20000 19000 18000 17000 16000 15000 14000 13000 12000 11000 10000 9000
0.000 0.100 0.200 0.300 0.400 0.500 0.600 0.700 0.800 0.900 1.000 1.100 1.200 1.300 1.400 1.500 1.600 1.700 1.800 1.900 2.000 2.100 2.200 2.300 2.400 2.500 2.600 2.700 2.800 2.900 3.000 3.100 3.200 3.300 3.400 3.500

Fig. 13c

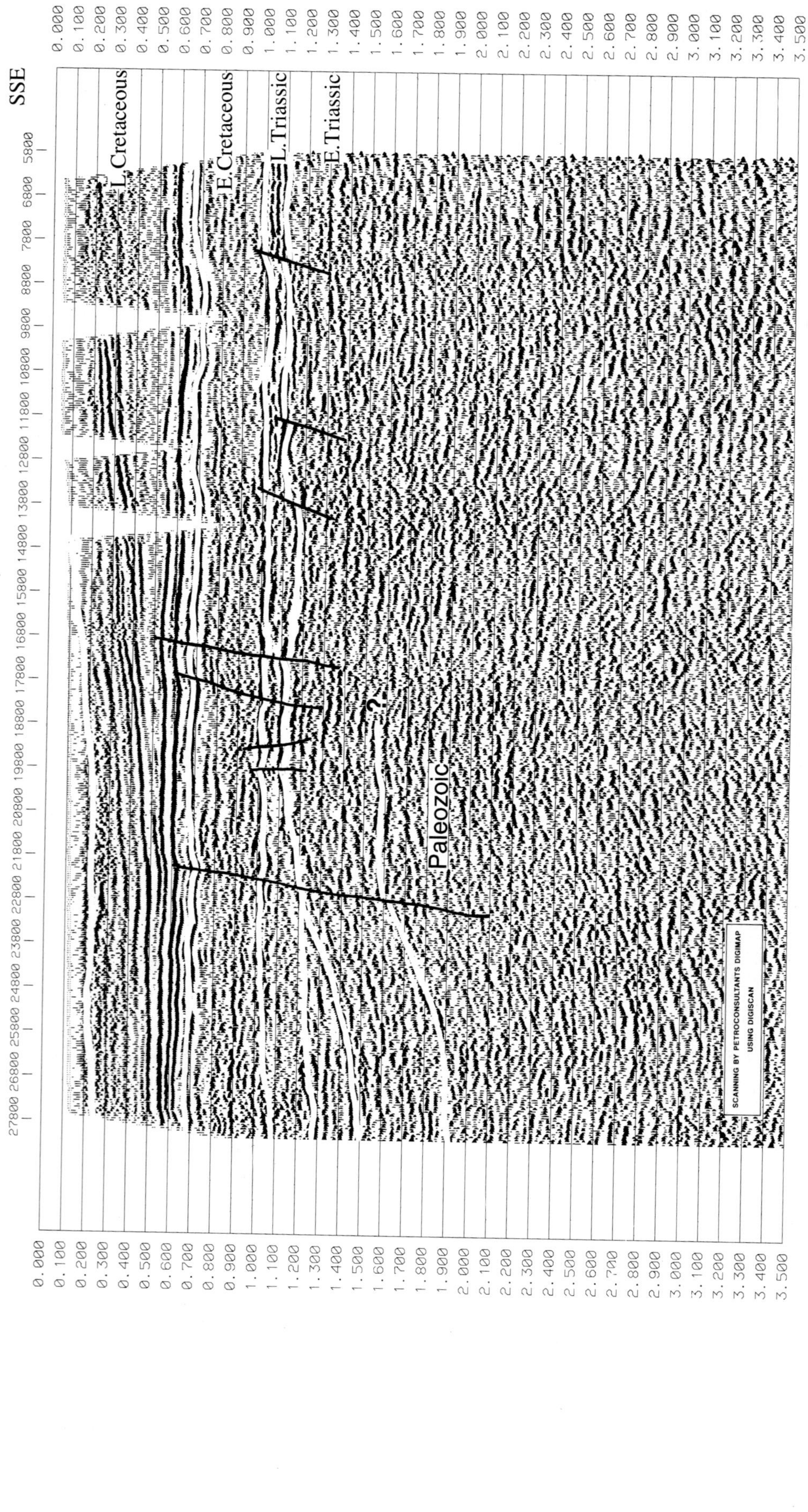
SSE
L. Cretaceous
E.Cretaceous
L.Triassic
E.Triassic
Paleozoic
?
SCANNING BY PETROCONSULTANTS DIGIMAP
USING DIGISCAN

Fig. 13d

explored areas being the central-southern and western regions.

Up to the present, more than 140 oil and gas fields have been discovered in the Devonian, Triassic, Upper Liassic, Dogger, Malm, Neocomian, Albian, Senonian, Sarmatian, Pontian and Dacian. The reservoir lithologies vary from limestones and dolomites in the Devonian, Upper Triassic, Malm and Cretaceous sequences to sandstones and sands in the Lower and Upper Triassic, Lower and Middle Jurassic, and Neogene reservoirs.

Geochemical analyses indicate that the following stratigraphic intervals contain important potential source rocks: Silurian, Mid-Devonian, Dogger, Albian and Neogene. The source rocks in the Moesian Platform are both clastics and carbonates. There is an immense variety of trap conditions, from structural to lithostratigraphic type (pinch-outs, unconformities, paleoreliefs).

The depth of discovered fields ranges from 350 m to 5000 m (Bibeşti, Brădesti zone). The deeper oil or/and gas fields are located on the northern part of the Moesian Platform (Fig. 14). In the western region, fields are mainly oil-bearing, while in the eastern part gas fields have been the main discovery. This situation could be attributed to the discrepancy which exists with the geothermal gradient from 6°C/100 m in the western part, to 1.5°C/100 m in the eastern part.

Despite the mature stage of exploration in the Moesian Platform, there are still some zones which could intrigue the petroleum geologist, as drilling often contradicts the generally accepted initial hypothesis. These zones are located in the Rosiori-Alexandria Depression, and in the western part of the Moesian Platform, where oil and gas shows have been identified in a number of wells, and where theoretically the conditions for hydrocarbon accumulations are adequate. However, so far, no commercial oil or gas fields have been discovered in these areas. Some encouraging results have been obtained by Bulgarian explorationists in Middle Triassic sequences south of the Danube, which improves the prospects for the above-mentioned areas, especially for the Paleozoic and Triassic.

4.2.2 *Birlad Depression*

1. Area: 19,700 km^2
2. Seismic lines: 1040 km (CDP)
3. Number of geological wells: 1262
   deeper than 4000 m: 55
4. Number of oil and gas fields: 14
5. Main source rocks: blackish shales (Neogene)
6. Main reservoir rocks: sands and sandstones (Paleozoic, Middle Miocene)

   Average parameters:
   - porosity: 15–20%
   - permeability: 5–400 md
   - saturated thickness: 60–150 m

7. Type of traps: normal and faulted anticlines, pinch-outs, paleoreliefs
8. Average size of structures:
   - area: 1 to 4 km^2
   - oil in place: 8 to 50 MMbl
9. Initial average daily production per well: 35–130 bbl
10. Objectives of interest: Paleozoic, Mesozoic, Middle Miocene

The Birlad Depression is a graben situated between the Moldavian Platform to the North and the North Dobrudgean Promontory to the South. The Birlad Depression plunges to the west under the Carpathian Foredeep, in the Adjud-Bacau region. In this area the North Dobrudgean Basement is at depths greater than 6000 m.

Practically unknown until 1950, the Birlad Depression was discovered as a result of systematic geophysical surveying, gravity, magnetic, electrical and especially seismic. Seismic lines have good to very good quality for Neogene formations (Fig. 15). The older Mesozoic and Paleozoic formations are less clearly defined in the seismic profiles. Otherwise, attention has centred on the Neogene formations where 12 oil and gas fields have been discovered in clastic reservoirs. The pelitic intercalations of the Neogene are considered as being the source rocks. However, the existence of some source rocks in the Middle Jurassic cannot be excluded. The traps are structural, lithostratigraphic, or in relation to the pre-existing paleoreliefs.

Interest in this zone has increased very much over the last few years, when the gas fields of Conteşti and Burcioaia were discovered in the Pre-Neogene paleorelief. The reservoirs are composed of hard quartzitic sandstones belonging to the Jurassic (Conteşti) and Paleozoic (Burcioaia). These gas fields have accumulated in the deeply buried areas of the Birlad Depression near the contact

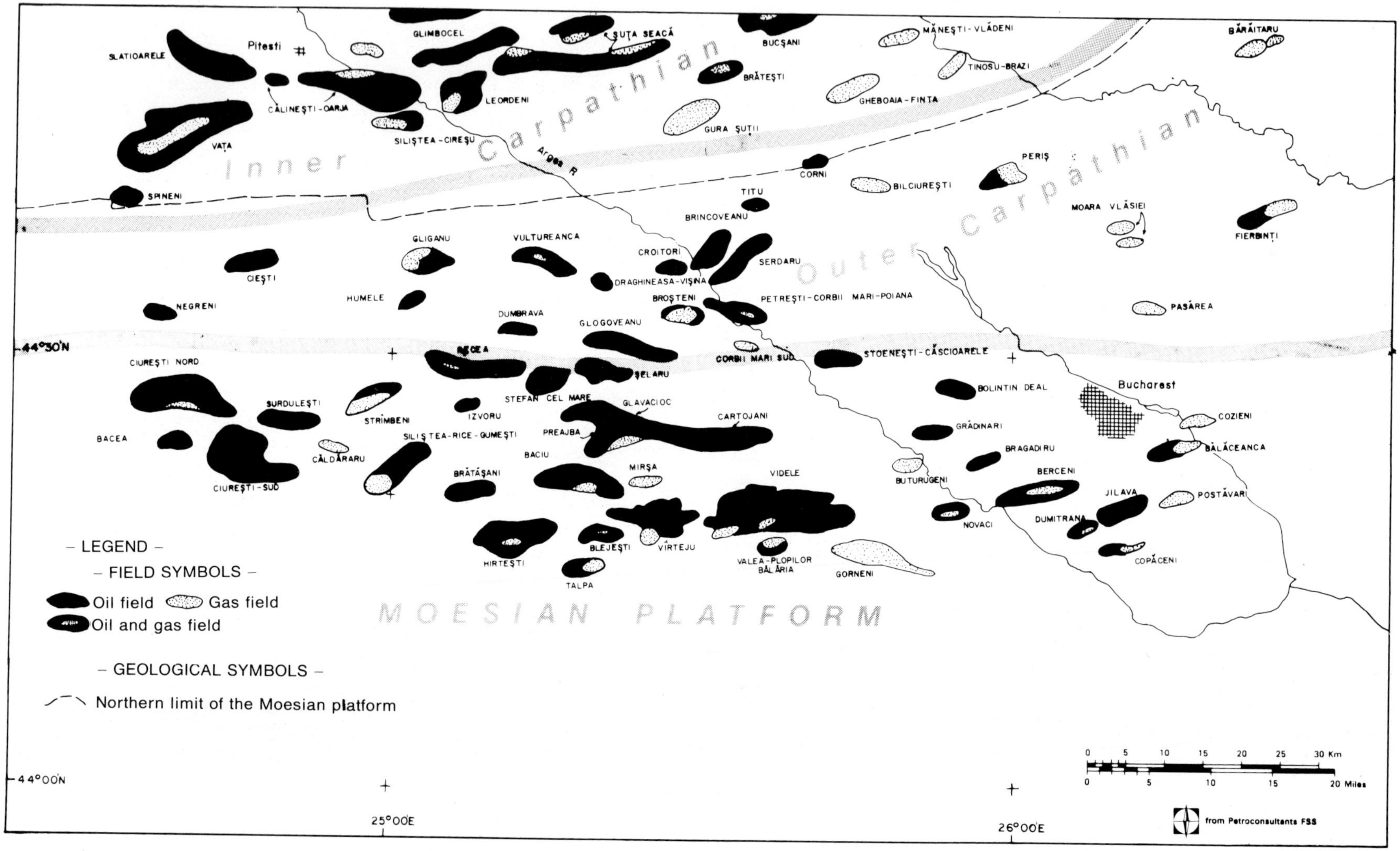

Fig. 14. Central Moesian Platform oil and gas field map (From Petroconsultants FSS map)

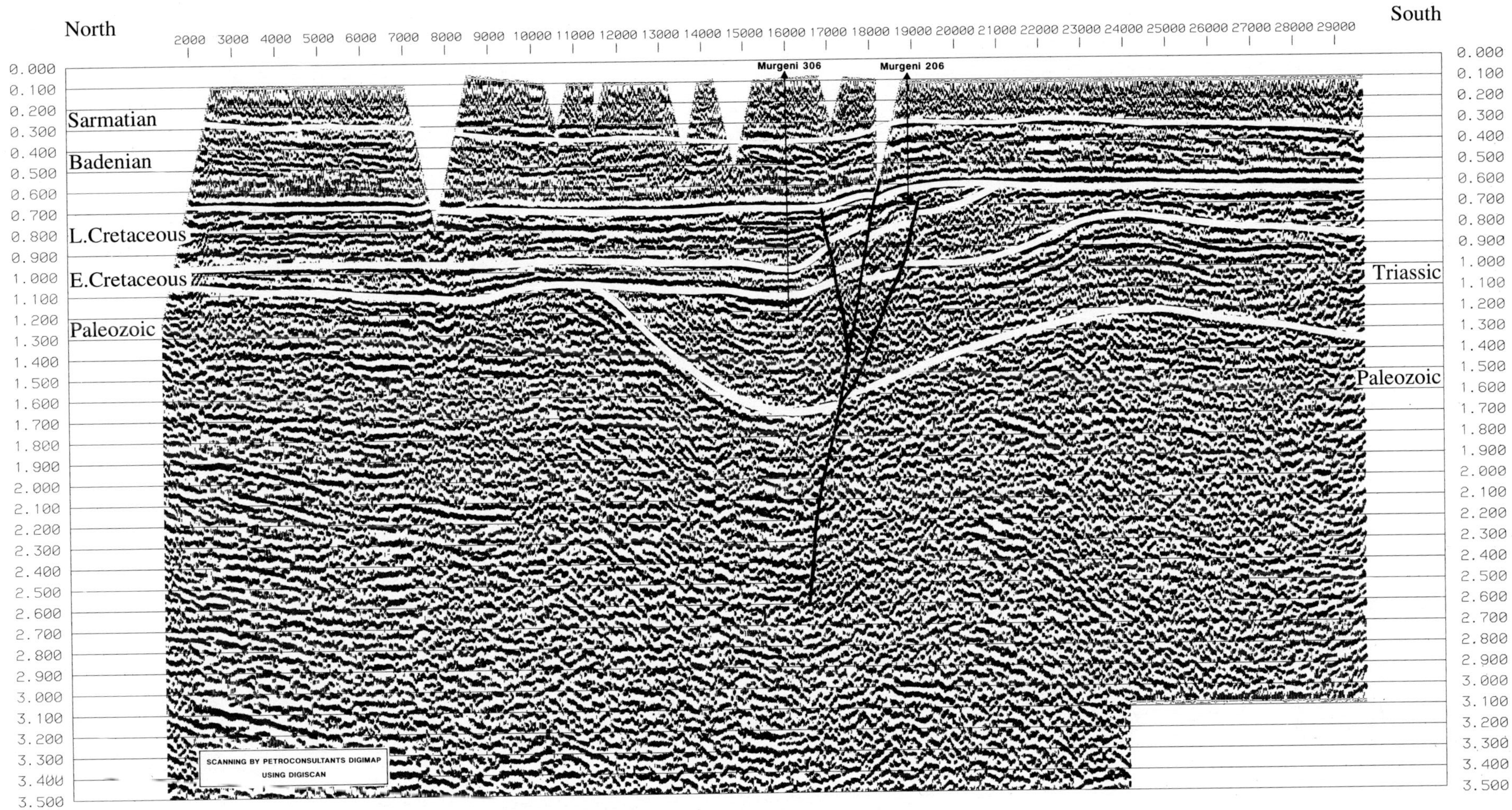

Fig. 15. Representative seismic line; Birlad Depression

with the Miocene zone in Moldavia. The average depth of these fields is 4200 m.

Although the present daily production of these fields is relatively small (100–150 m^3/well), due to the low parameters of the reservoir rocks (average porosity 4–9%, and average permeability 1–5 md), they prove the existence of favourable conditions for genesis, and preservation of hydrocarbons in this region. The relatively large, 60–150 m, thickness of the saturated interval is also worth mentioning.

The hydrocarbons are thought to be sourced by the Badenian or Sarmatian pelites. The accumulations in the Pre-Neogene paleoreliefs are the result of lateral migration process.

4.3 Intracarpathian Post-Tectonic Basins

Neogene molasse hydrocarbon basins develop over the inner, western part of the Carpathians. The Transylvania gas-bearing basin is the Neogene subsident area between Eastern and Southern Carpathians, the Apuseni Mts. The eastern Pannonian basin lies west of Apuseni Mts and Southern Carpathian; it largely covers Hungary, Southern Czechoslovakia and Northern Yugoslavia.

4.3.1 The Transylvania Basin

1. Area: 15,000 km^2
2. Seismic lines: 3909 km (CDP)
3. Number of geological wells: 2339
   deeper than 4000 m: 4
4. Number of gas fields discovered: 88
5. Main source rocks: blackish marls (Cretaceous), Dysodile Schists (Oligocene)
6. Main reservoir rocks: sands and sandstones

Average parameters	Permian to Middle Miocene	Miocene-Pliocene
– porosity (%):	5–15?	25–30
– permeability (md)	?	10–100
– saturated thickness	?	30–150

7. Type of traps: structural and lithostratigraphic and combined
8. Average size of structures:
   – area km^2 (gas): 10
   – oil in place (bbl): ?
9. Initial average daily production per well: oil: nil; gas: 1,500–5,000 MCFD
10. Objectives of interest; Mesozoic, Paleogene, Lower Miocene

The Transylvania Basin is known as the main gas-producing province in Romania, with two-thirds of the total gas reserves, and 80% of all the gas fields. The first industrial gas discovery was made in 1909, when the Sarmasell well produced 864,000 m^3 of gas a day at a depth of 302 m.

To date, 88 structures with 144 pools have been discovered. They are distributed as follows: Badenian 54 pools; Buglovian 43 pools; Sarmatian 45 pools; Pliocene 2 pools. Natural gas is 98% CH_4 and negligible amounts of CO_2 and N_2. The Badenian-Sarmatian shales are considered to be the source rocks. The reservoirs are represented by sands and silts grouped in 1 to 22 pay zones in the Badenian, Buglovian, Sarmatian and Pliocene. Fields are grouped in the central part of the basin, characterized by large development of the domes and of the brachyanticlinal and anticlinal folds, where genesis was controlled by the Badenian salt.

In this chapter we do not want to refer to the Neogene gas fields of the Transylvania Depression, which are very well known, but to the possibility of crude oil discovery. Points (5) to (10) above mainly refer to the crude oil estimated parameters.

Although up to the present, no oil accumulations have been discovered in the Transylvania Basin, some Romanian and foreign petroleum geologists remain optimistic. Their optimism is based on insufficient knowledge of the geological formations, situated under the salt sequences (older than Badenian) which have been considered as economic basement. This is despite the fact that all the geophysical methods have been used, and more than 90 wells have penetrated the salt. It is worth mentioning that the seismic lines shot up to the present do not show enough convincing information about Pre-Badenian formations, and the wells which penetrated the salt layers are shallow. The deepest well, 4502 Filitelnic, bottomed at 4533 m in Senonian limestones while only four other wells are deeper than 4000 m. With so few wells, the geological data obtained about Pre-Badenian (presalt) formations are widely dispersed over a large area.

The absence of good seismic data, contributing to the lack of correlating information between wells, makes the understanding of the spatial distribution of Early Neogene and Pre-Neoene formations difficult. In fact, no information is available to assess the potential of the local Pre-Badenian sedimentary basins.

One can conclude that the use of new seismic techniques could give a better insight into the Pre-Badenian formations, and therefore allow wells to be drilled in the most favourable locations. There are some more arguments that favour the existence of oil accumulations in the Pre-Badenian:

- Considering the thickness of the Pre-Badenian formations, as seen outcropping on the borders of the Transylvania Depression, or penetrated by wells, we can deduce that the total thickness of these sediments could be over 3000–3500 m in some areas. This conclusion is substantiated by some seismic profiles (Fig. 16).
- Inside these formations are numerous intercallations of sands and sandstones with very good petrophysical parameters. This has been confirmed by several production tests (significant salt water rate flow), which indicates good potential reservoir properties.
- The blackish grey marls of the Upper Cretaceous can be considered as source rocks. Also identified in some wells are the Oligocene bituminous marls of the Ileanda Formation that outcrops in the North Western part of the Transylvania Depression (Jibou subbasin). These are also very good potential source rocks.
- The geothermal gradient is relatively low (3 °C/100 m) in the deepest central areas.

4.3.2 Pannonian Depression

1. Area: 30,000 km^2
2. Seismic lines: 29,239 km
3. Number of geological wells: 1860
   deeper than 4000 m: 1
4. Number of oil and gas fields: 71
5. Main source rocks: marls (Miocene), marls and marly limestones (Pliocene)
6. Main reservoir rocks: sandstones (Pliocene), marly sandstones (Miocene-Tortonian), conglomerates (Miocene), weathered crystalline schists (Paleozoic-?), fractured geniss and granites
   Average parameters:
   - porosity: 8–27%(Miocene and Pliocene) 5–16% (weathered basement)
   - permeability: 5–650 md (Miocene and Pliocene) 0–130 md (weathered basement)
   - saturated thickness: 7–280 m
7. Type of traps: structural – normal and faulted anticlines
   stratigraphic – sandstone pinch-outs and sand lens
8. Average size of structures:
   - area: 3 to 9 km^2
   - oil and gas in place: 100 MMbbls; 400 BCF
9. Initial average daily production per well: 20–630 bbl
10. Cumulative production (end 1983):
    oil – 59 MMbbl
    gas – 150.1 BCF
11. Objectives of interest: Pliocene, Miocene, weathered basement

The Pannonian Basin lies in the westernmost part of Romania. It covers the eastern flank of the greater Pannonian Basin which rests against the Apuseni and Banat Mountains, forming several gulfs with eastward closures. The stratigraphic sequence is composed essentially of Pliocene and Middle-Upper Miocene clastics. Occasionally, in the deepest part of the basin and extending over small areas, older formations of Paleogene, Upper-Middle Cretaceous, Jurassic and Triassic age are found.

The structure is generally simple, with gentle anticlines sometimes faulted, located over buried reliefs (Fig. 17). Exploration activity started in 1942, and the first oil field was discovered in 1963 at Turnu. The quality of seismic surveys is good to very good and has contributed enormously to the identification of traps. At present, more than 70 oil and gas fields have been identified mainly in the Pliocene, Miocene and in the weathered zone of the metamorphic or igneous basement. The entrapement of oil and gas is most often the combination of tectonic, lithologic, paleogeomorphic and stratigraphic factors. More than 1860 geological wells have been drilled, one of which, Tomnatec, bottomed below 4000 m.

The main source rocks are the Lower Pliocene and Middle Miocene shales. Their maturation is directly connected to the geothermal gradient which is high (5–5.5 °C/100 m). The petrophysical parameters of the reservoirs are generally good: porosity 16–25%, permeability 5–600 md. A large variation of the petrophysical parameters is observed in the reservoirs belonging to the weathered zone of the basement or to Mesozoic buried reliefs.

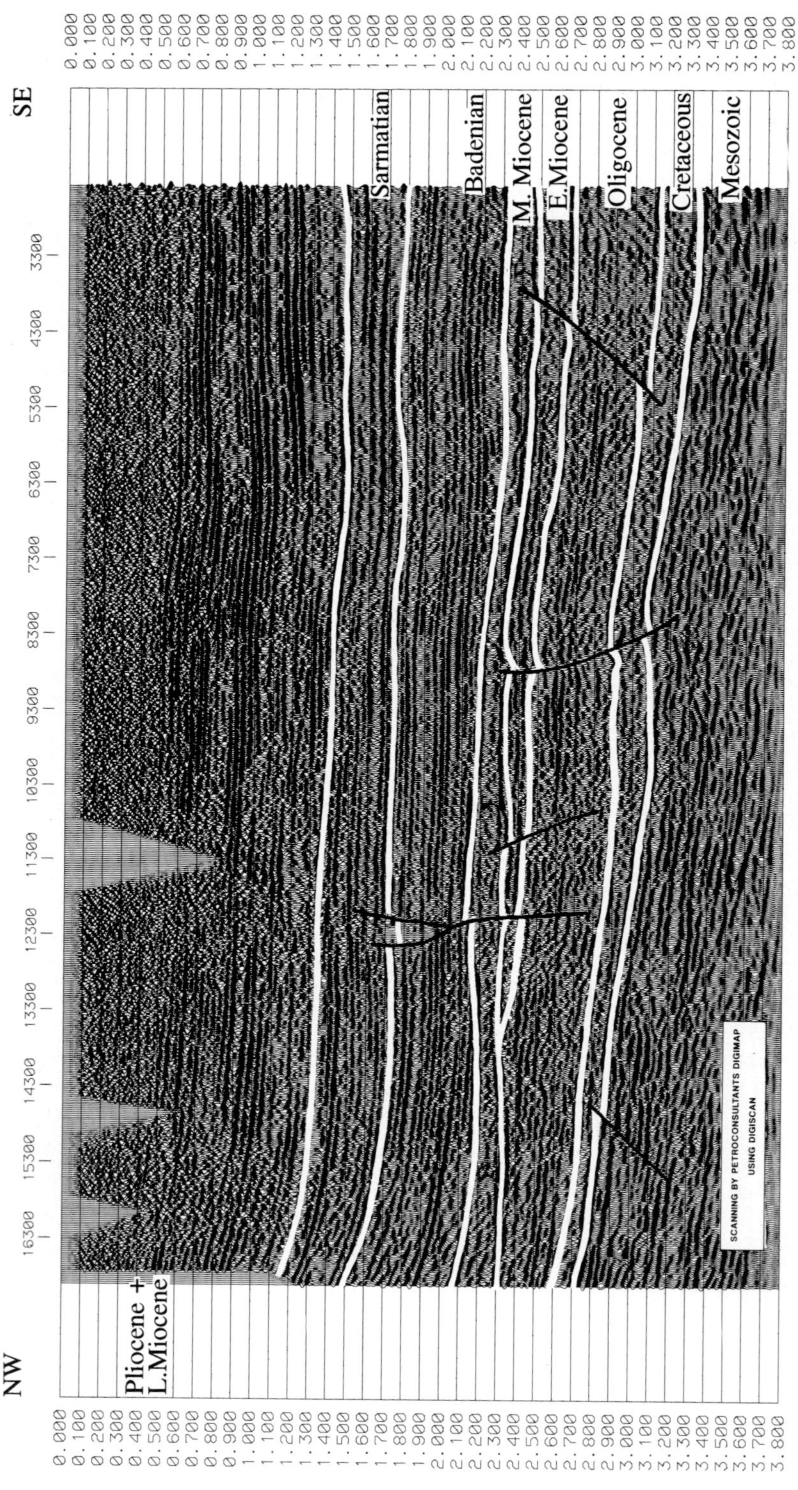

Fig. 16. Representative seismic line; Transylvania Basin

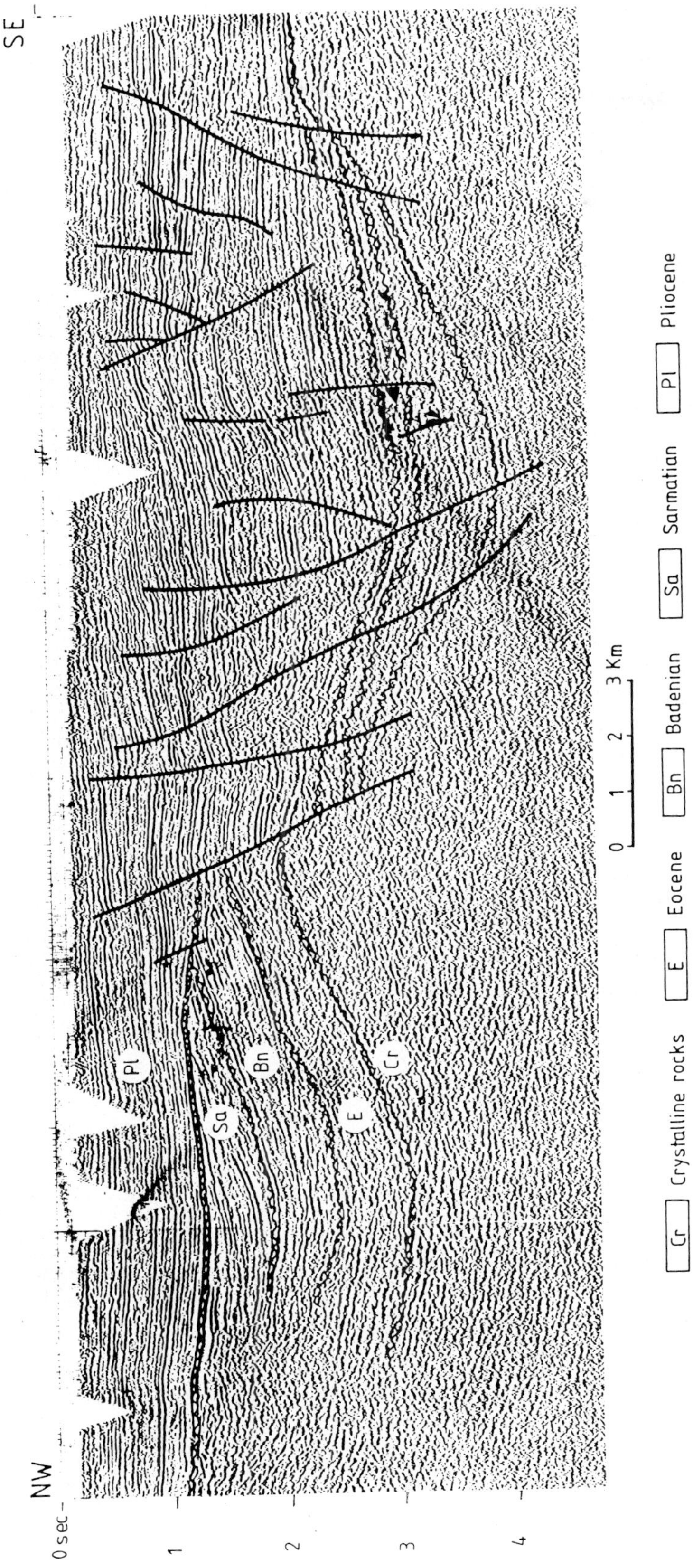

Fig. 17. Representative seismic line; Pannonian Depresssion

To help identify the potential prospects of the Pannonian Depression, a better understanding of the paleogeography and facies distribution of the Mio-Pliocene clastics and the resulting distribution of reservoir bodies needs to be achieved.

4.4 Maramures Depression

1. Area: 2500 km^2
2. Seismic lines: 170 km (CDP)
3. Number of geological wells: 36
   deeper than 4000 m: 1
4. Number of oil and gas fields: 1
5. Main source rocks: bituminous marls (Cretaceous), dysodile schists (Oligocene)
6. Main reservoir rocks: sandstones (Oligocene)
   Average parameters:
   - porosity: 1–5.6% (?)
   - permeability: 10 md
   - saturated thickness: 15–20 m
7. Type of traps: faulted anticlines
8. Average size of structures:
   - area: 10–14 km^2
   - oil in place: 2 MMbbl
9. Initial average daily production per well: 65–75 bbl
10. Objectives of interest: Upper Cretaceous-Oligocene

The Maramures Depression is a segment of the Transcarpathian Flysch on Romanian territory. The whole area is geologically mapped at a scale of 1:25,000. Gravity and magnetic surveys have been undertaken, and few seismic lines shot. The information obtained from wells, indicates the deposition of sediments in two cycles: (1) Upper Jurassic-Neocomian and (2) Upper Cretaceous-Neogene with a maximum thickness of about 6000 m. The deepest well was bottomed in the Eocene at 4600 m.

Like the Paleogene Flysch Zone in the East Carpathians, the structure of the Transcarpathian Flysch consists of nappes, overthrust faults and scale folds. The stratigraphic and structural evolution of the Transcarpathian Flysch offers favourable conditions for hydrocarbon genesis, accumulation and preservation. Some geochemical analyses indicate the presence of source rocks in the Paleogene whose organic carbon is over 6%, and the total hydrocarbon content, in the rock almost 5000 ppm.

The Upper Cretaceous, Eocene and Oligocene clastic sequences, with good porosity properties, could be very good reservoirs. The specific tectonic style, the lithofacies variations, and the geological evolution of the region have been favorable to the formation of a variety of structural and lithostratigraphic traps. There are numerous oil seepages at the surface, oil shows in wells, and a small oil field has been productive at Săcel. This oil field is located in Oligocene sediments. The petrophysical parameters of the reservoirs (Borşa Sandstone) are poor. In addition, on the Hungarian territory, larger deposits have been discovered in this series in the Szolnok Trough area. All these considerations allow us to expect, with an increase in exploration activity, the discovery of new hydrocarbon accumulations.

4.5 The Black Sea

1. Area: 22,000 km^2
2. Seismic lines: 42,441 km + 760 km^2 3-D
3. Number of geological wells: 19
   deeper than 4000 m: 2
4. Number of oil and gas fields: 2
5. Main source rocks: blackish clays (Cretaceous and Oligocene
6. Main reservoir rocks: sandstones and limestones (Cretaceous, Oligocene)
   Average parameters:
   - porosity :15–20%
   - permeability 5–1000 md
   - saturated thickness: 40–50 m
7. Type of traps: faulted anticlines and paleoreliefs
8. Average size of structures:
   - area: 4–4.5 km^2
   - oil in place: 150–300 MMbbls
9. Initial average daily production per well: 1150–1600 bbl
10. Objectives of interest: Cretaceous, Oligocene

The exploration activity on the continental shelf of the Black Sea was started in 1969, when the first seismic lines were shot. The density of seismic lines can be considered as very good for a sea bottom depth of 20–100 m, and as satisfactory for areas where the bottom of the sea is deeper than 100 m. No seismic line has been recorded between the shore and the 20 m isobath. The quality of seismic recordings can be considered as good only for the Tertiary formations (Fig. 18). The first well was spudded in 1976, and at present 19 geological wells have been drilled on 15 structures. Exploration

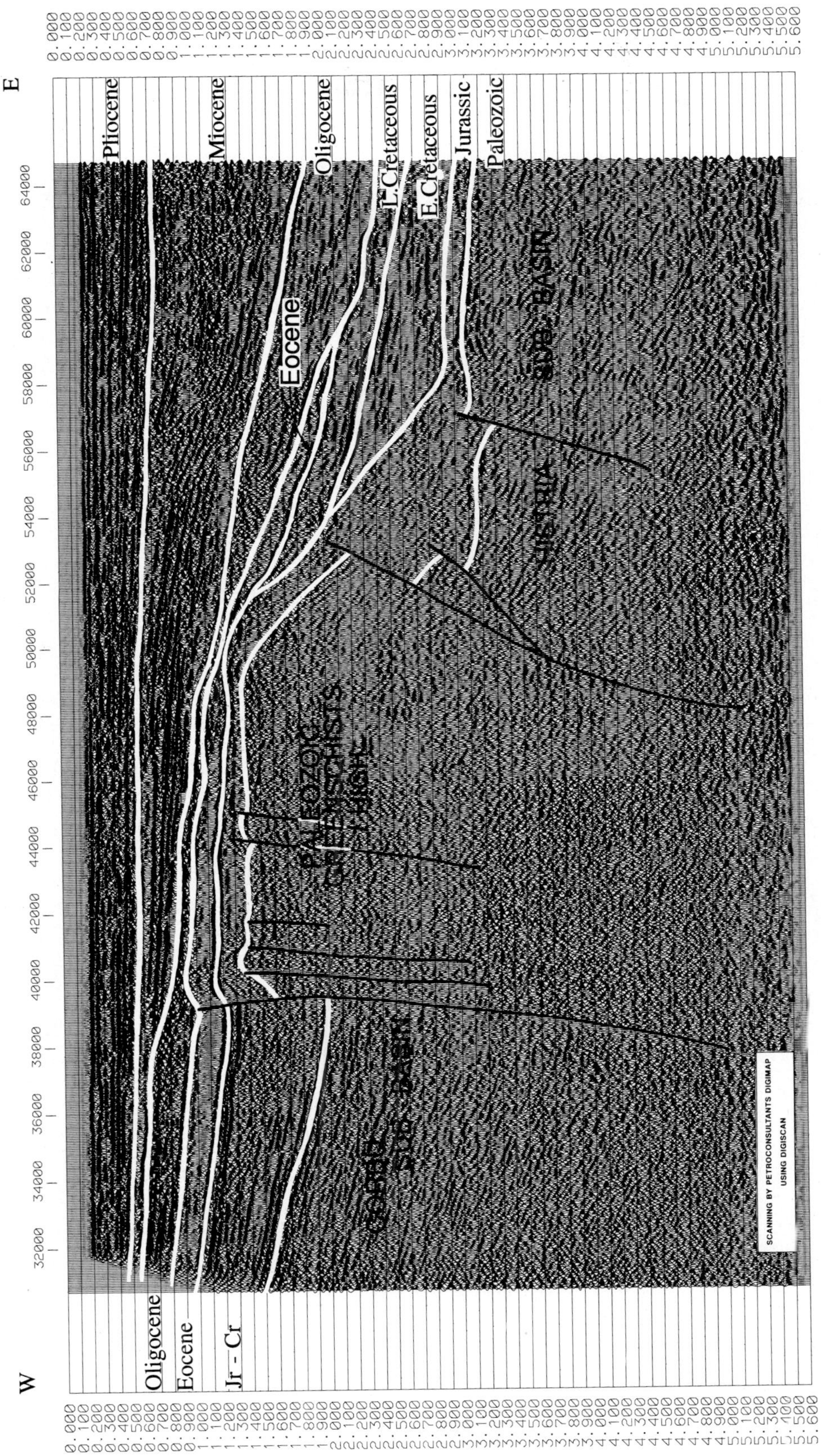

Fig. 18. Representative seismic line; offshore Black Sea

activity has proved that the North Dobrudgean Promontory and Moesian Platform plunge towards the East. Starting with the Lower Paleogene, this zone has been continually subsiding, and consequently the deposits of Tertiary formations exceed maximum thicknesses of 5000 m. The Tertiary formations are mainly detrital, and composed of intercalating sandstones, marls and clays. The sizes and shapes of the pre-Tertiary paleorelief have influenced the structure of the Tertiary formations, which is generally very simple. Exploration has identified four hydrocarbon fields two of which are commercial discoveries (Fig. 19), in Cretaceous sandstones and Oligocene limestones, at depths ranging from 1600–2500 m. Despite the fact that these two oil fields are producing, the structural-geological model of the accumulations is not yet fully understood.

Based on existing data, one can accept that the genesis of these hydrocarbons is related to the Paleogene clays, and the accumulations have been trapped structural and paleogeomorphic traps, above the positive forms of the pre-existing paleoreliefs. When considering the small number of wells, all drilled by jack-ups, and the quality of seismic that cover the total area of interest (22,000 km^2), exploration of the Romanian Black Sea offshore can be regarded as in its infancy. It must be added that the geological interpretation of the seismic data has focused on identifying the structural traps. Therefore, many stratigraphic elements of interest have been overlooked so far. More-over, data available is only representative for the area situated between the 20–100 m isobaths. This means that the tick sedimentary sequences developed in waters deeper than 100 m are practically unexplored. The main issues requiring attention are improvement of the seismic lines reprocessing, and some additional 2-D and 3-D seismic work in order to obtain a better knowledge of pre-Tertiary formations. The seismic grid must also be completed nearshore, and in the areas where the seafloor is deeper than 100 m.

5 Conclusions

For an exploration geologist, Romania presents some similarities with a few old oil-producing countries (e.g. USA, Indonesia). In Romania, the problem does not lie in proving the existence of favourable conditions for the genesis and preservation of hydrocarbons, but that undiscovered hydrocarbon fields still remain. No exploration geologist can agree with the concept that in a certain sedimentary basin all the oil and/or gas

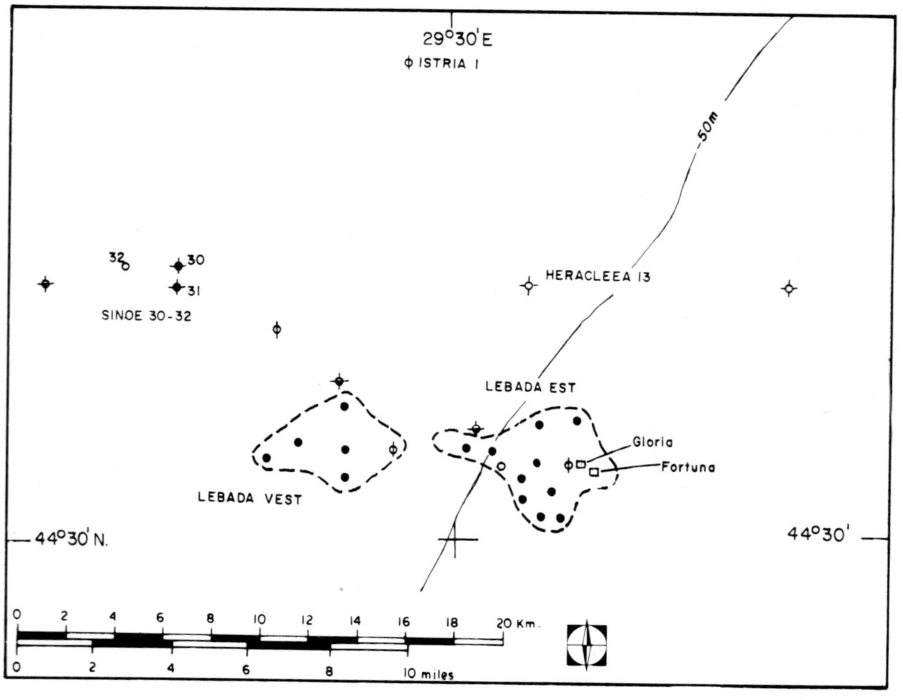

Fig. 19. Black Sea offshore oil field map (From Petroconsultants FSS map)

fields have been discovered. Current exploration activity in many countries worldwide, including Romania, leads to new field discoveries even in the most explored basins as a result of improving the technology, and changing theories.

The questions confronting the explorationist in Romania are:

1. where are these new fields located?
2. how can they be discovered?
3. would these be commercial discoveries?

To answer the first question, in this chapter we have supported the idea that new Romanian oil and gas fields exist in plays where the present exploration activity was nonconclusive. In a nutshell, the best opportunities are in deeper than 3,500 m seated reservoirs onshore and offshore in areas where an encouraging number of shallower hydrocarbon fields have been discovered. This is proved by the very good correlation between the ratio of drilling and discovered reserves for depths over 4000 m in recent years (Table 3).

New oil and/or gas fields could also be discovered at shallower depths in traps overlooked as a consequence of poor seismic results or insufficient research. The complex areas, need to be viewed in both geological and spatial prospective. Additional investigation is needed for the geological formations under the salt in the Transylvania Basin and to numerous stratigraphic events within the producing areas which have not been studied sufficiently, but which could have created good stratigraphic trapping conditions. The interest in exploring for structural traps has been dominant in Romania, as in many countries worldwide. However, the discoveries made in the last decades show that at this stage of exploration, attention must be focussed equally on the lithostratigraphic traps.

Data accumulated up to the present, the new possibilities offered by modern seismic techniques in the field of data acquisition and data processing, the improvement of log operations, and most importantly, the possibility of interpreting information acquired while drilling, allows us to anticipate successful subtle trap identification, regardless of their nature.

Regarding the other questions, the answer to how new oil and/or gas fields can be discovered and of what size belong to the exploration geologists who will analyze and then select one or more exploration areas. In our opinion, the extensive use of new acquisition and processing techniques of seismic and geochemical data, the reprocessing of some of the existing seismic profiles, the improvement of the quality of log operations, and consequently of a refined geological interpretation, can lead to new commercial discoveries in Romania.

Finally, the skill and the ability of future exploration scientists will play a prominent part in any forthcoming success. In this respect, despite the

Table 3. Results of the exploratory drilling below 4000 m

	1961–1965	1966–1970	1971–1975	1976–1980	1981–1985	1986–1990	Total 1961–1990
Total number of wells	4	31	43	82	131	76	367
Successful wells (oil and/or gas)	—	9	15	27	62	25	136
Success ratio %	—	29.0	34.9	32.9	47.4	32.9	38.4
No. of commercial discoveries	—	3	2	6	19	5	35
No. of deep wells from total number of geological wells (%)	No statistics		5.36	15.8	23.3	19.5	15.4[a]
Discovered oil in place: (deeper than 4000 m versus country total)							
Oil (%)	—	na[b]	3.4	9.8	23.4	17.2	10[a]
Gas (%)	—	na	—	6.3	9.4	12.4	5[a]

[a]Approximate.
[b]na, Not available.

fact that the statistics show Romanian oil and gas fields are small to medium in size, we consider there are great chances for new commercial discoveries. There are some specific elements:

- The existence of a well-developed infrastructure, in all the regions where permits have been selected. There are railways and access roads, dense oil pipeline (4600 km from 2- to 28-inch size) and gas pipeline (over 20,000 km from 2- to 40-inch size) network. Almost all facilities exist to make the transport of equipment, people, materials, and the evacuation of crude and/or gas as inexpensive as possible.
- There are specialized organizations for seismic surveys (15 crews), drilling (more than 250 active rigs, 150 of them having the capacity to drill up to 6000 m, 7 offshore platforms) and oil facilities.
- The operation costs are far less expensive in Romania compared to the average on the European market.
- In Romania the refining capacity (600,000 bbl/d) is almost three times greater than the actual domestic production so there is a local market for almost any share of crude belonging to the investor.
- Romania is a crude importer and an oil products exporter, so that any arrangement is possible to sell/exchange the crude with oil products, or with other crude coming from another market.
- There is a good local market for natural or associated gas.
- The development period is very short and the wells can produce immediately after completion, which is further eased by the short distances to the road/railway facilities or to the main pipelines.
- The oil and gas so far discovered in Romania is very clean, with no sulphur, CO_2 or other impurities. This means that no special treatment plants are necessary.

Acknowledgements. The data presented in this paper have been collected from several Romanian publications, and numerous unpublished reports. The complete list of these would be too long and irrelevant for foreign readers. However, comments and evaluation of petroliferous basins' prospectivity belong to the author.

I take this opportunity to thank colleagues from IPGG and Petroleum Department for fruitful discussions and generous supply of illustration and data. I also thank B. Popescu of Petroconsultants S.A., Geneva, for reviewing and greatly improving the text.

Author Index

Liz Lador, Petroconsultants, Geneva

Subject Index

Springer-Verlag and the Environment

We at Springer-Verlag firmly believe that an international science publisher has a special obligation to the environment, and our corporate policies consistently reflect this conviction.

We also expect our business partners – paper mills, printers, packaging manufacturers, etc. – to commit themselves to using environmentally friendly materials and production processes.

The paper in this book is made from low- or no-chlorine pulp and is acid free, in conformance with international standards for paper permanency.